Building Maintenance and Preservation
A guide for design and management

Second edition

Edited by
Edward D. M͏ E, Hon DLitt, FRIBA, FCSD, FRSA

Architectural Press
An imprint of Butterworth-Heinemann
Linacre House, Jordan Hill, Oxford OX2 8DP
A division of Reed Educational and Professional Publishing Ltd

℞ A member of the Reed Elsevier plc group

OXFORD   BOSTON   JOHANNESBURG
MELBOURNE   NEW DELHI   SINGAPORE

First published 1980
Second edition 1994
Paperback edition 1996

British Library Cataloguing in Publication Data
Building Maintenance and Preservation:
  Guide for Design and Management. –
  2Rev.ed
  I. Mills, Edward D.
  690.24

ISBN 0 7506 3398 0

Library of Congress Cataloguing in Publication Data
Building maintenance and preservation: a guide for design
  and management/edited by Edward D. Mills. – 2nd ed.
  p.   cm.
  Includes bibliographical references and index.
  ISBN 0 7506 3398 0
  1. Building – Maintenance.   2. Buildings – Conservation
  and restoration.   I. Mills, Edward David.
  TH3351.B84
  690'.24–dc20   93–45309
                                          CIP

Typeset by Vision Typesetting, Manchester
Printed and bound in Great Britain by The Bath Press, Bath

Contents

Foreword

It is often forgotten that buildings need looking after once they have been designed and constructed. Somehow we take for granted that our precious historic cathedral and churches deserve our care; like the Forth Bridge, medieval structures need constant attention to survive the elements. But what about more recent buildings? Here we tend to overlook the fact that some modern materials like concrete and plastic are liable to decay. And with buildings constructed in the last 40 years an increasing number of problems are now coming to light. Of course, architects and their clients are going to continue to specify new materials for buildings. That is the way that architecture advances.

The need for knowledge about materials and the remedies for defects is constant. This handsome new edition of *Building Maintenance and Preservation* will enhance the survival of buildings for future generations to enjoy.

The Lord Palumbo of Walbrook
MA, Hon FRIBA
Chancellor of the University of Portsmouth

Cover illustration

The De la Warr Pavilion, Bexhill, Sussex. Architects Mendelsohn and Chermayeff. Completed in 1935 and now a Listed Grade I building. The Pavilion is generally regarded as one of the most striking Modern Movement buildings completed in the UK before World War II.

The Rother District Council appointed Architects Troughton McAslan to prepare a report in 1991 on the repair and future use of the building. Considerable maintenance and restoration work has now been carried out under their supervision in conjunction with the De la Warr Pavillion Trust.

The cover photograph shows the very high standard of restoration work carried out to date, and the Architects have been retained for a continuing programme of restoration and improvement.

Lord Palumbo has been a strong supporter of the restoration programme which has received substantial Grant Aid from English Heritage.

Editor

Preface

The first edition of *Building Maintenance and Preservation* was published at a time when the building industry was fully occupied with a large volume of new building projects, but even then nearly £6 billion pounds was spent annually on the maintenance of the building stock of Britain.

The national building programme has suffered a serious decline since that time, and although the buildings and infrastructure on which we depend for our economic wellbeing have in many cases grown older the rate of maintenance and repair has not accelerated. We are now faced with a growing list of post-war buildings, in addition to our large stock of older ones, such as hospitals, schools and industrial buildings, which need urgent attention if their useful life is to be prolonged.

The rapid technological advances in medicine, industry and commerce means that many buildings which may be in a reasonably sound condition structurally are unable to provide the conditions demanded by modern technology. This means that major structural modification is often needed. This can be an economic and practical alternative to demolition and rebuilding.

Many older buildings are of historic and architectural value but the passing of time and sometimes carelessness and neglect has meant that such buildings need urgent attention. Recently a considerable sum of money was set aside by the government towards, essential maintenance work on some of our great cathedrals, many with a history of growth over a thousand years. A report by English Heritage, *Buildings at Risk* a sample survey, shows that of the 500 000 Listed Buildings in England, 7.3 per cent are at risk and 14.6 per cent are vulnerable. It is of interest to note that neglect can be related to lack of use, perhaps because the building has no viable use, and has been neglected and fallen into disrepair.

At the other extreme a number of early modern buildings (often Listed) also need restoration, because of the ravages of time, change of ownership or other causes. Fortunately the efforts of DoCoMoMo, the international organization set up in 1988, with representatives in twenty-nine countries, are already proving vital in the maintenance and restoration work on such buildings as the Bauhaus, Dessau; the Piamio Sanatorium, Finland, and the Penguin Pool at London Zoo.

The importance of planned maintenance, therefore, spans all ages and types of architecture, from the ancient historic buildings of the distant past to those of more recent years, whether they are well known or anonymous.

It is vitally important that all building owners and users should regard the regular, planned maintenance of their buildings as a matter of serious concern, for we cannot afford to allow our buildings, old or new, to decay and become unsightly and uneconomic through neglect. It is both impractical and undesirable to replace all our older buildings. It is therefore essential that they should be kept in good repair. Everyone concerned with buildings as designers, constructors, owners and users must continue to take the vast problem of building maintenance seriously, and to that end this new edition of *Building Maintenance and Preservation* has been planned.

The general pattern of the book follows the first edition, and several of the authors have updated their original chapters. Where the original authors were no longer able to do this, new contributors have either rewritten or updated the particular sections. The scope of the book has also been enlarged by the inclusion of completely new sections and readers will find much valuable material in Chapter 11, 'The conservation of modern buildings', Chapter 15, 'Rehabilitation and re-use of existing buildings', and Chapter 16, 'Euro legislation'.

All the contributors, old and new, are experts in their particular fields, with a wide practical experience spanning many years, and acknowledgement of their cooperation and assistance is gratefully recorded by the Editor. Thanks are also due to all who have loaned photographs and other material. Individual acknowledgement is made in the appropriate place in the case of photographs that have been reproduced.

Particular thanks are due to Lord Palumbo (former Chairman of the Arts Council of Great Britain) for his valuable support and Foreword, to the Building Centre Trust and Pauline Borland for revising the important Directory of Organizations and the staff at the publishers for their continued practical help.

It is hoped that this new edition will prove to be of value to many people as an introduction to a vast and important subject. The Editor would welcome constructive comments and suggestions from readers for future consideration.

Edward D. Mills

Contributors

John Allan, MA (Edin), BA Hons, Dip Arch
MA (Sheffield)
Registered Architect, Avanti Architects Ltd.
Author of *Lubetkin, Architecture and the Tradition
of Progress*, RIBA Publications (1992)

Jacob Blacker, BArch, RIBA, FRSA
Chartered Architect, Jacob Blacker Architects.
Author of the first and second editions of the
Building Centre Maintenance Manual

Alan Blanc, FRIBA
Chartered Architect. Alan and Sylvia Blanc
Architects

Sylvester Bone, BA, RIBA, AA Dip, Dip TP,
FASI
Chartered Architect. The Camden Consultancy

William T. Bordass, MA, PhD
William Bordass Associates

Pauline Borland
Information Officer. The Building Centre,
London

Hugh Clamp, OBE, VRD, FRIBA, FLI,
FCI Arb, BA Arch, Dip LA
Chartered Architect and Landscape Consultant

Richard Dyton, MSc
Kennedys Solicitors (Construction and
Commercial Department)

John Earl, FSA, ARICS
Chartered Building Surveyor. Director of The
Theatres Trust

John Field
Target Energy Services Ltd

Alastair R. T. Gardner, BA (Edin)
Registered Architect. Lecturer, School of
Architecture and Planning, University of
Nottingham

Alan Johnson, BA (Hons), Arch, BArch, MA,
RIBA
Chartered Architect. Alan Johnson Consultancy.
English Heritage

Margaret Law, MBE, BSc, MSFSE, MIngF
Fire Engineer, Consultant, Ove Arup and Partners
(Arup Fire)

E. Geoffrey Lovejoy, CEng, FICE, FIStructB
Chartered Engineer. Senior Construction
Engineer, Health and Safety Executive

David M. Lush, OBE, BSc (Eng), CEng, MIEE,
FCIBSE
Chartered Electrical and Building Services
Engineer
Technical Director Ove Arup Partnership

Edward D. Mills, CBE, Hon DLitt FRIBA,
FCSD, FRSA
Chartered Architect. RIBA Bossom Research
Fellow, Churchill Fellow. Editor, *Building
Maintenance and Preservation*

Norman Sheppard, RIBA, Dip Arch, Dip
Urban Valuation
Chartered Architect. The Camden Consultancy

Douglas L. Warner, BA, FRICS
Chartered Building Surveyor. Joint Partner
Cluttons Chartered Surveyors

John T. Williams, FRICS, MIBM
Chartered Surveyor. Partner Hugh Knight, Lomas
Associates Quantity Surveyors

1
Design and building maintenance
Edward D. Mills

1.1 Introduction

1.1.1 Much has changed in the world of building since the first edition of this book in 1980, but the relationship of Design to Building Maintenance is as important today as it was in the past. The efficiency, convenience, life span, economic viability and appearance of any building can be affected by decisions taken and actions performed at any time in the history of a building project, from its initial conception to its final demolition. Designers should be involved in all these decisions, and their relationships with the other participants is of vital importance.

(a)

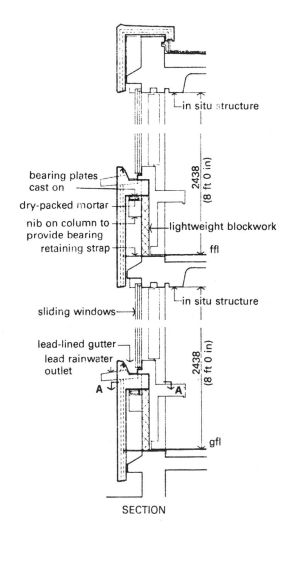

Figure 1.1 (a) Precast concrete cladding units to the Structures Laboratory Extension at the C&CA Wexham Springs complex. Erected in 1966, the building is still in excellent condition due to careful detailing and maintenance.

(b) Section through the external cladding. (Architects: Casson Conder and Partners; photograph and details; C&CA)

Figure 1.2 May & Baker pharmaceutical building. An example of high-technology construction which receives regular maintenance to maintain its condition. Maintenance is greatly assisted by the built-in cleaning facilities (see also *Figure 1.8*). Architects: Edward D. Mills & Partners. (Photograph: William Rawles)

(a)

(b)

Figure 1.3 The Daily Telegraph building, London, (a) showing the effect of over 30 years of weathering and pollution on the Portland stone facade, (b) the facade after cleaning and restoration in 1992. (Photograph: Permanens Restoration Ltd)

1.1.2 There are four main categories of people concerned with all buildings;

Owners
Designers
Constructors
Users.

Sometimes these groups are separate and sometimes they merge together, but ultimately they form the total building team. If any one group is missing or inactive, the final result is less than satisfactory. Each group may have many component parts, but unless they work together as a team, problems will arise.

1.2 The universal problem

1.2.1 Modern buildings are designed to meet more complicated needs than those of previous times; improved space standards, higher environmental standards, and new patterns of use all affect their design and construction. The natural result of these changes is higher building standards and more complicated buildings. This means that the design influence on the maintenance of all buildings is greater than ever before. As the dividing line between different building types, for example factories, offices and laboratories, is rapidly disappearing the problem of building maintenance is a universal one and the consideration of the problem at the design stage is of vital importance.

1.2.2 All buildings start to deteriorate from the moment they are completed and from that time begin to need maintenance in order to keep them in good condition. A building is attacked from all sides, from without and within. The structure settles down, materials swell and shrink, expand and contract. Air and waterborne chemicals and dirt attack surfaces causing discoloration and corrosion. Normal wear and tear soon have an effect on all parts of the building: floors and walls lose their original sheen, fittings work loose through constant use, the efficiency of light sources decreases, and machinery begins to wear out.

1.2.3 This process of gradual deterioration is inevitable, but the speed at which it proceeds can be regulated and the ultimate failure of the building, in whole or in part, can be avoided or accelerated according to the way in which it is maintained. This, in turn, is to a large degree conditioned by the amount, of maintenance required and the ease with which the essential work can be carried out. The problems of building maintenance concern client, architect and builder alike, and each can play an important role in minimizing the maintenance needed in a completed building.

1.3 Design decisions

1.3.1 In the early stages of any building programme the building owner must decide, in addition to the amount of space he or she needs, the amount of money available and the length of life required for the building. At this point in the preparation of the design brief, decisions can be taken which will have a vital effect on the amount of maintenance the completed building will need. The lowest initial cost is not necessarily the most economical in the end, for cheaper materials often require more frequent maintenance and may have a shorter working life than the more expensive alternatives. This is particularly true of wall and floor surfaces, moving parts of machinery, ironmongery and many other parts of a building, especially those which are subject to constant use and wear.

1.3.2 Building owners often place undue emphasis on initial costs and ignore the annual cost of repair and replacement. In the UK it must be admitted that the existing pattern of taxation encourages this approach to building finance, as tax allowances are given on building maintenance but not on capital expenditure.

1.3.3 The preparation of the original design brief is of great importance, and decisions must be made by the client concerning capital expenditure in relation to annual repair and replacement costs, and on the ultimate life to be expected of the completed building or its constituent parts. How long should a building last? How soon will it become obsolete because of sociological, technical or scientific developments in particular fields of activity? These problems face manufacturing industries in particular, where the rate of technological change is rapid, new manufacturing methods, new types of machines and new products all tend to result in early obsolescence of buildings.

1.4 Briefing

1.4.1 We are told that most of the characteristics of a human being are decided at the moment of conception, the colour of our hair and eyes, whether we will be tall or short, practical or artistic, introvert or extrovert, etc., for 'it's all in our genes'. This theory can be applied equally to a building. At the moment of conception – the preparation of the brief – the decisions made and the parameters established determine for good or ill the physical characteristics, the life span and the aesthetic qualities of the final completed work. It is, therefore, vital that this stage in the life of any project should be the subject of very careful study and preparation. This is often not the case. In these days of increasing bureaucracy the designer of a building often works for a 'faceless' Client, a Committee, a Government Department or Developer, with no personal interest or involvement in the project. In many cases the Client or prime mover is not the final user of the building, and sometimes has little understanding of the problems of the particular building in use. Their only concern is too often the possible financial return.

1.4.2 Perhaps the reason for the inadequacy of much of British post-war housing has been due to the fact that the people who would ultimately live in dwellings were seldom consulted about their needs. How often does a Local Authority involve the head teacher in the development of the brief for a new school, or old people in determining the facilities to be provided in the home where they will spend their later years? Too often the fundamental decisions are made by a Welfare Committee, a Housing Officer or an Education Officer or even a Company Accountant, well meaning, but under pressure and often remote from the real problems.

1.4.3 It is hardly surprising that mistakes are made and wrong decisions reached. If we are to design buildings that are technologically and aesthetically satisfactory, the importance of the original brief must be fully appreciated.

1.4.4 There are a number of basic principles to be observed. The physical function of the building must be understood, and the way in which it should work to achieve the greatest measure of efficiency and economy of operation. This applies equally to a factory, a school, a community centre or any other building.

1.4.5 Functionalism as a creed is no longer fashionable, but that does not mean that in its broadest interpretation it is less valid. Mies van der Rohe's dictum 'Form follows function' is as true today as it has always been, for a building that is not based on the functional needs of its user is merely a 'folly'.

1.4.6 However, we live in a rapidly changing world where we have changes of function and developing technologies which can drastically influence the buildings we design. Many industrial, commercial, educational and even medical techniques which appeared to be firmly established only a few years ago have already undergone radical changes, so that in some cases the buildings in which they are used are now uncomfortable and inconvenient.

1.4.7 The lesson to be learned is that, whatever buildings we build, the brief must be adaptable and flexible enough to allow for future change if their full economic potential is to be realized and premature redundancy avoided. Alex Gordon has defined this very aptly as 'loose fit'.

1.4.8 The economics of any building are important. One of the major contributors to construction defects and other unsatisfactory features of many building erected in post-war years, has been the application of unrealistic 'cost budgets'. Both public and private clients have laid down unreasonable cost restrictions, demanding cheap buildings which have proved very expensive as a result of increasing maintenance costs, caused by the initial use of cheap or short life materials in order to reduce first costs.

1.4.9 This can be false economy particularly in an age of rapidly increasing buildings costs and general inflation. Cheap building is usually expensive in the long run. Realistic budgeting at the briefing stage is usually a long-term economy, the time spent on careful and comprehensive briefing before any design work is started is a wise investment, which will pay valuable dividends by ensuring a long life and trouble-free building.

1.5 The importance of details

1.5.1 Detail design is of paramount importance in all buildings and many examples can be seen of buildings of considerable architectural merit which, because of inadequate detailing at critical points, have failed to withstand conditions of weather and normal use. These buildings have either required extensive maintenance to maintain their original appearance or have been allowed to deteriorate to a point of no return because of the high cost of repair.

1.5.2 Uncontrolled weathering and even normal use can lead to physical decay and deterioration, resulting in the need for an abnormal amount of repair and renewal, and often leading to a change in the appearance of the building, usually to the detriment of the original design. This can be avoided by the exercise of greater care in detailed design at critical points of the structure, and a better understanding of the nature and behaviour of materials, as well as their proper application. This problem of deterioration and weathering of buildings is as old as the art of building itself. Today not only are traditional materials used in non-traditional ways, but architects are constantly faced with new materials and building techniques with little information about their behaviour and characteristics. Chapter 5 deals with the maintenance of a wide range of building materials, old and new, in common use.

Figure 1.4 Traditional method of solving weathering problems by means of cornices, copings and similar details. (Photograph: Edward D. Mills)

1.5.3 The traditional builder used eaves, overhangs, cornices, drips and many other devices to solve the problems of weathering and structural and material movement. These traditional details had a functional as well as an aesthetic purpose. Such means were used to provide protection to wall surfaces, to allow building movement without showing surface cracks and provide protection for openings, etc. With the development of new building materials and techniques, these traditional means of protection for a building have largely been abandoned and have not always been replaced by adequate contemporary counterparts. Failures have therefore occurred in modern buildings which have required expensive reinstatement or repair, which could have been avoided had the basic detail been properly solved.

1.5.4 If the traditional solutions to the detailing problems do not fit in with the contemporary conception of architectural form, new solutions must be devised, for the problems still remain. The natural forces that cause deterioration in all building structures will always be with us and experience has shown that the climatic conditions in which building in Britain is carried out, are among the worst in the world.

1.5.5 The contribution of sound detail design to the reduction of maintenance can be considerable, and here again higher initial cost may sometimes be involved. A few of the important considerations in the design of a building which can have a bearing on annual repairs and renewals must be mentioned.

1.6 Detailed design considerations

1.6.1 Of the factors which contribute to the deterioration of a building, the following are among the most important:

1. Moisture
2. Natural weathering
3. Corrosion and chemical action
4. Structural and thermal movement
5. User wear and tear

Many of these are interrelated; for example, thermal movement can cause moisture penetration through a light cladding, and natural weathering can produce chemical action on an external surfacing material. Careful detailing and the choice of the correct material and structural technique, can, however, eliminate avoidable deterioration and reduce the extent of the wear and tear on the fabric and fittings of a building.

Figure 1.5 Contemporary solution to the problem of weathering with metal flashings and sills. (Photograph: Edward D. Mills)

1.6.2 Moisture can attack many parts of a structure. The principal means are

1. Rain
2. Condensation
3. Leakage from pipes or drains
4. Rising damp.

Penetration of moisture through a structure due to weather conditions is perhaps the most important of these problems, particularly where curtain walling or other light cladding is concerned.

1.6.3 The problem of weatherproof joints is one which cannot be over-emphasized. Properly designed damp-proof courses, drips, weepholes, sills and copings, can eliminate the trouble in traditional construction, together with the use of properly detailed cavity walls, gutters and eaves overhangs. Not all of the moisture problems originate outside our buildings.

1.6.4 With more airtight structures, better thermal insulation, less natural ventilation and a greater use of thin, impervious claddings, condensation has become a major source of trouble in many buildings. This causes damage to structure and finished surface and often needs extensive maintenance work. Greater consideration of problems of ventilation, the use of vapour barriers,

Figure 1.6 The waterproofing of joints in heavy panel construction is a major problem. One solution is the self-draining joint designed by the BRE which permits the water entering the joint to discharge harmlessly to the outside. (Photograph: courtesy of the Building Research Establishment)

and a more scientific study of thermal insulation are all needed.

1.6.5 Leakage from pipes and drains can be related, among other things, to poor workmanship or building movement causing fractures in pipes, or joint failure. The latter can be controlled to some degree at the design stage by the provision of flexible joints. Rising damp can be prevented by adequate damp course design at ground level and below, and the inclusion of over-site damp-proof membranes of a suitable material. Failure to give these matters attention at the design stage can lead to rapid deterioration of both structural members and visible surfaces, which can be difficult to cure permanently.

1.6.6 All buildings face the hazard of natural weathering. Rain, wind and sun all help to shorten the life of building materials. The use of an unsuitable material in an area exposed to frequent salt-laden winds, for example, can mean a never-ending maintenance problem. In recent years the effects of high winds on unusual roof shapes of light construction has been only too painfully apparent, and it is now widely appreciated that sun dries out some types of mastic rapidly, causing failure and opening-up paths for moisture penetration. This is of particular importance where curtain walling, infill panels and other lightweight techniques are employed.

Figure 1.7 An eighteenth century historic country house in Turkey now derelict, will its future be demolition or preservation and reuse? (Photograph: Edward D. Mills)

Figure 1.8 Access gantry for the maintenance of curtain walling. This was built into the structure and, because it was used for the erection of the curtain walling, paid for its original cost (see also *Figure 1.2*). (Photograph: William Rawles)

Table 1.1 Thermal movement of building materials

Material	Coefficient of thermal expansion per $°C \times 10^6$	Approximate increase in length of 30 m for a 28°C rise in temperature (mm)
Concrete	11–14	10.00
Mild steel	11–13	9.90
Aluminium alloys	23	19.30
Brickwork	5–7	5.33
Limestones	2.4–9	4.82
Sandstone	7–16	9.90
Granite	8–10	8.12
Slates	6–10	6.86
Glass	9	6.86
Asbestos cement	12	9.90
Wood:		
along grain	6	4.82
across grain	46	35.50
Plastics (glass reinforced as in moulded panel)	12	9.90

Table 1.2 Moisture movement of building materials

Material	Expansion (% of original length)	
	Minimum	Maximum
Clay bricks; limestones	0.002	0.01
Hollow clay blocks; terracotta	0.006	0.016
Expanded clay; concrete, fly ash concrete	0.018	0.04
Sandstones; sand-lime and concrete bricks	0.01	0.05
Foamed slag concrete 1:6	0.04	0.05
Cast stone; dense concretes; mortars	0.02	0.06
Autoclaved aerated concrete; clinker concrete	0.03	0.08
Wood:	0	– – –
Douglas fir, European spruce	Tangential av. 1.6%[a]	
Western hemlock	Radial av. 0.8%[a]	

[a] Movement of timber with relative humidity change from 90% down to 60% at 25°C.

1.6.7 The weather is often responsible for chemical action on building materials. Atmospheric pollution associated with rain causes rapid deterioration of some materials and regular cleaning is essential; if this is to be practical, then it must be considered in the design stage of the building.

1.6.8 Structural and thermal movement have already been mentioned. It has been said that 'if we don't make provision for thermal movement in our buildings, nature will do it for us, and usually where we don't want it'. This is distressingly true and of all the detailing failures, this unwillingness to accept the fact that all structures are continually moving results in the greatest outlay in maintenance. This movement will not only open up paths for moisture penetration but also produce cracked wall and floor surfaces, ill-fitting doors and windows, and a host of other troubles, some of which can never be properly cured. It must be realized that expansion joints and other devices for accepting the inevitable building movement are vital for the success of any design. Buildings with light structural frames and large spans are particularly subject to thermal and structural movement. Chapter 8 explores the problems of joints and jointing materials in detail.

1.6.9 At the design stage of a project the building must be visualized in use, and materials and finishes chosen capable of withstanding everyday wear and tear. This will vary with different types of building but some parts of every building receive constant wear. In particular, poor-quality ironmongery, sanitary fittings and floor finishes are a false economy in any situation. Wall, floor and ceiling surfaces all need regular maintenance to preserve their appearance, but the extent and cost of such maintenance, and the frequency of replacement will depend to a large extent on their original ability to withstand the daily use to which they will be subjected. The choice of the right material – not always the cheapest – at the design stage is therefore of great importance to the building owner, long after the building is completed.

Figure 1.9 Thermal movement in brickwork. (Photograph: Edward D. Mills)

1.7 Access for maintenance

1.7.1 Chapter 12 of this book is devoted to the question of accessibility in relation to maintenance, repair and cleaning allied to the subjects of safety and security, but it is appropriate for brief reference to be made in this section to accessibility as a design criterion.

1.7.2 Much of what has been said so far is basically common sense and should be standard practice, for it is important to remember that *all* buildings, however well designed and conscientiously built, will require repair and renewal as they get older. The maintenance-free building, like the maintenance-free car, will never exist.

Figure 1.10 The Taj Mahal, Agra, built by Shah Jahan in 1660. One of the world's most famous architectural monuments, the marble of which is now being destroyed by fumes from nearby industrial development. (Photograph: Edward D. Mills)

1.7.3 It is important, however, that essential maintenance should be capable of being carried out easily, quickly and economically. Vital decisions can be made at the design stage of a building which will ensure that this is the case. Major replacement can often be avoided if regular cleaning and minor repair work can be carried out without difficulty. All buildings are increasingly dependent on extensive and complicated services. Adequate provision must be made for these and the necessary plant and equipment. Service zones are desirable when large-scale ducting, pipework and cabling are needed. These should be easily accessible and provide adequate working space, well lit and ventilated. Where the scale of the operation only necessitates services duct ways these should be adequate in size, with an allowance for expansion with access panels at regular intervals. Service pipes must never be buried in the structure or floors; where they are exposed, they should be clearly marked using the British Standard colour code.

1.7.4 Where electrical supplies are buried these must be in rewirable conduits, and provision must be made for easy access to lighting fittings for regular cleaning and relamping. Electric motors, starters and other machinery do not have an

Figure 1.11 Repairs to a London building. Access for maintenance to historical buildings is often a serious and expensive problem. (Photograph: Edward D. Mills)

Figure 1.12 Travelling access gantry to the roof of the Old Hall, Royal Horticultural Society, London (built 1904) making maintenance to patent glazing a simple operation. (Photograph: Cluttons)

unlimited life and replacement cost will increase unnecessarily if the building structure has to be disturbed. Adequate working space, access traps and lifting points must therefore be considered at the building design stage (Chapter 6).

1.7.5 If provision for easy access for maintenance purposes is important inside a building, it is vital outside. All major buildings should be designed so that access to the whole of the building face and the roof is readily available by ladder, mobile tower or permanent cradle gantry for cleaning, repair and maintenance. This is of particular importance in relation to high structures and where light wall cladding and roofing is used. With light roofs, proper crawl-ways to provide safe access is essential. If easy access is not available minor failures soon develop into major ones through neglect. Obvious examples are infill panels, mastic joints and sealing gaskets which should be inspected regularly. A planned inspection and cleaning programme not only maintains the appearance of a building but also preserves the material of which it is constructed.

1.8 New materials and construction techniques

1.8.1 Today the building designer is faced with an ever-increasing range of materials, often artificial, sometimes relatively untested and often without adequate guarantees. Our ancestors had a limited choice, brick, stone, timber, metals, glass and some jointing materials, and most of these were natural materials with a long history and high performance record.

1.8.2 The use of such materials made building a labour-intensive industry which relied heavily on a large community of craftsmen devoted to the art of fine building. This has all passed away and the construction industry is becoming increasingly industrialized, relying more on 'assemblers' who fit together factory-made components to form an enclosure for an ever-increasing complex network of internal services.

1.8.3 Architects must choose materials and building techniques that will enable them to design a building that first, meets their clients' functional needs, second, meets the budget constraints laid down and third, can be maintained in good working order for a reasonable time at a reasonable cost. The second and third requirements are, as we have already seen, not necessarily compatible.

1.8.4 It is a regular complaint that the demarcation between design and construction is one of the causes of the problems of building industry. Many alternatives have been proposed and just as many discarded.

1.8.5 The increase in the use of factory-made building components will bring special problems in the field of maintenance, and at the design stage of any building these need special consideration. Wherever possible, components should be standardized throughout a building so that spare parts and replacement units can be either kept in stock or be readily available. Standardization in the building industry is of vital importance, otherwise replacement may be impossible if the original supplier has ceased to manufacture. A spare-parts bank for essential equipment is a valuable means of reducing delays and replacement costs in the event of a breakdown. Chapter 17 deals with this issue and the important question of Statutory Inspections made necessary by ever-increasing legislation in Britain relating to buildings.

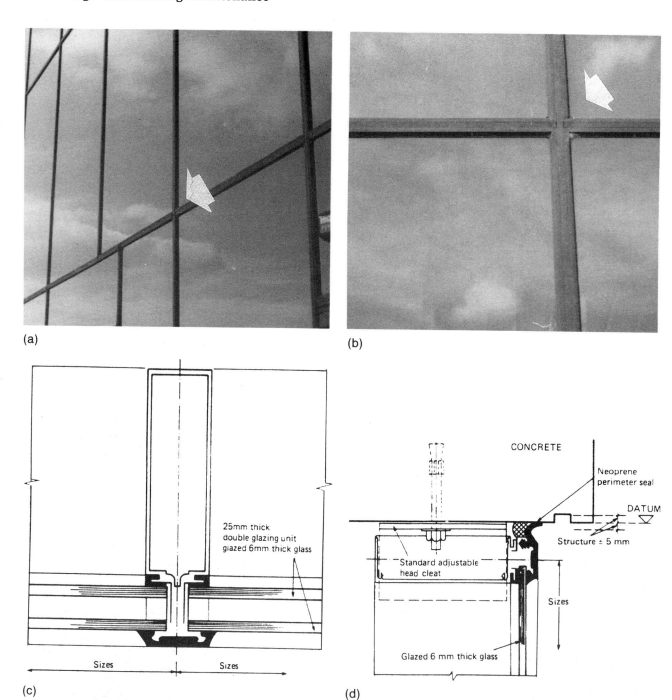

(a)

(b)

(c)

(d)

Figure 1.13 Modern building cladding is often assembled and the Astrawall system is a typical cladding system relying on factory-made components using industrial techniques to obtain adequate weather protection. In this case the nophene gasket replaces putty for holding the glass in position as these sectional drawings show. The photographs indicate the system in use. (Photographs: Edward D. Mills; drawings reproduced by kind permission of Astrawall Ltd)

1.8.6 Interchangeability, standardization and replaceability of components and machinery parts needs detailed study and could be the subject of very valuable research.

1.9 Regional considerations, location and climate

1.9.1 Much of the material in this book has been prepared in relation to the background and problems of building in the UK, and considerable reference is made to legislation and practice in Britain. However, the majority of the information is of general application to buildings anywhere in the world. This is particularly true of Chapter 6 on 'Service design and maintenance' and is becoming increasingly the case in respect of Chapter 3, 'The efficiency of energy utilization', as the world's energy crisis deepens as the result of ever-decreasing stocks of fossil fuels.

Figure 1.14 Modern well-constructed and maintained buildings resist the forces of nature more effectively than those which may have been inadequately maintained. The illustration shows a modern reinforced concrete building still standing after an earthquake while the mud-wall dwellings have completely disintegrated. (Photograph: Edward D. Mills)

1.9.2 Fire safety and means of escape (Chapter 16) is another subject with universal application. For although this chapter has been prepared with UK legislation in mind, the principles will apply anywhere, particularly as British legislation tends to be the basis of building law, especially in developing countries.

1.9.3 Nevertheless, a number of specific design factors which have a direct bearing on building maintenance do arise, and these can be briefly mentioned. There are special problems in hot and dry climates, earthquake zones, and areas where typhoons and hurricanes are regular events. All these need particular attention, and the Bibliography includes reference material that should be especially valuable.

1.9.4 Earlier in this chapter, paragraph 1.6.6 referred briefly to natural weathering. In hot, dry climates the effect of almost continuous sunshine can accentuate the problems of thermal movement in both buildings and building materials, and if adequate design precautions are not taken in the form of carefully designed expansion joints, considerable maintenance problems will be encountered. In such climates an 'ant-proof

Figure 1.15 Light roof construction on a housing scheme in the UK destroyed by exceptionally high winds. (Photograph: courtesy of Building Research Station)

course' may be of greater importance than a 'damp-proof course' in order to prevent the ravages of invading termites. A careful study of local detail design problems and an understanding of the special problems involved is essential, and special reference must be made to the work of the Building Research Establishment and its Overseas Division. The work of these departments includes research and development programmes devised to help the construction activities, particularly in developing countries. The BRE issues a list of books and Overseas Building Notes which are invaluable as design aids for building work outside the UK.

1.9.5 Building in earthquake zones of the world has its special problems and many natural disasters have resulted in a greater loss of life and damage to property than need have occurred if basic design considerations had been observed. A BRE publication *Small buildings in earthquake areas* (1972) by A. Daldy provides much design information which, if properly applied, can prevent loss of life and minimize structural building damage.

1.9.6 Similarly, publications and Codes of Practice are available dealing with wind, including design precautions which can minimize the damaging effects of tornados, hurricanes and typhoons. Even in the UK problems have arisen through inadequately detailed light roof construction, which did not take into account the possibility of exceptionally high winds.

Figure 1.16 The restored facade of the Library of Celcius in the ancient city of Ephesus, Turkey. (Photograph: Edward D. Mills)

1.9.7 The effect of severe climatic conditions on building materials and finishes such as paintwork can cause considerable maintenance problems, and BRE Overseas Building Notes are available on a wide range of materials, including concrete, plastics, metals, brickwork and surface coating. These notes are specially prepared in relation to tropical and similar climatic situations.

1.9.8 Finally the need for the conservation of buildings of historic and architectural importance is not confined to the Western world. In many countries there is a growing awareness of the value of many famous buildings which are irreplaceable. 1975 was designated 'European Architectural Heritage Year'. Chapter 10 of this book, which deals with the maintenance of older buildings, contains much that is as relevant to the Taj Mahal as to Canterbury cathedral. In many developing countries, and in the Middle East, rapid industrialization is often having a seriously damaging effect on historic buildings that have successfully withstood the natural elements for centuries. This is a problem that requires immediate and urgent attention if the international architectural heritage is to be preserved.

1.10 Maintenance planning

1.10.1 Maintenance planning should start at the design stage of any building project and should continue throughout the life of that building. In this the building owner, and or user, must play an active part. Chapter 2, 'The economics of maintenance', discusses the balance to be struck at the design stage between first costs and maintenance costs.

1.10.2 Chapter 13 deals with maintenance planning policy and relates this to the important issue of information feedback. This latter subject is not one that arises only when the building work is completed. Feedback should be planned at design stage, and any extra professional cost involved in such activity can show increasing dividends in later years.

1.10.3 When a building is handed over to the user – the client – those who have designed and built it are often forgotten, unless something goes wrong. It is of interest to recall the story of Charles Garnier, the blacksmith's son, who built the Paris Opera. They forgot to invite him to the opening night and so he paid 20 francs and bought a ticket for a box on the second tier. They did not even know that he was there!

1.10.4 This lack of continuity is unfortunate, for much useful information for future reference can

Figure 1.17 The State Tobacco Factory, Linz, Austria. Designed by Peter Berhens and Alexander Popp, 1931–4. The buildings are still in use for their original purpose and have been carefully maintained and extended. They are still in perfect condition after 60 years of continuous use. (Photograph: Michael Gilman)

be obtained if a proper feedback system can be devised. The tragedy is that many buildings are destroyed not by outside forces, such as wind and weather, but by their owners, through bad housekeeping, inadequate maintenance, and even outright neglect. A recent school inspection produced complaints of water penetration through a wall, resulting in damaged decorations. A quick investigation and the removal of two tennis balls, a soft drinks can and a large quantity of leaves from a gutter cured the problem instantly. An insignificant example, but typical of the sort of minor problem which can grow into a major one through neglect, and which can be multiplied a thousandfold in many buildings.

1.10.5 All new buildings, as a matter of course, should be provided with a maintenance manual, so that a building owner can look after and service a building in use just like a car. The Building Centre maintenance manual and its use is discussed in Chapter 18. A proper maintenance routine should be established and *followed* so that repairs and replacements are carried out in proper sequence and before they develop into major problems. The designer must visualize future maintenance situations and build into the building structure the means of access and other facilities needed to make redecoration, repair and replacement as simple and economical as possible. In many instances a 'replacement bank' of spare parts (Chapter 17) for essential equipment should be established when a building is completed to avoid delays and cost increases at later dates.

1.10.6 Building owners, including public sector clients, often pay too little attention to keeping their buildings in good working order and are surprised when they fail to give the service they expect. It has already been emphasized that buildings start to deteriorate from the day they are

completed, just as human beings start to die the day they are born. The length of time this process of decay takes depends on the care taken in first the design, second the construction, and third, the upkeep or maintenance.

1.10.7 An important matter related to the question of maintenance concerns alterations and modifications to a building during its life, without proper consultation with the original designer. Not

Figure 1.18 The boiler house of the State Tobacco Factory, Linz (see also Figure 1.17). (Photograph: Michael Gilman)

only does this usually impair the integrity of the architecture but it can also damage the structure and give rise to serious defects. Overloaded services; floors loaded beyond the designed capacities; factory roof trusses used for hoisting heavy machinery, an endless list of similar cases of misuse could be compiled. The cost of repairing the resulting damage can be monumental.

1.10.8 Most architects regard the building they design as part of themselves, and many are horrified by the way buildings are treated by careless or thoughtless owners, who certainly treat their cars with much greater respect. If a pattern could be established where the building design team was involved with the lifetime health of a building – like a family doctor – perhaps more buildings would enjoy a longer trouble-free life and give greater service to the owner and the community. Architecture is concerned with buildings that are practical, economical and beautiful for the whole of their life span, and modern economic conditions require this to be as long as possible.

1.11 Summary and conclusions

1.11.1 To summarize the important issues relating to design and its relationship to building maintenance the following points need to be emphasized, remembering that all buildings require regular maintenance to ensure a useful and economic life, and that it is essential that users of any building should have a planned approach to this work.

- The first point at which the design of a building can influence the cost of the inevitable maintenance programme is the formulation of the design brief. First, cost must be equated with running costs. The life cycle of any building is automatically determined before it is built.

- The second point of influence is the choice of materials and structural technique which must be carefully considered in relation to foreseeable patterns of use and possible changes in those patterns.

- The third point is the importance of the detailing of the building which can be the cause of failure or success in any building design.

- The fourth point relates to the design of access, both inside and outside a building. This is of vital importance if maintenance, repair and replacement work is to be carried out economically and without interference with the occupants of the building.

- The fifth and final point of importance is the availability of replacement parts for building components, machinery and equipment when they have reached the inevitable end of their useful life.

Figure 1.19 The Brynmwr Rubber Co factory in Wales, now a Listed Building widely recognized as one of the outstanding industrial buildings of the twentieth century. This photograph, taken in 1953, should be compared with the more recent one in Chapter 11. The building has been unused for many years owing to the problem of adapting it for present day manufacturing requirements, unlike the Linz Tobacco factory (see Figures 1.18 and 1.19). (Architects: Co Partnership)

1.11.2 The following paragraphs from a leader in the *Architect's Journal* of 23 June 1993 [by permission of the Editor] sums up the fact that 'Maintenance is key to the life of a building' and emphasizes not only the importance of detail design but also that continuing maintenance to all parts of any building is essential for a long life span.

'God, according to Mies van der Rohe, was in the details. By that he meant the design, craftmanship and workmanship. What he never said was that God was in the maintenance.

It's very easy to hand over a building, publish and then forget about it. But designing and detailing are increasingly not just about good practice. Planned maintenance, from the inevitable everyday kind, to that required after a number of years, is increasingly becoming an integral part of the design brief, and the economic life of the building.

Overhauling and servicing a building at regular intervals, and even its eventual recladding and refurbishment, are fast becoming essential considerations for developers, institutions and tenants.'

Acknowledgement

Tables 1.1 and 1.2 are reproduced by kind permission from the chapter on 'The effect of design and maintenance' by Dr N. G. Marsh in *Developments in Building Maintenance – 1* (Ed. E. J. Gibson) published by Applied Science Publishers Ltd.

2
The economics of maintenance
J. T. Williams

2.1 Introduction

2.1.1 Maintenance is defined in BS 3811: 1984 as 'The combination of all technical and associated actions to retain an item in or restore it to a state in which it can perform its required function'.

2.1.2 This definition recognizes that, subject to a minimum set by the statutory authorities, there is no single 'state' which will be equally applicable to all the different types of building and circumstance likely to be met in practice. This flexibility, which is admirably suited to an industry whose products are so varied and long lived as those of the construction industry, nevertheless raises two questions.

2.1.3 First, to whom does 'a state in which it can perform its required function' have to be acceptable? The easy answer 'To the client' raises the question endemic to many problems in the industry: 'Who is the client?' Obviously there may be differences of opinion between the occupier or tenant of a property and its owner. From an economic standpoint these differences are reconciled in the rent payable; for example, a standard of condition which may be acceptable to tenants at one rent may not be so at a higher one. Conflicts are resolved in the market, inability to let a property indicating that putting considerations of obsolescence on one side, either the rent must be lowered or the standard of condition improved.

2.1.4 This leads into the second problem, that of improvement. There is, of course, an element of improvement inherent in almost any maintenance operation, if only because the original materials and fittings are no longer made and have to be replaced by better ones, but unavoidable improvement of this nature creates no difficulty. The problem arises when we consider changes of acceptable standards over time.

2.1.5 The Committee on Building Maintenance[1] argued that, given the long life of buildings,

acceptable standards of amenity and performance will rise over their lifetimes. They therefore defined maintenance as:

Building maintenance is work undertaken in order to keep, restore or improve every facility, i.e. every part of a building, its services and surrounds to a currently acceptable standard, and to sustain the utility and value of the facility.

2.1.6 It is doubtful if this definition should be accepted. The resources invested in the construction of a building give rise to a capital asset of a particular value and life expectancy. Subsequent maintenance expenditure prevents the value of the asset being eroded by wear and tear, accidental damage and the like. The installation of a bathroom in a house built without one, however, changes the nature of the asset; the investment would have been greater and the asset more valuable if the house had been built with a bathroom initially.

2.1.7 The fact that with the passing of time and the improvement of living standards it has become impossible to let a house without a bathroom so that some part at least of the original investment would have to be written off earlier than was foreseen, or that the return on the investment is less than was anticipated, indicates an error in the original investment decision. However, we should not confuse maintenance and improvement merely to correct past errors, even if they were justifiable at that time. Improvement expenditure is properly a new investment and should be appraised financially as such.

2.1.8 Maintenance is not the only recurrent expenditure which arises from the decision to construct a new building. In this chapter the following hierarchy of cost definitions will be used:

- *Maintenance cost* The cost of any actions carried out to retain an item in or restore it to an acceptable condition but excluding any improvements other than those necessitated by inability to replace obsolete materials or components.

- *Running cost* The cost incurred in occupying a building or other facility whether paid by the owner or occupier. These include national non-domestic rates or council tax, maintenance costs, fuel, light and water supplies, and cleaning cost, but not the cost of fuel used in manufacturing processes.

- *Cost in use* The costs incurred in owning and occupying a building or other facility whether paid by the owner or occupier. These therefore represent running cost plus the initial cost of the site, construction and associated fees.

- *Operational cost* The cost of the staff, machinery and materials used in the building or facility.

2.1.9 M. E. Burt[2] estimated that for non-domestic buildings the average annual running costs are about the same as the annual cost of repaying at current interest rates the original capital cost as a mortgage spread over the life of the building. For clerical and administrative staff operational costs may be up to four times the cost in use, and will be even higher when expensive material or equipment is used.

2.1.10 Flanagan and Norman[3] estimated that for an office building, for every £100 spent annually in repaying the original site and building costs, £138 will be spent annually on running cost of which £38 will be on rates, £29 on maintenance, £48 on cleaning and the remainder on energy over a period of 40 years at a discount rate of 2 per cent.

2.2 Maintenance demand, supply and price

2.2.1 Contractor's output of repair and maintenance work, during the 1990–93 recession fell 17 per cent from £3600 million to £3000 million. Present low interest rates are intended to revive the market.

2.2.2 The total expenditure on maintenance in the UK in 1990 is estimated at £27 763 million. At constant prices this represents a slight fall compared with 1989 and a rise of 25.5 per cent since 1980. The estimate of total maintenance expenditure is built up from government figures for contractors' repair and maintenance output, an estimate of output by private sector direct labour organizations and figures for householders' expenditure on DIY materials taken from the UK National Accounts.

From 1982 to 1989 the growth in maintenance spending outstripped the rate of general growth in the economy but fell back in 1990 and now represents 5.04 per cent of the UK Gross Domestic Product.

The replacement value of the stock of buildings which represents the size of the estate to be maintained has risen even faster than the growth in maintenance spending. Consequently, maintenance spending expressed as a percentage of the Gross Capital Stock of Buildings and Works fell from its peak of 1.69 per cent in 1987 to 1.47 per cent in 1990.

2.2.3 Maintenance output has, however, tended to rise and fall at approximately the same times as new works output, indicating that while there is a basic, fairly steady maintenance workload, the factors that influence demand for new work also affect maintenance demand.

2.2.4 It would be expected that in a recession as developers turn away from new build that the output of maintenance would increase but this has not happened. Changes in the average age of the stock could obviously affect demand, but in view of the small proportion of the stock demolished each year and the small proportion represented by new building such changes should be minimal.

2.2.5 Data on the numbers (and ages) of buildings and other works are not readily available. The most suitable proxy to illustrate trends over time is probably the value of the gross capital stock of the nation at constant replacement cost. Indices of the values for dwellings and for other buildings and works are shown in the CSO Blue Book,[4] together with the annual outputs of repairs and maintenance. Even when allowance is made for the probable growth of DIY over the period it is evident that repair and maintenance output is much more affected by the general economic situation that it is by the size of the stock. Neither is there much support for the frequently made suggestion that present-day building technology is condemning the nation to devote a steadily increasing proportion of its Gross Domestic Product to maintenance.

2.2.6 It is also to be expected that firms undertaking maintenance work do so under conditions of diseconomies of scale. A high proportion of the cost of minor repairs and maintenance is the time and expense of workers travelling to and from the work. As a firm expands it can at first seek to gain an increasing proportion of the work available within a convenient distance from its yard or base. Beyond a certain point it can expand only by either enlarging its radius of operations, so increasing the cost of travelling time, or by setting up a new establishment further afield. Entry to the maintenance industry is also easy; many of today's successful building firms have their origins in the traditional man and a boy with a barrow and a ladder in the maintenance field.

2.2.7 Economic theory therefore postulates that the firms engaged in maintenance should be predominantly small. In fact, despite the frequently criticized 'fragmentation' of the construction industry, there is not the sharp division between firms engaged in new works and those engaged in maintenance as is found in many other industries, such as the car industry. Even the largest building firms tend to have a minor works department which may carry out significant amounts of maintenance work, especially when the demand is concentrated in a particular area, as it is, for example, in the Measured Term Contracts used by the Property Services Agency for the maintenance of Government buildings.

2.2.8 In the past despite the participation by firms of all sizes maintenance work remained to a considerable extent the domain of the small firm. However, with the rise of Facilities Management and the demise of directly employed labour it is seen that maintenance work is increasingly carried out by large Management Contracting organizations.

2.2.9 It is frequently argued that as maintenance work is labour intensive its costs tend to rise faster than those of new work. However, that argument does not hold and many other factors affect the market prices of new and maintenance work.

2.3 Techniques of economic appraisal

2.3.1 The economic theory of production makes a useful distinction conceptually between the short and the long runs. The short run is that period within which the firm cannot significantly add to or reduce its stock of fixed capital – its buildings and machinery – and has to make the best use of what it has. In the long run it can increase or decrease its fixed capital and its range of options is correspondingly widened.

2.3.2 The same distinction can usefully be made when considering the economics of maintenance. In the short run the owners of buildings or estates of buildings and other works have a fixed capital stock which cannot increase, reduce or change quickly. Their problem then is to determine the appropriate amount of resources to devote to

maintaining it. In the long run they can change their holdings, and the problem then, if they decide to build new buildings, is to determine the optimum balance between initial capital expenditure and subsequent running and maintenance cost.

2.3.3 Both these problems involve choosing between alternative courses of action which will result in different cash flows over a period of years, and the technique used to resolve them is discounted cash flow.

2.3.4 There is a large and growing literature on the theory and practice of discounted cash flow. A small selection of the more useful works is given in the bibliography, and no more is attempted here than to outline the principles involved and to discuss briefly some of its applications and the more important points likely to cause difficulty in practice.

2.3.5 The difficulty which arises when choosing between alternative courses of action which result in different cash flows over time is that, even without inflation, money in the future is worth less to an individual than money today. Offered £100 today or £100 next year most people will choose the £100 today, not merely because of the risk that it might not materialize next year but also because money received today can be invested and so, again ignoring inflation and risk of loss, will be worth more next year. It follows that payments to be made in the future must be 'discounted' or reduced in some way to enable valid comparisons to be made between present and future payments, or between payments to be made at different times in the future.

2.3.6 If the assumed rate of interest is 10 per cent per annum then £100 invested today will yield £110 next year. Conversely, £100 to be received next year has a 'present value' (or is worth today):

$$£100 \times \frac{£100}{£110} = £90.91$$

In other words, £90.91 invested today at 10 per cent interest will enable a payment of £100 to be made next year. Similarly:

$$£100 \times \frac{£100}{£110} \times \frac{£100}{£110} = £82.64$$

invested today will enable a payment of £100 to be made in two years' time.

2.3.7 In essence, therefore, discounted cash flow is no more than the listing of all cash flows which will arise as a result of some course of action, together with the year in which they will occur, discounting each payment or receipt to its present value and summing the total. If choice is to be made between two or more alternatives the calculations are repeated for each alternative and the totals compared.

2.3.8 Manual calculation of the present values is simplified by the use of published tables which show, for different rates of interest, the present value of £1 to be received at different periods in the future and the present value of an annual payment to be paid or received for a number of years in the future.

2.3.9 Simple calculations are also readily made on many pocket calculators, and there are simple computer programs available for more complicated problems. The example below illustrates the application of discounted cash flow to a simple problem and shows why summation of the (undiscounted) cash flows may lead to a faulty decision.

2.3.10 In practice, of course, the problem is rarely so simple, and many other factors have to be considered – not least in the example below the probable effect of the frequent disruptions caused by the major renewals of components for Design B.

2.3.11 An alternative method of comparing different cash flows is to calculate their 'Equivalent Annual Values'. The equivalent annual value is the constant sum which is paid annually over the life of the project would have the same present value as the actual cash flow. In the example below the equivalent annual value of Design A is £11 376 (i.e. £11 376 paid each year for 60 years has a present value of £11 376 × 9.967 = £113 381). The equivalent annual value of Design B is £10 475.

2.3.12 Before discussing the difficulties of applying discounted cash flow in practice it is worth considering its applications a little deeper; by so doing some confusion can be avoided.

2.3.13 The simple cost-in-use study illustrated below has three characteristics:

1. The client is choosing between the mutually exclusive courses of action; if he or she decides on Design A he does not want Design B as well.
2. The cash flows being considered are all negative. There is an underlying assumption that any receipts such as rent, etc. will be the same whichever design is chosen. Incidental

A client has the choice of two alternative designs for a proposed building project. Design A costs £100 000 to build and will require annual running costs of £1000 and major renewals of some components each costing £20 000 and occurring at years 20 and 40. Design B will cost £80 000 to build, and will need annual running costs of £1200 and the same major renewals of components every 10 years. The expected life of both designs is 60 years and the discount rate to be used is 10 per cent. The comparison is made:

Design A

Cash flow	Year	Discount factor[a]	Present value
£100 000	0	1	£100 000
£ 1 000	1 to 60	9.967	£ 9 967
£ 20 000	20	0.1486	£ 2 972
£ 20 000	40	0.0221	£ 442
£200 000			£113 381

Design B

Cash flow	Year	Discount factor[a]	Present value
£ 80 000	0	1	£ 80 000
£ 1 200	1 to 60	9.967	£ 11 960
£ 20 000	10	0.3855	£ 7 710
£ 20 000	20	0.1486	£ 2 972
£ 20 000	30	0.0573	£ 1 146
£ 20 000	40	0.0221	£ 442
£ 20 000	50	0.0085	£ 170
£252 000			£104 400

[a] Discount factors taken from published tables.

Therefore an investment of £104 000 now will pay the total cost of Design B over its entire life, whereas £113 381 would need to be invested to pay for Design A despite the fact that the undiscounted costs of B are greater than those of A.

receipts such as the scrap value of the renewed components can be subtracted from the renewal cost so that all cash flows are payments made by the client.
3. The client is seeking to minimize his costs. Other things being equal, he or she will select the design with the lowest present value or equivalent annual value.

2.3.14 In the wider and more general application of discounted cash flow to Investment Appraisal these characteristics will be different:

1. The client may have several investment opportunities which may not be mutually exclusive; if he or she has sufficient capital the fact that he builds an office does not stop him deciding to build a factory as well. If he has several investment opportunities, all profitable, but insufficient capital or borrowing power to finance them all he is said to be 'capital rationed'. The choice of the best combination of investments then raises difficulties beyond the scope of this section.
2. The cash flows are both positive and negative – there are receipts as well as payments. Each cash flow, positive and negative, is then discounted separately and the 'net present value' is calculated for comparison.
3. The client is seeking to maximize the return on the capital invested. Other things being equal he or she will select the project, or combination of projects, with the highest net present value.

2.3.15 Whereas private investors making an investment appraisal need consider only the cash flows to and from themselves or their company, government and other large public sector organizations may need to take account of the social costs and benefits to the community in general. This will involve Cost Benefit Analysis. The term is often misapplied to any private investment appraisal where only private costs and benefits are being taken into account. Cost Benefit Analysis raises difficult economic problems of public goods, the valuation of social environmental and other intangible costs and benefits, and the social time preference rate or social opportunity cost. It will normally be undertaken by a multi-disciplinary team, one of which will be an economist, and is not further considered here.

2.3.16 A term often encouraged in connection with both investment appraisal and cost benefit analysis is the 'internal rate of return' or the 'discounted cash flow return'. This is the discount rate at which cash flows must be discounted so that the present value of a project will be zero. Given a choice of projects, the one with the highest rate of return is normally that to be selected.

2.3.17 The attraction of the discount rate is that it depends only on the cash flows generated by the project and is inflation-free.

2.3.18 We turn next to the difficulties associated with the use of discounted cash flow in normal cost-in-use or investment appraisal studies. All calculations must compare like with like and the difficulties concern principally the problem of forecasting the future, inflation, the discount rate to be used and taxation.

2.3.19 *Forecasting the future*

Discounted cash flow compels the decision maker to forecast the future cash flows over the life of a project, which for building works is commonly taken up to 60 years and for road bridges up to 120 years. All forecasting is inevitably an error-prone process, and neither building owners nor their advisers are likely to be 100 per cent right in their predictions over such a long time span.

2.3.20 Fortunately, the risk of error is normally greater, the further in the future the payments occur; as seen in the example on page 19, the effect of discounting is to reduce the present value of payments more, the further in the future that they are made, so that the effect of errors a long way ahead is very small indeed. If the £20 000 to be paid in year 50 for Design B was 100 per cent in error so that the actual payment was £40 000, the effect would be to increase the present value by only £170.

2.3.21 There is, nevertheless, a considerable risk of serious error in forecasting, and it is sometimes argued that this is best allowed for by increasing the discount rate by a 'risk premium'. The method is really only satisfactory if every cash flow in the study is subject to the same risk of error, and if the risk increases uniformly over the life of the project.

2.3.22 A better method is to use sensitivity analysis. This is to examine the basis of forecasting of each of the cash flows, and the assumptions, i.e. rate of inflation, the discount rate, the period of analysis and life expectancy of the options, ranking of cost options and taxation underlying them. It will commonly be found that only a few of the forecasts are critical to the final result. These can then be considered in more depth before any final decision is made.

2.3.23 *Inflation*

In pure terms, inflation means a fall in the value of money relative to other goods, and as such can simply be ignored. If the best available option is selected on today's prices, i.e. inflation-free, it will still be the best if the price of everything doubles or quadruples.

2.3.24 What cannot be ignored however, is what is sometimes called differential inflation. Even in non-inflationary times the prices of some goods will be rising and those of others will be falling, for reasons such as the exhaustion of cheaply available raw materials, improved methods of production, etc. During inflationary periods these differential movements will be superimposed on the general rate of inflation, so that the prices of some goods and services will rise faster than others.

2.3.25 The simplest way to deal with such differential price movements is to continue to use inflation-free discount rate and increasing the forecast cash flows of the faster-rising items, or, if the rate of increase is thought to be constant over the period considered, by discounting those items at a lower discount rate.

2.3.26 Discount rate

Choosing the appropriate rate of discount raises difficult problems and is of crucial importance. If, in the example above, the costs of the alternative designs had been discounted at 5 per cent instead of 10 per cent the present value of Design A would have been £129 308, and of Design B £131 744; Design A would now be the lowest-cost option and not Design B. Because money invested at 5 per cent would increase more slowly than at 10 per cent, the client would need to invest more money now to make the future payments on both schemes, and there would be less interest available on the £20 000 saved initially on Design B to pay its higher (undiscounted) future costs.

2.3.27 The above discussion illustrates the point that the higher the discount rate, the lower the present value of future payments and the greater weight attached to initial capital costs. It may also be noted that at higher rates of discount, payments occurring after 60 years are reduced to such an extent in present value terms that it is immaterial whether the life of the building is taken at 60 years or more. Lower discount rates have the reverse effect by increasing the relative importance of future costs.

2.3.28 In the private sector the appropriate discount rate is the cost of capital to the client. In a normal situation a firm may obtain capital for investment from a variety of sources – short- or long-term borrowing, share issues, depreciation provisions, retained profits, etc., each of which will have a different cost to the firm by way of interest payments, dividends, or other profitable investment opportunities foregone. The average cost of capital to the firm is the weighted average cost of each source, the weights used being the proportion which each source contributes to the total capital of the firm. The subject is extensively discussed by Merrett and Sykes.[6] It is to be expected that for most cost-in-use or investment-appraisal studies in the private sector the discount rate to be used will be provided by the client or the firm's accountant. It is essential that cash flows and discount rates are chosen on the same basis. If cash flows are expressed in today's prices then inflation-free discount rate must be used.

2.3.29 In the public sector more difficult theoretical problems of the social time preference rate and social opportunity costs arise. fortunately for the average architect or surveyor, the Treasury determine a test Discount Rate (7 per cent in 1978) to be used in this type of analysis in most of the public sector.

2.3.30 Taxation

It is important in cost-in-use or investment-appraisal studies to take account only of the cash flows resulting directly from the project under consideration, and to do so on a net of tax basis. This means that if a project will produce, for example, a rent of £100 in years 1 and 2, on which £50 tax will have to be paid in each of years 2 and 3, then the cash flows to be discounted are +£100 in year 1, +£50 in year 2 and −£50 in year 3. Other cash flows such as investment grants, VAT, etc. should be similarly taken into account.

2.3.31 It has been argued that as expenditure on maintenance is allowable as an expense against tax liabilities whereas expenditure on new capital projects is not, private firms are encouraged to build at lowest initial cost, ignoring any future higher costs of maintenance etc. Since any money saved on the initial capital cost could be invested by the company in some other project with a comparable rate of return on capital and therefore a comparable tax liability, the taxation system does not discriminate against maintenance-reducing expenditure. However, fiscal considerations on plant, machinery and industrial buildings are relevant to building decisions, and the awareness of fiscal aspects are expected to increase.

2.4 Cost-in-use in the design of new buildings

2.4.1 The preceding section has outlined the techniques to be used in cost-in-use studies and illustrated the need for them to be made.

However, it is still the exception rather than the rule for formal studies to be made during the design of new buildings, due to doubts as to whether the benefits from such studies are worth their cost to the developer.

2.4.2 Data

Cost-in-use studies need data on the life of materials and components, and on the cost of their maintenance, etc. Reliable data on these matters are difficult and costly to acquire and to keep up to date. Organizations such as the Royal Institution of Chartered Surveyors (RICS) are able to collect and analyse the cost information and reduce it to a form in which it can be used. See the Building Cost Information Service by the RICS.

2.4.3 The Building Maintenance Information Service Ltd, sponsored by the RICS and the Building Research Station and CIOB, collects and disseminates some information on a cooperative basis.

2.4.4 The Royal Institution of Chartered Surveyors publishes a set of life cycle costs papers giving worked examples of the lives of some common materials and components and their cost-in-use. Other information is also assembled and published by the BRE and CIOB.

2.4.5 Costs and benefits of cost-in-use studies

The careful consideration of alternative design options and the calculation of their cost-in-use costs time and money. Professional fee scales are based on traditional design practice when design decisions were made more on the basis of accumulated experience, and for all but the largest projects do not allow much margin for the more systematic methods now developed. The question must arise as to whether the additional benefit is worth the cost.

2.4.6 It is undeniable that cases will occur in which the full benefit will not be obtained. Changes in technology and the economic situation will often nullify the most careful study, and many examples could be cited. Calculations as to the optimum level of thermal insulation made before the considerable increase in energy prices were falsified by that event; modern formulations of plastics materials used in buildings have far better performance characteristics and lower repair costs than earlier ones.

2.4.7 Calculations are also falsified by the incidence of obsolescence. Components and entire buildings may become obsolescent for a variety of reasons before they are worn out. Furnishings and fittings, particularly in buildings such as shops, hotels, entertainment buildings, etc., are liable to be replaced due to changes of fashion before the end of their useful life. Buildings can become functionally or positionally obsolescent.

2.4.8 Finally there is also the problem of defects. All forms of construction are liable to prove defective, sometimes in design, sometimes in construction, necessitating expensive remedial work. Examples abound in recent cases where innovatory materials or methods have been used. By their nature these are largely unpredictable and again they can nullify the most careful cost-in-use study.

2.4.9 Nevertheless, designers would be patently failing in their duty if they did not at least make their clients aware of the effects on the probable cost-in-use of their major design decisions on their larger projects. The above difficulties may be regarded as illustrations of the subjects to be considered in sensitivity analysis described earlier. It remains to outline the results of some cost-in-use studies to assist designers in concentrating their efforts where they are likely to be most effective.

2.4.10 The breakdown of annual running costs found in office buildings in a study by Flanagan and Norman[3] was approximately:

Rates	28%
Cleaning	34%
Annual Maintenance	12%
Other Maintenance	9%
Energy (fuel and light)	17%

2.4.11 Much of maintenance expenditure is spent on a wide variety of minor repairs, etc. and no correlation between maintenance cost and type of construction is found, except that smaller offices are relatively more expensive to maintain. Rates are largely beyond the control of the designer and it follows that cost-in-use studies should, in general, concentrate upon energy costs, cleaning costs, and those major items of cyclical maintenance such as redecoration, the renewal of short-life materials and components, services installations, etc. whose frequency can be predicted with some degree of certainty.

2.4.12 Low initial cost does not necessarily lead to high running or maintenance cost or vice versa. In some cases this is self-evident – a building with a sophisticated air-conditioning system will cost more both to erect and run than one simply heated and naturally ventilated. In other cases it is not so obvious. Arguments of the kind that an increase in the capital cost to provide double glazing will be more than saved in reductions in the cost of heating ignore the fact that the same result may often be achieved more cheaply by reducing the glazed area and increasing the thermal insulation in other parts of the structure.

2.4.13 Energy costs are high and likely to increase faster than other costs. They depend to a large extent on the strategic design of the building – its shape, orientation, wall/floor ratio and window/wall ratio, and they should be considered in the very earliest stages of design, preferably on a multidisciplinary basis. It is important to allow for the differential increase in the cost of fuel in any study.

2.4.14 Many parts of the building, particularly the structural frame and floors, foundations, etc., usually incur no cleaning or energy costs, and for some forms of construction are virtually maintenance-free. Furthermore, the statutory requirements are such as to ensure that even the cheapest form of construction will often have a life of 60 years or more. Unless there are significant differences in the cyclical maintenance of the structures being compared therefore, design decisions can be based solely on initial capital cost. Finishes and components such as roofing, cladding, sanitary fittings, fittings and fixtures, windows, etc. tend to incur high maintenance and cleaning costs and should be the subject of costs-in-use studies which should take account of their frequently shortened lives due to changes of fashion or occupier. Before selecting any material or component which may be thought possibly liable to premature failure it should, of course, be subjected to a cost-in-use study on both 'normal' and pessimistic assumptions as to its life.

2.5 Maintenance, modernization and conversion of the existing estate

2.5.1 The economic problems faced by the owner of an existing building or estate are first, to determine the amount it is worth spending on maintenance, and second, when and to what extent he or she should change the estate by improvement, conversion from one use to another, modernization or rebuilding.

2.5.2 When buildings are leased or rented these problems are solved relatively simply. The economic amount of maintenance is that necessary to maintain the rental value of the estate, to maximize the return on the capital invested, or to meet any other objective of the client. Both costs and rents can be estimated on the basis of market prices or forecasts of them in the future and normal methods of investment, conversion, and rebuilding are similarly solved.

2.5.3 When making investment appraisals to decide between demolishing and rebuilding an existing building or converting or modernizing it it is important to ensure that all the factors which affect cash flow are considered and taken into account. They include the following:

1. The initial cost of demolishing and rebuilding, or converting/modernizing it.
2. *Differences in rents* The differences in rents likely to be received for a new as compared with an existing building.
3. *The different time scales* It will normally be found that rebuilding takes longer than conversion or modernization, affecting the period over which negative cash flows (payments) and positive cash flows (receipts) occur.
4. *Professional and other fees and charges* It may be necessary to include management expenses, especially in the case of Local Authority housing.
5. Any improvement, rehabilitation or investment grants or tax allowances receivable, and any profits or other taxes payable.
6. *Council tax or national non-domestic rates* Possible differences in the rates which will be levied on a new as compared with an existing building.
7. *Running costs* Possible differences in the running costs, especially of energy, cleaning and cyclical maintenance.
8. *Possible differences in the efficiency in use* It will often be found that for any given gross floor area it is possible to achieve a higher rentable or usable floor space in a new building than in an existing one. Similarly, there are sometimes significant differences in operational costs.
9. The expected future life of the existing building after modernization or conversion.

2.5.4 Obviously, it will not always be necessary to consider every one of these items in particular cases. A client proposing to demolish and rebuild an existing building and then to sell it will not need to consider differences in rates or operational costs, for example. Some examples of investment criteria appear in reference 3.

2.5.5 Needleman[8] gives a useful formula for setting an upper limit to the amount it is economic

Table 2.1 Design guide: floor finishes[7]

		Capital	Cleaning	Cyclical and renewal	Defects	Total
				Costs in £/m²		
1	50 mm Granolithic (1:1:2) paving laid on concrete and finished with carborundum grains trowelled in, including 15 × 75 mm granolithic skirting.	16.10	126.80		0.30	143.20
2	150 × 150 × 15 mm Red quarry tiles to BS 1286 type A with 10 mm cement and sand (1:3) bed, on and including 25 mm cement and sand (1:3) screeded bed and 90 mm high quarry tile skirting.	33.70	126.80		2.60	163.10
3	25 mm Iroko blocks fixed with adhesive to and including 25 mm cement and sand (1:3) floated bed complete with 70 mm high hardwood skirting and wax polish all hardwood surfaces.	46.60	97.90	25.40	3.80	173.70
4	Dunlop Semflex vinyl sheet flooring fixed with adhesive to and including 50 mm cement and sand (1:3) trowelled bed complete with 70 mm high softwood painted skirting.	21.50	91.50	6.10	2.00	121.10
5	Low loop pile tufted carpet including integral underlay on and including 50 mm cement and sand (1:3) trowelled bed complete with 70 mm high hardwood polished skirting.	27.00	53.00	13.20	0.90	94.10

to spend on improving an existing building for any predicted future life. It is based on the premise that the money saved by improvement rather than rebuilding, if invested over the period of the future life of the building, should be sufficient with accrued interest to pay for its eventual demolition and rebuilding. Account would need to be taken of many of the other factors listed above.

2.5.6 It would appear to be probable that annual maintenance costs increase with the age of the building, and if so, this would need to be taken into account. The evidence, however, is rather conflicting.

2.5.7 A study of Local Authority housing maintenance[9] showed that annual maintenance costs rose slowly until they were double their initial value by the time a house is 50 years old, but part of this increase may be due to the inclusion of some element of improvement. On the other hand, variations in day-to-day maintenance costs over a period of years appear to be due to local organizational or budgetary factors. No significant correlation between annual maintenance costs and age of building was found.

2.5.8 Finally, we need to consider the problem of determining the economic level of maintenance, improvement, etc. in buildings for which no market rent or suitable analogy for it exists. Such buildings are common in the public sector and include, for example, hospitals, libraries, defence establishments, etc. Here it must be admitted that no method which is both theoretically sound and reasonably practicable to operate appears to exist. However, with government policy of privatization of organizations and services the methods detailed previously in the chapter may apply.

2.5.9 Cost benefit analysis is usually much too costly for the scale of expenditure under consideration, except possibly for very large Local Authority housing estates, for which a rent or rent analogy exists in any case, and some simpler development of cost benefit analysis which can be applied relatively cheaply appears to be needed. As it is, decisions are usually based on some combination of the availability of finance and similar budgetary considerations, local politics, and the (often conflicting) pressures exerted by the managers of different departments. It is to be hoped that the greatly increased attention being given to cost-in-use techniques for energy, environment and maintenance management by many major organizations will lead to a better solution to this problem.

Table 2.2 In-use budget data: floor finishes[7]

| | | | | | Costs in £/m² | | | | |
| | Cleaning | | | Cyclical and renewal | | | Defects | | |
	Work	Annual cost	Freq-uency ratio No:Yr	Work	Cost	Cycle in years	Remedial action includes:	Cost	Occurs in year
1	50 mm Granolithic paving, skirting, carborundum finish.	Scrub and rinse floor, skirting.	9.00	254:1	–	–	–	Relay, repair cracks	0.60	3
2	Quarry tile paving, bedded on screed,	Scrub and rinse floor, skirting.	9.00	254:1	–	–	–	Relay and repair floor.	3.50	3
3	Iroko block floor, laid on screed, hardwood skirting, wax polish.	Sweep and mop floor. Dust down skirting	6.10 0.35	254:1 254:1	Polish floor. Polish skirting. Sand floor.	1.90 0.30 5.80	1 8 30	Repair floor.	4.80	3
4	Vinyl sheet floor, laid on screed, painted skirting.	Sweep and mop floor. Dust down skirting	6.10 0.35	254:1 254:1	Renew floor. Paint skirting.	13.20 0.15	15 8	Renew floor, repair screed.	1.60	3 18 33 48
5	Pile carpet, integral underlay, laid on screed, polished hard-wood skirting.	Vacuum carpet. Shampoo carpet. Dust down skirting.	3.50 0.26 0.35	254:1 1:2 254:1	Renew carpet. Polish skirting.	15.10 0.30	10 8	Relay carpet, repair screed.	0.70	3 13 23 33 43 53

2.6 Conclusions

2.6.1 Maintenance is a significant part of the output of the construction industry, and running costs account for a large part of the budget of building owners and occupiers. Economic techniques for determining the optimum levels of expenditure are well developed and soundly based. In the field of maintenance much data have now been published by the Building Costs Information Service (RICS), BRE and CIOB to use with the techniques which have been developed.

2.7 Explanatory notes for cost-in-use tables

2.7.1 Cost-in-use Tables e.g. published by PSA are in two parts. A typical example appears above in Tables 2.1 and 2.2. The first provides the costs-in-use of selected building elements, components, etc. The second part provides the underlying data such as cleaning frequencies, renewal and redecoration cycles on which the costs-in-use have been calculated. The costs shown include overheads and profits, and are based on prices updated to March 1991. They assume a building life of 60 years, and the discount rates used are 7 per cent for cleaning costs and 8 per cent for cyclical renewals and defects costs.

2.7.2 The cleaning costs shown indicate the cost of normal office cleaning and window cleaning services derived from current standards and practice. They do not include anything for cleaning external brick or stone, etc. elevations, nor for general cleaning such as dusting furniture, changing towels, etc.

2.7.3 The cyclical and renewal costs includes all maintenance; the requirement for which is known to the designer when selecting a component and which will vary depending on his choice of material etc. Where components have an expected life of less than 60 years renewal has been allowed at the original cost plus the cost of taking down the old work. No allowance is made for any costs arising from premature obsolescence arising for reasons other than normal wear and tear, nor for accidental damage such as reglazing broken glass, clearing blocked wastes, etc.

2.7.4 While it is expected that sound workmanship and materials will be used in conjunction with good design, there will inevitably be some failures, and some forms of construction are more prone to failure than others. The defects costs shown will not, of course, arise on every project on which a particular form of construction is used, but if that form of construction is employed on a number of projects then the average cost on all of them will be the amount shown, even though most of them may have no defects at all. It has been assumed that defects will come to light within three years of first installation, and that where a component has an expected life of less than 60 years there will be a similar defects risk on each replacement.

2.7.5 Most maintenance activities entail a certain amount of disturbance to the occupants of buildings. The degree of disturbance can have a significant and adverse effect on productivity, particularly where rooms or even entire buildings have to be temporarily vacated. No allowance has been made for likely disturbance costs, which will, like the cost of accidental damage, depend much more on the nature of the building's occupants and use than on the original design selection. Where likely to be significant, however, as in the case where frequent cyclical renewals are required, they should be taken into account in any assessment of design options.

References

1. *Report of the Committee on Building Maintenance*, HMSO.
2. Burt, M. E., *A Survey of Quality and Value in Building*, Building Research Establishment.
3. Flanagan, R. L. and Norman, G., *Life Cycle Costing* (1989).
4. *United Kingdom National Accounts 1992.* (The CSO Blue Book).
5. *Housing and Construction Statistics*, (March 1992) HMSO.
6. Merrett, A. J. and Sykes, Allen, *The Finance and Analysis of Capital Projects*, Longman, Harlow.
7. *Cost-in-use tables*, Property Services Agency.
8. Needleman, J., *The Economics of Housing*.
9. Clapp, M. A., Cost comparisons in housing maintenance, *Local Government Finance*, **67**.

3
Energy utilization, audits and management
W. T. Bordass and J. W. Field

3.1 Introduction

3.1.1 Almost exactly half the energy the UK consumes is used in buildings, and nearly every building uses more energy than it should. The reasons are both technical and managerial, and often fall into the following groups:

1. The building has an intrinsically high demand. For instance, it could be poorly insulated or admit little natural light.
2. The building services are inherently wasteful, perhaps having inefficient lamps and light fittings, or air conditioning with excessively high fan power.
3. The control systems are inherently wasteful, for instance with too many lights on one poorly positioned switch.
4. Buildings and systems are poorly maintained, for example having damp walls, damaged window catches, and leaky boilers with loose insulation, poorly adjusted burners and dirty heat exchangers.
5. The systems are poorly operated and managed, with controls incorrectly set and perhaps with plant running unnecessarily at night and weekends.
6. The systems are wastefully utilized by the occupants, for example, heating full up and windows wide open.

Excessive energy consumption not only wastes fuel and money, it is also a major contributor to environmental damage through air pollution and greenhouse gas emissions. Money, although not energy, can often also be saved by reviewing fuel purchasing and tariffs.

3.1.2 In new designs, and sometimes in major refurbishments, one has the opportunity to work through the above list from top to bottom. However, when seeking cost-effective opportunities in existing buildings, the list is best read backwards, a point not always appreciated by designers. Better operation and management can cost little or nothing to improve, while building alterations of all but the simplest kind tend not only to be expensive but can also prove disappointing if not seen in the context of the overall operation. For instance, if the management cannot maintain the existing systems properly, will new technologies ease their problems, or add to them? Similarly, if heat losses are reduced – say, by double-glazing – but supply, management and control of heating and ventilation are not altered accordingly, then savings may be smaller than anticipated, with system inefficiencies continuing and the windows being opened to control the resultant overheating. Changes to heating, ventilation and insulation may also affect the moisture balance of the building (see Chapter 4).

3.1.3 A wide range of techniques can be used to help one to understand how energy is used and to improve energy performance. The six main sections of this chapter discuss them in the following groups:

Comparing annual energy consumption and cost with norms.
Reviewing patterns of energy-consumption behaviour.
Understanding utilization and identifying scope for savings.
Selecting appropriate projects.
Taking action.
Maintaining and improving performance.

Ideally one would progress logically through all the stages from top to bottom. In practice, this is not always essential: one does not need to know everything in order to start doing something useful.

Table 3.1 Conversion of fuel and energy units to kWh

Fuel type	Billed unit	To get kWh multiply by
Natural gas	Therm	29.31
Natural gas (variable)	Cubic foot	0.303
Liquefied Petroleum Gas (varies slightly with the gas)	Tonne	13 800
Gas oil (35 sec)	Litres	10.6
Light fuel oil (290 sec)	Litres	11.2
Medium fuel oil (950 sec)	Litres	11.3
Heavy fuel oil (3500 sec)	Litres	11.4
Coal and coke (variable)	Tonne	7 800
Anthracite (variable)	Tonne	9 200

Notes: Gross calorific values are used here and conventionally in the UK. 1 kWh (kilowatt-hour) = 3.6 MJ (megajoules).

3.1.4 Energy units

Fuel and energy are measured in many different units, which can make comparisons confusing. Data here are normally given in kilowatt-hours, the unit in which electricity and now natural gas are billed. Typical conversion factors from other units are shown in Table 1; better figures may be printed on bills or available from suppliers. Osborn[1] gives a wide range of useful data required for energy-related calculations generally.

3.2 Comparing annual energy consumption and cost with norms

3.2.1 Energy indices, performance assessments, and targets

Comparing a building's annual energy consumption and cost with figures for others of a similar type can be helpful, although even a 'very good' assessment seldom means that nothing more remains be done. Making genuine comparisons is not always quite as straightforward as it may appear, and so it is worth considering the steps involved, which are usually as outlined below. Terms commonly used are shown in italics.

1. A building has a certain annual *energy consumption* of individual fuels, sometimes assigned to different end uses.
2. These are then converted to *energy consumption or cost indices*, usually of consumption per square metre of 'treated' (= heated or air conditioned) floor area, though other units (for example treated volume, number of hotel bedrooms, or number of school pupils) are also used.
3. The indices may or may not be *normalized* for weather, hours of occupancy, volume or product, etc.

4. The indices, normalized if appropriate, are then compared with *yardsticks*, which may be either fixed or adjusted for circumstances.
5. The comparison may lead to qualitative *performance assessments*, such as 'very good', points systems, or quantitative *labels* or *ratings*.
6. Numerical *targets* may also be set, either for the label or for the energy consumption or cost index.

To make realistic comparisons, the factors employed in indexing must be reliable and in the same units as the yardsticks: this applies particularly to floor area. Normalization, particularly for hours of occupancy, can be very useful but it can also introduce further errors, cloud understanding by the non-expert, and conceal opportunities in intensely used buildings which, even if relatively efficient, offer greater returns on capital investment owing to their longer operating hours. It is therefore often worth employing a range of different indicators – for instance, total energy consumption and cost (to identify high consumers); energy consumption and cost per unit area (to identify the more energy-intensive buildings); and normalized indices (to identify the highest consumers in relation to their level of utilization).

3.2.2 Methods of comparing performance with yardsticks

Two methods and sets of yardsticks are widely used in non-domestic buildings: Normalized Performance Indicators (NPIs), which are discussed in CIBSE[2] and applied in more detail to individual building types in Audit Commission[3] and EEO;[4] and Energy Consumption Guide yardsticks, EEO.[5] Domestic buildings have other procedures such as the National Home Energy Rating (NHER) which are offered commercially. EEO[4] is aimed primarily at energy managers and uses flowcharts to normalize for weather, occupancy, etc. and derive a single number which is then compared with yardsticks to give an assessment from 'very good' to 'very poor'. Figure 3.1, from Audit Commission,[3] shows the range of NPIs for a selection of local government buildings. The energy consumption of any building to the left of the appropriate bar is regarded as good and to the right as very poor. Unfortunately, NPI assessments usually tend to be of total delivered energy only and do not differentiate between different fuels, even though electricity typically costs around four times as much as fossil fuels such as gas, coal and oil.

Energy Consumption Guides tend to have specific target audiences, from main board directors to caretakers, advising them on what to

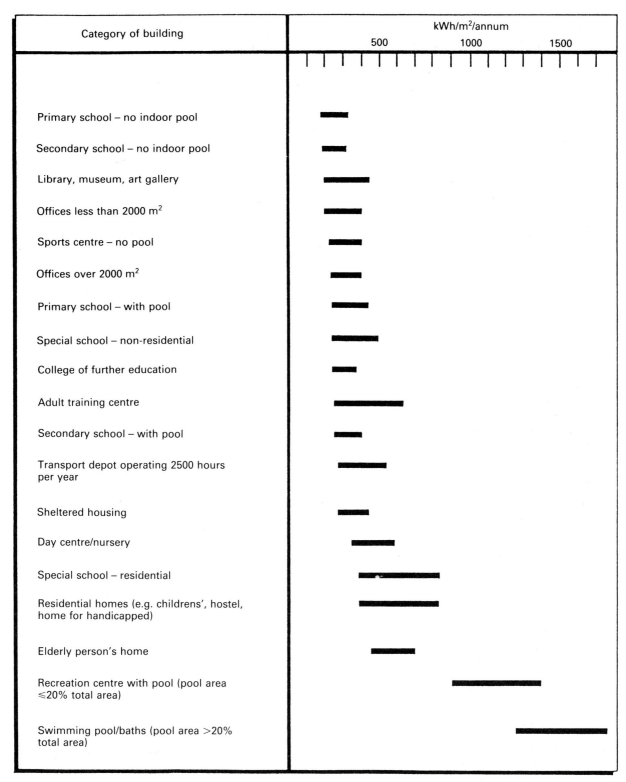

Figure 3.1 Range of normalized performance indicators (in kWh/m² gross floor area per year) for different types of building. (Source: *Saving Energy in Local Government Buildings*, Audit Commission)

look out for and the type of action they may wish to take. They include fewer standard corrections but more detail on how and where energy is consumed and how much it costs. For example, Figure 3.2, from EEO,[6] shows typical and good practice energy costs for four different types of office building. Notable features include:

1. The major variations in electrical cost between building type, showing the importance of the right yardstick against which to make comparisons, and that one should aim not to have an intrinsically high-energy building type if one does not really need it.
2. The much smaller variation in fossil-fuel

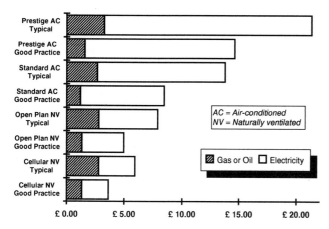

Figure 3.2 Typical and good practice energy costs in offices (1991 £ per square metre treated floor area per annum)

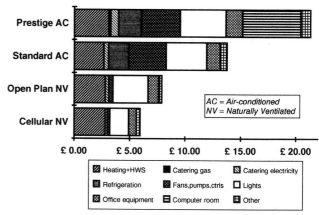

Figure 3.4 Annual energy costs of typical offices (1991 £ per square metre treated floor area per annum)

consumption and cost. In fact, in many sectors (houses, schools, offices, and light industrial units) typical heating energy consumption usually falls within a narrow band of 150–250 kWh/m², though variations between individual buildings can nevertheless be large.

3. The predominance of electrical costs over fossil-fuel costs in all but the Naturally Ventilated Cellular offices. In spite of this, many people still think that energy efficiency is primarily about heating.

4. The major differences between typical and good practice. Typical is only the average: many offices use more, and sometimes far more. Occasionally such high consumption is unavoidable, due to a large computer suite, for instance, but usually it is not.

Some of the Consumptions Guides also attribute energy consumption and cost to various end uses, as in Figures 3.3 and 3.4, based on EEO.[6] This helps to give a sense of proportion and assists comparisons with more detailed data, if available. The following points are of interest:

1. Fossil fuel for heating and hot water is much less important in cost terms than in terms of delivered energy.

2. The importance of lighting energy costs, particularly in the open-planned and sealed air-conditioned buildings where lights tend to be left on more often.

3. The relatively high energy costs for fans, pumps and controls in the air-conditioned buildings. With all-air systems most of this energy tends to be used by the fans.

4. The importance of the mainframe computer room in the Prestige Air Conditioned building. This, of course, varies widely with the installation. Sometimes over the whole year computer suites can use far more electricity than anything else, as they tend to be on continuously while most other building services and equipment typically run for less than one-third of the time. Computer room air conditioning can also run very wastefully, but seldom is there any management information on this as only rarely is the computer equipment and its air conditioning separately sub-metered.

3.2.3 Unusual features

Sometimes buildings will have unusual features not included in the yardstick for performance assessment, with either:

1. Very high energy consumption (for example, a swimming pool, a computer room or a large catering kitchen) or

2. Very low (for example, unheated storage or covered parking) or

3. No relation to the building area, such as illuminated signs or electric vehicle recharging.

If either the energy consumption of these items or their floor area is significant (say, more than 5 per

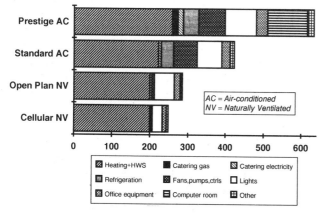

Figure 3.3 Annual energy consumption of typical offices (kWh per square metre treated floor area per annum)

cent of the total) then comparison with standard yardsticks could be misleading. Where possible, their area and energy consumption should therefore both be separately identified and eliminated in the calculation of the energy index. It follows that the high-energy area's should also be separately metered, and subject to separate performance assessments although at present this is rare. As a general rule, sub-metering of buildings, special areas or individual items of plant may be justifiable if the item consumes more than £1000 worth of gas or electricity or £5000 worth of hot water or steam per year. The difference occurs because metering steam and hot water is relatively expensive and the equipment tends to require more maintenance.

3.3 Reviewing patterns of energy-consumption behaviour

3.3.1 Introduction

Without even visiting a building, studying the patterns of energy consumption with time and season can highlight issues which need investigating. As dated records of actual consumption for the periods concerned, monthly gas and electricity bills are particularly useful, and simple month-by-month graphs and cumulative annual totals can be revealing. Quarterly bills for small buildings are too infrequent, and the suppliers now tend to estimate alternate ones, in any event. For stored fuels, such as coal and oil, consumption may bear little relationship to deliveries and useful information may be difficult to obtain unless there are careful stock-control procedures. Sometimes maintenance staff and caretakers keep diligent records of meter and sub-meter readings and stock levels, which can give a wealth of useful information. Frequently these data have been collected but never reviewed, and reveal years of undetected waste. In principle, any data that are collected should also be reviewed routinely.

Differing meter reading dates from month to month can substantially distort individual monthly figures. Where possible, one should therefore obtain the actual dates from the bills or records and calculate the average energy consumption per true month (or per working day) for the period concerned. The data are most easily handled on a computer: special-purpose software is available, but many tasks can be quite easily done using standard spreadsheet programs.

3.3.2 Heating performance – degree-day plots

Degree-days measure the coldness of a period of time (usually a month or a year) in relation to a certain base, usually 15.5°C (see EEO[7]). Annual averages vary with location (see Figure 3.5) and also with height, microclimate and so on; a typical UK annual is 2462 degree-days (see EEO[4]). Individual years tend to be within ±10 per cent of the average figure, while individual months can vary by 25 per cent or more. Monthly degree-day figures are published for standard regions in the UK, whose boundaries are also shown in Figure 3.5. More information, often now on disc, may be purchased from the Meteorological Office and others.

For heating fuel, a plot of monthly fuel consumption against degree-days is usually a straight line, known as the *performance line*, with three main characteristics, as shown in Figures 3.6 and 3.7. Further information can be found in Levermore and Wong[8] and Harris.[9]

1. *The gradient*, which measures the *weather-related demand*, the amount by which energy consumption rises for each degree-day. Sometimes this flattens off at high degree-days as either people close windows more tightly or the heating runs out of capacity and cannot meet demand. Figure 3.7, however, shows an example with not only a large scatter but also an uplift in cold weather which was traced to a control fault. It is not unusual to find sophisticated controls doing the wrong thing and abnormal energy consumption can be a powerful indicator.
2. *The intercept*, where the line cuts the energy axis at zero degree days. This represents the *weather-unrelated demand* (such as standing losses, domestic hot water, humidity control and other items such as gas cooking if metered with the heating). A high intercept usually indicates an inefficient system, although for certain buildings (such as hospitals) which are kept warmer than usual it may also indicate that the degree-day base temperature chosen was too low. Conversely, negative intercepts may occur in well-insulated buildings where too high a degree-day base has been used. A relatively flat slope in relation to the height of the intercept indicates poor control.
3. *The scatter* of points about the line (see Figure 3.7), indicating the *consistency* of heating control in relation to the weather. In most buildings a high scatter indicates scope for improvement in control, although sometimes it may have another direct influence – for example, changing product volume in a factory. CIBSE,[10] appendix A2, discusses how

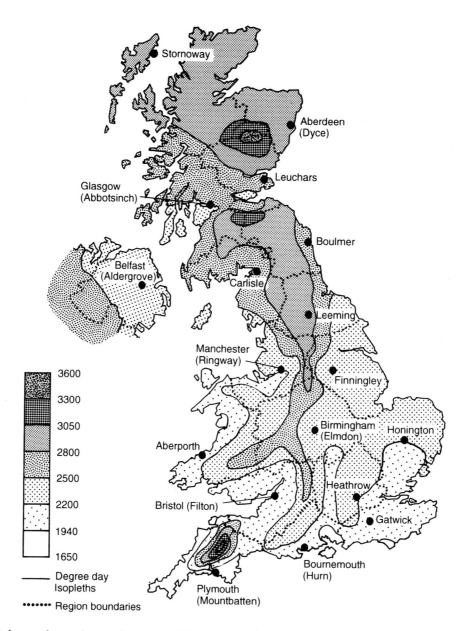

Figure 3.5 UK degree-day regions and contours. (Source: *Fuel Efficiency. 7. Degree Days*, Crown Copyright © 1989)

influences of multiple variables can be determined and allowed for, if appropriate.

3.3.3 Electrical consumption – monthly plots

Unless the building is electrically heated, monthly electricity use is only weakly related to degree-days and histograms of energy consumed per month may be easier to interpret. In most UK buildings, electricity consumption tends to be lower in summer than in winter, due largely to less artificial lighting, and with heating pumps, burner fans, etc. also off or at least less heavily used. Conversely, many air-conditioned buildings show peaks in summer due to the additional cooling load; if control systems are poor, high consumption and sometimes peaks may occasionally occur in spring and autumn when heating and cooling systems are fighting each other hardest to maintain control.

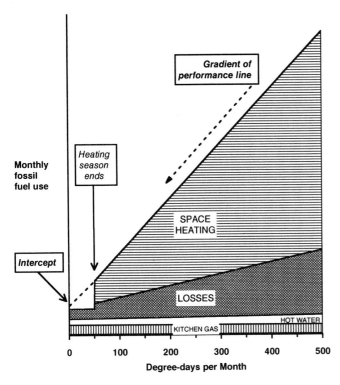

Figure 3.6 An idealized degree-day plot with its components

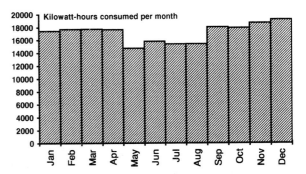

Figure 3.8 Monthly electricity consumption of a low-energy office building. Consumption is not much lower in summer than in winter because office equipment, extract fans, etc. still run, and most of the lights away from the perimeter remain on

3.3.4 Moving totals

When consumption is erratic or meter reading dates are not known, three-month moving totals (for example, January to March, February to April, etc.) can help smooth out irregularities in the data. Twelve month moving totals make underlying trends in consumption more apparent (see Figure 3.9). Growth in the use of information technology and electrical appliances over recent years has tended to create a rising trend in electrical consumption and a falling trend in heating: if both are rising something may be wrong.

3.3.5 More detailed fuel consumption monitoring

Frequent meter readings can give more information on load patterns. Sub-meters will give even more but on multi-building sites one often finds buildings without a single meter! For quick checks, people can be asked to read the meters frequently (say, every hour or each morning and evening). The date and time of the reading must also be recorded as people will seldom maintain a completely regular visit schedule.

For more details, electronic monitoring can be installed, and the information plotted, printed out or transferred direct to a computer for analysis. For temporary monitoring, a range of portable equipment is available and can be hired. Where an electronic Building Management System (BMS) is fitted it can monitor meters directly, normally by counting pulses, say for each kWh. However, counters can be corrupted and many users prefer systems where the meter transmits its actual reading to the BMS – this is particularly important if consumption is charged out to others, who will wish to audit their readings by inspecting

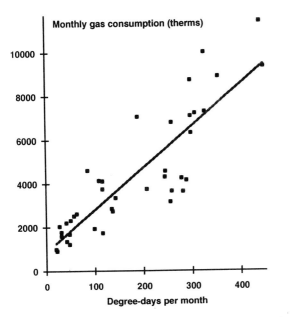

Figure 3.7 A degree-day plot for an air-conditioned office, revealing energy wastage. The intercept is fair, but well above zero. The scatter is not only high but the points are nearly all below the line in the mid-portion (200–300 degree days) and above it in colder months. Here the cause was a wrongly calibrated control unit for the main plant. The device had been intended to save energy by admitting extra outside air for 'free' cooling in mild weather only. It worked best in the mid-range but in colder weather the dampers opened wide, admitting excessive amounts of air and raising heating fuel consumption significantly

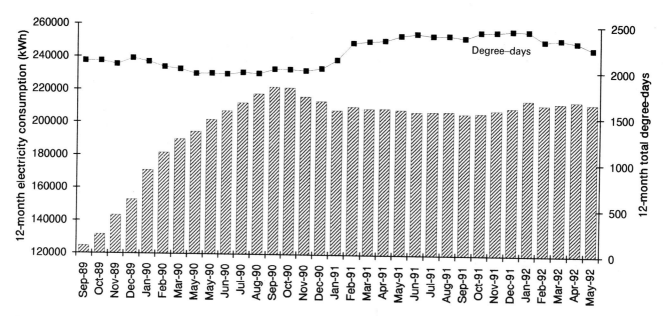

Figure 3.9 Annual moving total energy consumption and degree-days for an all-electric office extension. Up to September 1989 12-month energy consumption (vertical bars and left-hand scale) in the original building was stable at around 120 000 kWh per year. The extension was handed over in October 1989. Until October 1990 12-month consumption rose as the effect of the new building accumulated; it then fell back because the extension required less heat than at first when it was drying out and not yet properly commissioned. Through 1991 consumption was fairly stable, in spite of an overall rise in degree-days (the line at the top and the right-hand scale), indicating effective control and management and providing a good baseline for setting targets. In winter 1991–1992 energy consumption began to rise even though degree-days fell. Investigation revealed that the ventilation system's temperature setting had been turned up in an unusually cold snap and was not subsequently reset

their local meters. Typically, recordings are made every half-hour, the measurement period used by the electricity authorities. A few days' consumption, ideally including a weekend, gives a wealth of information (see Figure 3.10, where hourly intervals are shown for clarity). Often one finds unexpectedly high consumption at night and at weekends, with electrical equipment left on unnecessarily and heating and ventilation systems operating due to mis-set controls such as overridden time programmes and faulty frost-protection thermostats. Sharp peaks in daytime consumption can also occur, leading to expensive maximum demand charge penalties in winter.

3.4 Identifying the scope for savings

3.4.1 Energy audits and surveys

A well-known way of identifying energy flows and savings is to undertake or commission an energy audit or energy survey. A full survey tends to cost typically 3–5 per cent of the annual energy bill (less on large sites, more for small buildings) and, on average, recommends measures saving some 20 per cent, with an average simple payback period (SPP = cost of measure/annual energy cost saving) often of around 2 years or less. Survey techniques are outlined in Field,[11] with fuller guidance in CIBSE,[12] which also defines:

1. An *energy audit* as establishing the quantity and cost of each form of energy input and its breakdown between end users or users, and
2. An *energy survey* as a site investigation of all or selected parts of the flow, control and management of energy to identify and evaluate opportunities for savings.

In practice, the terms are often used interchangeably. In commissioning work, it is therefore very important to make clear what you want and to make sure what you will be getting: the references above give guidance on establishing the brief.

Seldom will there be sufficient information to set up a full auditing system without also doing at least some survey work. However, once an energy accounting system has been set up, auditing can and should continue regularly and can form part of the Monitoring and Targeting process discussed in Section 3.7.

Full energy surveys will only be required occasionally: they can be an invaluable way of taking stock and giving management a strategic understanding of energy utilization, the scope for savings and how they would best be achieved.

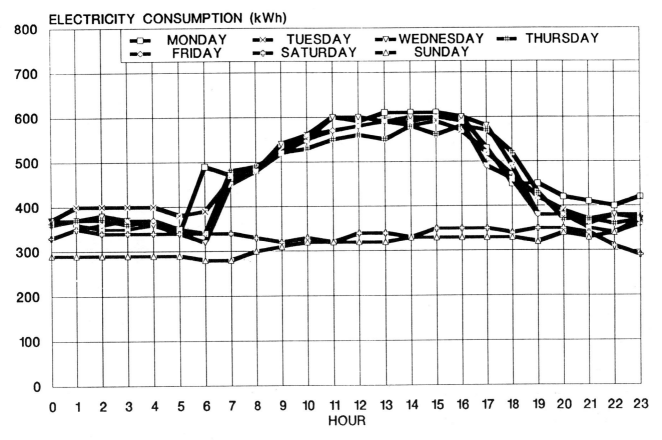

Figure 3.10 Typical hourly electrical demand profile one summer week in an air-conditioned office building. The computer room gives a high baseload at nights and weekends. Separate monitoring of the computer room supply would reveal what else is running at the same time: some things are clearly on in the small hours on Saturday but not on Sunday. In working hours the consumption rises, first as people arrive and switch things on, then as the cooling load increases. There is no sharp mid-day peak because there is no restaurant in this building; neither is there much of a trough, suggesting that lights and office equipment stay on over lunch. The peak at 6 a.m. on Monday occurs because the air-conditioning comes on early to pre-cool the building after the weekend off

Shorter surveys can be useful prior to major refurbishment, alteration or maintenance work, to ensure that energy-related opportunities are grasped: a guide for surveyors is available from the RICS.[13] Subsequently, detailed surveys may be undertaken as necessary into particular items (for example, catering, boiler plant, lighting in Block B) to establish feasibility of new initiatives or projects, and a re-survey or at least a review every 5–10 years may also be worth while. Government grants may be available to assist with the costs of energy surveys, implementing their recommendations, and setting up Monitoring and Targeting schemes.

3.4.2 Key issues

A survey should develop an understanding of energy conversion and consumption, produce a breakdown of energy used for the principal purposes in each building, to meet operational and comfort requirements, and identify items needing attention. Typical problems identified and issues raised are summarized in Table 3.2. The most important energy conversion item tends to be the boiler plant, whose efficiency should be considered: not only thermal (from fuel to heat) but also operational (are the boilers on too much?). Often one finds that all the boilers in a plant room are keeping themselves warm when there is little load to meet, and similar problems occur with refrigeration and compressed-air plant.

For all building services, one should consider what service is expected (temperature, ventilation rate, light level), whether it is being provided, if it is really needed and whether the actual requirements can be met more efficiently. However, it is also important to remember that the most influential surveys tend to be practically-orientated ones which establish a clear strategic understanding, capture the attention of management, provide a framework which encourages decision making, identify viable cost-effective projects and suggest a clear action plan.

Table 3.2 Measures commonly identified in energy surveys

	Typical payback period
Management	
Review use of building	Variable
Review and alter control settings	Instant
Improved staff and maintenance training	Rapid
Publicity to raise occupants' energy awareness	Rapid but perishable
Review fuel purchasing and tariffs	Rapid
Monitoring and targeting systems	3 months or less
Building fabric	
Restore missing insulation	Rapid
Simple draughtproofing	1 year
Better loft insulation	2–3 years
Injected or blown cavity-wall insulation (where appropriate)	3 years
Upgrade flat roof insulation when replacing roof covering	5 to 10 years
Replace unwanted glazing with insulated panels	5 to 10 years
Double-glazing (when windows are being replaced)	10 years
Heating systems	
Maintain boilers and burners	Rapid
Replace burners	1 to 3 years
Improve central time and temperature controls	Up to 3 years
Insulate pipework which does not provide useful heating	1 to 5 years
Separate zones for large parts of building with shorter use	1 to 5 years
Separate heating for small areas with extended usage	1 to 5 years
Replace boilers	2 to 5 years
Major controls replacement	2 to 5 years
Hot water systems	
Insulate poorly insulated cylinders	1 year or less
Include simple controls where necesary	1 year
Point-of-use heating for remote users	1 to 3 years
Electric heating in summer where central system is wasteful	1 to 2 years
Separate water heaters where system is generally wasteful	3 years
Ventilation and air conditioning	
Avoid overventilation	Rapid
Avoid extended use of entire system	Rapid
Reduce conflict between heating and cooling	Rapid
Shorten operating hours and seasons for chillers and pumps	Rapid
Separate zones for large parts of building with shorter use	2 to 5 years
Separate cooling for small areas with extended usage	2 to 5 years
Major controls replacement	2 to 5 years
Variable-speed fans and pumps	3 to 5 years
Lighting	
Fit higher-efficiency direct-replacement lamps next maintenance	Rapid
Label light switches and modify use of lights	Rapid
High-efficiency external and floodlighting	1 to 4 years
Automatic control of external lights which tend to be left on	1 to 2 years
Replace tungsten lights with mini-fluorescent fittings	2 to 3 years
New controls (including automatic) for internal lighting	2 to 5 years

3.4.3 Avoidable waste

In many buildings avoidable waste (for example, liberal running hours, poor operation and maintenance practices, elderly and inefficient lighting) occurs but nothing is done about it. While all these items will be picked up in a good energy survey, many can be identified and tackled by asking simple questions such as:

1. Why do we use only four times as much heating fuel in January as in July, when the heating is supposed to be off? Probably poorly controlled massive boilers are cycling inefficiently to heat the main circulation system in order to produce small amounts of hot tap water which would be best heated independently.
2. What is the insulation like in that roof? It may be missing, or have been cast aside by the pest-control people 5 years ago.
3. Why are the lights on all the time? Perhaps the switches are in the wrong place.
4. Why do we still have tungsten light bulbs in corridors and toilets?
5. What is the current time program for the heating?
6. Why does the air conditioning always seem to be on when I come in at weekends?
7. Why does no-one even mend the window catches?

Often such avoidable waste occurs and persists because it is no-one's problem. If management assigns responsibility for it, then things can happen. For instance, maintenance staff often merely maintain the *status quo*, but do not see it as their task to improve it. Fitters may have visited the roof space to repair a leak but did not see it as their concern to report or to put right the missing insulation: for a certain type of lamp no-one thought to put the better equivalent in the parts bin, or the spares were obtained at the lowest price without any regard to subsequent energy costs. Contractors are frequently appointed to maintain the plant but not to check control settings regularly and report on unusual changes. Staff or cleaners may leave the lights on in the knowledge that ultimately the security guard or automatic controls will turn them off. Often creative management and small changes to existing work descriptions will do the trick.

3.5 Selecting appropriate projects

3.5.1 Classes of project

Suitable projects tend to fall into three groups:

1. *'No capital cost'* involving better operation and management, fuel purchasing and tariff reviews,

Monitoring and Targeting (which may also require investment in consultancy, software, metering, etc.), better maintenance, upgraded maintenance replacement (for example, more efficient lamps) publicity, training, etc. These can often produce rapid savings of 5–10 per cent. However, their effects can wear off equally fast if the management impetus is not sustained, and the costs of maintaining awareness can mount unless good behaviour can be made automatic and habitual. Properly handled, no-cost measures can establish initial credibility for an energy-saving programme, and some organizations may permit the cash savings on the utilities budget to be devoted to further energy-saving measures.

2. 'Low cost', typically paying for themselves in 3 years or less, generally consisting of better time and temperature controls, catching up with maintenance replacement backlogs, simple draughtproofing, injected cavity-wall insulation (where appropriate), replacement or upgrading of damaged or missing pipe and loft insulation, and wasteful situations in which building services are grossly inefficient for the task in hand (for example, with the whole building left on regularly so that one area can be used out of hours). Together, these can typically save around 15 per cent of heating energy and 5 per cent of electricity. Usually they are undertaken as individual projects in their own right.

3. 'Higher cost', typically paying for themselves in 3 years or more, and consisting typically of major disruptive alterations to systems, controls and buildings. When these are undertaken, they are often not exclusively financed from energy savings, but are combined with something else, such as inevitable major maintenance. For example, the boilers may be due for a major overhaul but with an additional energy-related contribution it is decided instead to upgrade, replace or supplant them, or obsolescent lighting and controls are replaced by something more efficient.

Radical building fabric measures – for instance, double glazing, external wall insulation (other than injected cavity fill) and roof insulation (other than loft insulation) – tend to be expensive and have long payback periods. However, when there is an opportunity to upgrade them (for example, in reroofing or major refurbishment) it should normally be seized.

3.5.2 Assessing feasibility

Programmes are always resource-limited, so one needs to prioritize options. Low-cost projects are usually sorted in terms of simple payback period (SPP), while for high-cost projects discounted cash flow (DCF) techniques are often used (see CIBSE[12]). Other criteria, often neglected, include:

1. *The life of the measures proposed.* Some things may wear out quite soon after payback has been reached. DCF usually takes account of this, SPP does not.
2. *Knock-on energy cost implications*, particularly where equipment to save heat uses electricity, which is considerably more expensive. Such 'parasitic' energy consumption – such as extra fan power or running hours – often appears trivial but can swiftly eat into savings. On the other hand, a local project may generate knock-on savings elsewhere. For example, double-glazing a few rooms which tend to be cold – although expensive for the rooms concerned – may allow the heating system for the entire building to start up later and run at a lower temperature.
3. *Management and maintenance implications.* These are often ignored, but when properly costed some energy-saving projects cost more to look after than the energy they save. Conversely, other measures save both energy and maintenance time – for example, low-energy lamps which also last longer and automatic control and monitoring which may save on manpower, or allow staff to be re-assigned to more productive tasks.
4. *Robustness of energy-saving estimates.* Savings are usually quoted as single numbers but are in fact subject to uncertainties which are much greater for some measures than for others. For example, by replacing a lamp with a more efficient one having similar operational characteristics, one can 'lock-in' the energy savings with little uncertainty. If, however, the old lamp came on instantly while the new one takes several minutes to warm up, people may tend to leave the new ones on, increasing the running hours and eroding the energy savings.
5. *'Nuisance' technologies.* The sole purpose of some devices – for instance, many heat-recovery or 'free cooling' systems – is to save energy. If they do not work, the service still continues as before, but with a higher energy bill, sometimes higher than if the system was not present due to residual 'parasitic' losses, or unintended behaviour as seen in Figure 3.7. Systems of this kind are very likely to fail unless standards of on-site management are high and alarms need to be incorporated to warn if systems are not operating as intended.
6. *Convenience.* Many different small projects can be very difficult to get right and to organize. A

single project or standard package of measures may be easier to undertake and have a greater chance of giving a reliable result, even if its SPP or DCF does not seem to be quite so good.

7. *Disruption*. All things being equal, a project that can be implemented behind the scenes while maintaining continuity of supplies will be preferable to one in which access, noise, service interruptions, etc. disturb people.

8. *Opportunity points*. Other work – for instance, maintenance, refurbishment or alteration – allows energy-saving measures, even items with relatively long payback periods, to be carried out at the same time. There are usually four complementary reasons for this: the projects are often cheaper as part of ongoing work than by themselves, they are less disruptive, are administratively easier, and are sometimes the only chance there is. Unfortunately, such opportunities are often missed, frequently because the two types of project are handled by different people or paid for from different pockets.

9. *Too good to be true*. While most buildings contain many excellent opportunities for reducing energy consumption, excessive claims are sometimes made, particularly for imaginative pieces of technology, and occasionally energy-saving measures may reduce the life or the reliability of an item of plant. If in doubt, get a second opinion.

Sometimes one finds that major savings can be made by undertaking only part of a measure. For instance, in EEO[14] heating energy consumption was halved without replacing the boilers as originally proposed. Instead – and among other things – a new high-efficiency boiler was added to operate in the region of performance where the original boilers were at their least efficient.

3.5.3 The best combination of measures

Selecting the right combination of measures is thus rather more than picking the bargains off a shopping list. It is important to take a strategic view, or actions taken in the wrong order may become obstacles to longer-term improvement. Difficulties can arise if too much money is spent on upgrading inherently inefficient systems which really should have been replaced: this either delays replacement or means that new investment is scrapped prematurely. A similar problem is mutually undercutting projects: for example, boilers are replaced one year and next year the insulation is improved, making part of the new capacity redundant.

3.6 Taking action

3.6.1 Who decides?

Often measures which are cost-effective and well proven do not happen because their benefits are not properly appreciated by the real decision makers. Organizations need to establish mechanisms by which energy-saving projects receive the same general management attention as other investments, as often they are more cost-effective. EEO[15] gives some guidance on this.

3.6.2 What are the implications?

It is tempting to see energy-efficiency measures as technological and 'fit and forget': once the investment decision has been made, the result will work fine: this fantasy must explain why some buildings claimed to be energy-efficient do not even have meters. However, most things are actually 'fit and manage the consequences'. The management requirements vary:

1. Permanent insulation may need little or none, provided it is properly specified, installed and detailed.
2. Energy-efficient conventional items (such as lamps, pumps and boilers) may require very similar maintenance to the traditional item, though different procedures, contractors and spare parts may be required.
3. Extra items will require additional knowledge, training and attention: can this be provided or will they become neglected nuisances? Are there appropriate facilities and space for maintenance? Are there automatic means of alerting management to deteriorating performance?
4. Central control systems need to be checked and adjusted from time to time: are they within the capacity of the occupier and their on-site staff? If control systems are too complicated for the expertise available on-site they may act as obstacles rather than aids to effective energy management, particularly where the managers have no building services background.
5. Local controls need to be tailored to the needs of occupants and users. If not, the systems they control may run wastefully as it will be easier to leave them on.

3.6.3 Are the solutions appropriate?

Often one finds that a relatively small number of measures save most of the energy. The rest may in theory make things even better, but if they require attention they may also add complication which can easily prove unmanageable, at least within the resources available. For maintenance and management-intensive items, 'keep it simple' and 'if in doubt, leave it out' are good principles to adopt, provided that the simplicity is a result of clear thinking, not oversimplification, and the items left in are done well. If sophisticated features are advocated, then occupiers must understand and accept their consequences – some will welcome them, many will not. Aim to use no more technology than the minimum necessary to meet the objectives effectively and efficiently. Include meters to allow performance to be monitored and sustained.

3.6.4 Take the staff along with you

In occupied buildings with on-site maintenance staff, energy-saving measures designed by consultants or contractors or at head office can appear to have been imposed from outside. Where possible, site staff must be kept informed and involved in the decision making and in the installation, or certainly alongside it, so that they develop an intimate knowledge of what is happening. Their contributions are sure to improve the outcome and to save money by spotting potential problems, and the outcome is more likely to be greeted with pride, not hostility. Where what is being done affects occupants directly (for example, new lighting, better local control and switching or changes to the windows) preliminary pilot installations can be very rewarding.

3.6.5 Ensure an effective handover

In an occupied building, ensuring that new systems and control work satisfactorily is not a task that can be left entirely to third parties. The occupant needs to devote sufficient time to the systems well before handover, and to be involved in (and probably to specify and agree) all commissioning work and acceptance tests.

3.7 Maintaining and enhancing the savings achieved

3.7.1 The importance of management

For building managers, handover is only the beginning. Energy efficiency in operation must first be achieved and then sustained. This is seldom (if ever) a passive exercise: some unforeseen problems are likely to arise initially and will need to be diagnosed and corrected, and

subsequently good levels of service will need to be sustained with a minimum of energy consumption. Effective maintenance must be organized and control systems used to their best effect.

There is a (somewhat outdated) view that low energy consumption is only achieved at the expense of occupants. While this can occur under dogmatic 'switch off' regimes, recent survey evidence (for example, Bromley et al.)[16] suggests that buildings with proven low-energy performance (and these are often relatively simple straightforward ones) tend to have above-average levels of health, comfort and productivity. The link is not a direct one but via good management: of building procurement, building stock, general and facilities management, staff, and energy efficiency. The right management culture, if supported by effective and responsive local control facilities, can also encourage occupants to develop energy-saving habits.

3.7.2 Monitoring and targeting

Energy management cannot only operate in hindsight: one must be in command of the situation. This means understanding the building's energy performance, making forward projections, and reviewing actual consumption against them, using techniques such as those described in Sections 3.3 and 3.4, together with more powerful techniques of waste detection such as the CUSUM method.[9] The sophistication and frequency of the system will depend on the size and complexity of the building or estate. For many purposes, a monthly review when the bills come in will be sufficient, but for major sites continuous computer-based monitoring and review may be justified. Effective understanding and management of energy consumption not only saves energy but also provides a firm basis for negotiation with fuel suppliers to obtain the most competitive prices and tariff structures.

3.7.3 Maintaining the momentum

Seldom (if ever), even in the best and best-managed buildings, do no cost-effective opportunities for saving energy remain. Opportunities and standards should therefore be kept under constant review and maintenance operations should be organized so that, where possible, they do not merely maintain the *status quo* but improve it.

3.8. Conclusions

This chapter has been able to outline only some of the essentials of energy utilization, audits and management and has exposed the intimate links between design, technology, maintenance, control, management and occupants. The general message is that low energy consumption is an essential indicator of good design and effective management, and that any technical measures which are not genuinely 'fit and forget' must be considered in their management context and indeed designed for manageability.

Editor's note

The following extract from Scorpio's column in *Building Design* (28 September 1979) and reprinted by permission of the Editor is relevant comment on the necessity for a check on energy consumption (see paragraph 3.4.1):

Fuel for thought There may be something in these energy surveys after all. One factory owner recently stirred his stumps and got one of the aerial survey people to take infra red photos of his plant. They are those photos which show buildings in different colours according to how much heat is escaping.

Poring over the prints the owner was puzzled by a very high heat emission from a playing field on his land. After his lads had dug around a bit on the field they discovered that it had been laid over a large concrete slab, the floor of an old wartime factory.

No, it wasn't acting as a heat sink.

The old slab had been heated with embedded electric cables which no one had got round to turning off these 30 or so years.

References

1. Osborn, P. D., *Handbook of Energy Data and Calculations*, Butterworths, London (1985).
2. Chartered Institution of Building Services Engineers (CIBSE), *CIBSE Building Energy Code Part 4: Measurement of Energy Consumption and Comparison with Targets for Existing Buildings and Services*, London (1982).
3. Audit Commission, *Saving Energy in Local Government Buildings* (1985).
4. Energy Efficiency Office (EEO), *Energy Efficiency in Buildings* (various dates 1987–1993). A series of booklets, typically of 20–30 pages each, covering a range of building types and including step-by-step flowcharts for calculating NPIs.
5. Energy Efficiency Office (EEO), *Energy Consumption Guides* (various dates 1990–1993). A series of booklets, typically of 2 to 6 pages each, giving practical annual energy consumption and cost targets for a range of building type and identifying typical ways in which savings can be made.
6. Energy Efficiency Office (EEO), Energy Consumption Guide 19, *Energy Efficiency in Offices: A Guide for Owners and Single Tenants* (1991).
7. Energy Efficiency Office (EEO), Fuel Efficiency Booklet No. 7, *Degree Days* (1989). Other booklets in this series may also be of interest. They tend to have an industrial bias.
8. Levermore, G. J. and Wong, W. B., Performance lines and energy signatures, *Building Services Engineering Research and Technology*, **10**, 105–114 (1989). A further paper by GJL on this subject appeared in the same journal in 1993.
9. Harris, P. J., *Energy Monitoring and Target Setting Using CUSUM*, Cheriton Technology Publications, Cambridge (1989).
10. Chartered Institution of Building Services Engineers (CIBSE), *CIBSE Applications Manual AM6, Contract Energy Management*, London (1991).
11. Field, A. J., Energy audits and surveys, BRE Information Paper 12/92, Building Research Establishment, Watford (1992).
12. Chartered Institution of Building Engineers (CIBSE), *CIBSE Applications Manual AM 5, Energy Audits and Surveys*, London (1991).
13. Royal Institution of Chartered Surveyors (RICS), *Energy Appraisal of Existing Buildings: A Handbook for Surveyors* (1993).
14. Energy Efficiency Office (EEO), Best Practice Programme, *Case Study No. 16: Heslington Hall* (1991).
15. Energy Efficiency Office (EEO), General Information Report 13, *Reviewing Energy Management* (1993).
16. Bromley, K., Bordass, W. T. and Leaman, A., User and Occupant Controls in Buildings, *Building Services Journal*, April (1993).
17. Chartered Institution of Building Services Engineers (CIBSE), *CIBSE Building Energy Code Part 3: Guidance towards Energy Conserving Operation of Buildings and Services*, London (1979).

Note: References published by the Energy Efficiency Office are normally available free of charge. For these and further information on energy efficiency in buildings, apply to the Enquiries Bureau, BRECSU, Building Research Establishment, Garston, Watford WD2 7JR.

4
Thermal standards, methods and problems
Norman Sheppard and Sylvester Bone

4.1 Introduction

4.1.1 This chapter is concerned with insulation, shading and the control of ventilation. It concentrates on the thermal performance of the building fabric rather than on the services installations that use energy for heating, cooling, lighting, transport, cooking or other processes. Chapter 3 deals with the energy utilization, audits and management of the services installations. The service installations and the building fabric should always be considered together, they must in fact be balanced, if the internal environment of the building is to be what the occupants require – and can afford.

4.1.2 Poor thermal performance of the building fabric not only wastes resources but can also severely restrict the use of the building and even cause extensive damage. A building that is too cold, too hot or too 'stuffy' is not fully usable and one that suffers from condensation may need major repair and modification before it can be used at all (see Figure 4.1). Thermal performance of the fabric can deteriorate if a building is poorly maintained – for example, if insulation is displaced or becomes saturated from a leaking roof or pipe. However, a more common reason for the need to improve the thermal performance is that the use of the building is changed. This may be a radical change such as from an unheated warehouse to an office or it may be a simple relocation of workspaces that brings them closer to unshaded windows. It may even be a change of timetable or a more intensive use of an existing facility – for example, if a naturally ventilated art gallery has an unusually popular exhibition in winter when crowds in damp clothing can raise the humidity to a level that threatens the exhibits. When either the use of a building or the uses of space within it are changed the need for a modification of the fabric may have to be considered, perhaps by such

simple measures as making some windows openable or by draughtproofing the external doors.

4.1.3 The temperate climate of the UK can produce shade air temperatures between $-24°C$ (Scottish highlands) and $37°C$ (London). Normal temperatures are in a narrower range. Materials exposed to the midday sun or to clear night skies reach much higher and lower temperatures than the shade air temperature. Heating systems in most parts of the UK are designed for external temperatures of $-3°C$, air-conditioning and cooling systems for an external temperature of $28°C$. In tropical climates and in climates with long cold winters different levels of thermal performance are required for buildings. The control of ventilation and the sealing of the building envelope becomes increasingly important in colder climates or when mechanical cooling is used in hot ones. Shading is essential for windows exposed to tropical sunlight and is often needed to obtain good working conditions in temperate climates.

4.2 Standards

4.2.1 Existing buildings, built before 1970, make up the bulk of the UK building stock and the rate of replacement is hardly enough to change this, particularly as conservation policies retain many of the older buildings indefinitely. To make major savings in the 50 per cent of the national energy use that relates to building attention will need to be focused on buildings that are uninsulated and were not designed to use energy efficiently. The efficiency of these buildings can be improved by installing more efficient lamps and heating equipment but to achieve the maximum benefit from such measures, modification of the building fabric is also needed: to increase insulation, modify

Figure 4.1 Inadequate insulation contributes to condensation and mould growth in a bathroom

thermal capacity, control ventilation and optimize solar heat gain.

4.2.2 In the UK, since 1972, the Building Regulations have set the minimum standards that should be achieved in different building types. Initially, the Regulations only included a comprehensive set of requirements for thermal insulation and heat loss through windows in new dwellings. Standards for new buildings other than housing were introduced in the 1976 Building Regulations and subsequent amendments to the Regulations have progressively increased the insulation required. The 1990 revision of Part 'L' of the Regulations for England and Wales did not set standards for existing buildings (although extensions over 10 m² were covered). The revision of the regulations proposed for 1994 does include requirements for insulation to be added when there is a change of use of a building or an alteration that provides an opportunity to improve insulation standards. Part J of the Scottish Building Standards has, however, applied to

changes of use of existing buildings for some years. For example, when a house is converted to flats the Scottish Building Standards may require additional thermal insulation. Figure 4.2 summarizes the requirements for thermal insulation of the building fabric in the Building Regulations. The revision of the Regulations in 1994 that is expected to extend the application of the England and Wales Regulations to certain categories of work in existing buildings is also noted in Figure 4.2 but readers should consult the final version of the regulations when they are published.

4.2.3 The minimum temperature in which employees are required to work (for more than short periods) is 16°C (Workplace (Health, Safety and Welfare) Regulations 1992). If the work involves severe physical effort, 13°C is accepted. The 1992 Regulations also give a minimum fresh air supply rate per person of 5 to 8 litres per second, which has a bearing on energy use. There is also guidance, published by government

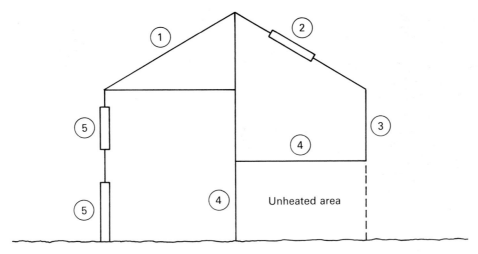

Maxima for new buildings

Key No.	Dwellings	Residential, Offices Shops and Assembly	Industrial, Storage and Other Buildings
①	0.25 W/m²K (0.35 with double-glazing)	0.45 W/m²K	0.45 W/m²K
②	Rooflights counted in with windows	20% of roof area	20% of roof area
③	0.45 W/m²K (0.6 with double-glazing)	0.45 W/m²K	0.45 W/m²K
④	0.6 W/m²K	0.6 W/m²K	–
⑤	15% of floor area (more if multiple-glazed)	25% of wall area for residential 35% for others	15% of wall area

EXTENSIONS less than 10 m² are exempt
EXISTING BUILDINGS are exempt in England and Wales (but this may be changed)

PROPOSALS FOR REVISION OF REGULATIONS (January 1993 draft)
Double glazing to be normal standard. Applies to both windows and doors
New calculation methods introduced to take account of mortar joints and timber studs and increased requirements for windows and door surrounds
Solar gains and more efficient plant can be used to justify lower insulation
Air leakage should be controlled by reducing gaps in the fabric
Non-domestic building guidance given on air conditioning and lighting
Material alterations and conversions should now involve extra insulation
Home energy rating now required for all new dwellings (scale of 1–100)

Figure 4.2 Summary of UK Building Regulation guidance on thermal insulation of the building fabric

departments, on maximum and minimum temperatures for different activities. The courts have ruled in a number of cases brought by Local Authority tenants complaining of inadequate heating to their dwellings. In 1964 it was recommended that the heating and insulation of a dwelling should be to enable a temperature of 18°C to be maintained in the living rooms and 13°C in circulation spaces when the outside temperature is −1°C. However, failure to meet this standard has still not been accepted as evidence that a house is unfit for habitation.

4.2.4 In 1978 grants for loft insulation, pipe lagging and cylinder insulation were introduced for dwellings. The grant covered a proportion of the cost of insulation in domestic roofs. However, due to financial restrictions, these grants are now only available from certain Local Authorities and then only in very specific circumstances – for example, for elderly tenants or where without insulation the dwelling is considered to be unfit because the heating is inadequate.

4.2.5 One part of the government's programme of research into energy efficiency has been the development of 'energy targets' for different building types. Energy targets for offices and public houses have been published by the Department of Energy (now part of the Department of the Environment) and more building types will be covered in due course.

Targets for other building types such as schools and hospitals are published by other Departments (see Figure 4.3). Targets for existing buildings are designed to be achievable by balancing energy-efficient service installations with proved methods of increasing the energy efficiency of the building fabric, such as weatherstripping and loft insulation. Unlike the application of the Building Regulations, the application of energy targets is still entirely voluntary but it is progressively being linked to environmental auditing schemes, such as the Building Research Establishment's BREAM scheme that awards more points to buildings that use less energy as well as points for other environmentally friendly features. At present these schemes are mainly for promotional purposes – to assure prospective buyers and tenants that the building is environmentally friendly and meets certain standards of energy efficiency. Future extension of these schemes may be linked to European legislation or voluntary policies and may possibly involve tax or other incentives to

encourage their adoption. The proposed amendment to the England and Wales building regulations includes a requirement that all new dwellings should have an energy rating.

4.2.6 Solar gain through windows exposed to direct sunlight can provide a valuable source of heat. Double-glazing and low-emissivity glass can increase the efficiency of windows as solar heat collectors (see Figure 4.4). In summer the heat collected through unshaded windows may need to be dispersed by additional natural ventilation or mechanical cooling. Orientation affects the solar radiation of heat falling on a vertical surface in different latitudes (see Figure 4.5 for the south coast of England) Even in the UK the energy required to cool a fully glazed air-conditioned building in the hottest weather is likely to be more than that required to heat it in winter and shading should be considered (while remembering that planning permission will be needed for alterations to elevations of most buildings).

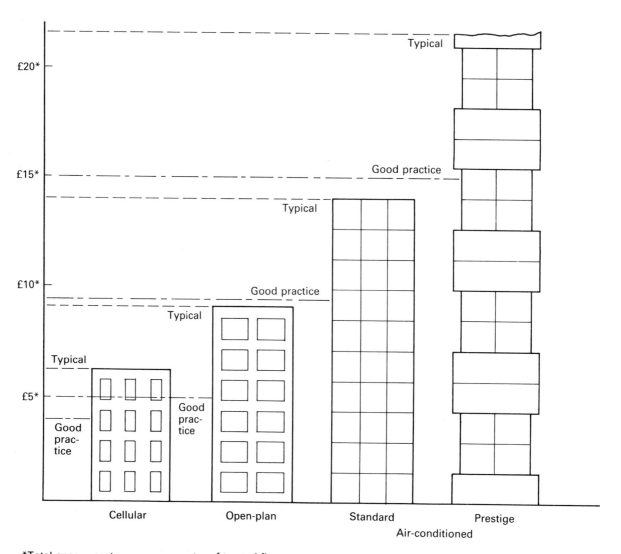

*Total energy costs per square metre of treated floor area per annum (1991)

Figure 4.3 Target heat loss figures for offices

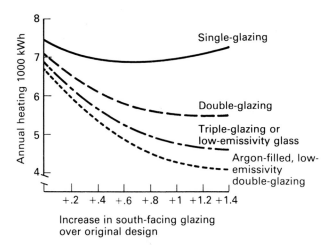

Figure 4.4 Savings through using different types of glazing (based on *Renewable Energy 3*, Department of the Environment (1988))

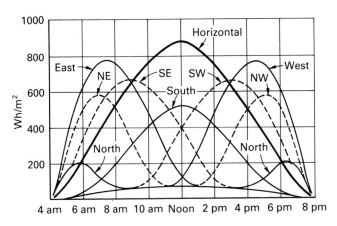

Figure 4.5 Solar heat falling on differently orientated building surfaces (based on ISO 9492, TR 1987)

4.2.7 In hotter climates cooling is the main user of energy and energy efficiency must concentrate on shading, insulation, ventilation and natural cooling from water and the mass of the ground or the structure. The thermal capacity of the structure is particularly important. If the structure requires a long heat-up period it will keep the interior cool when the temperature peaks outside. Equally, if the structure is well insulated and takes a long time to cool down it may retain high indoor temperatures when it has become cooler outside. Warm climates have been categorized by BRE (Digest 302) under six headings each requiring a different design strategy (see Figure 4.6). Different comfort conditions are accepted in different climates. Figure 4.7 shows the comfort zones aimed at in different countries and the limits within which human beings can function.

4.3 Use of the building

4.3.1 For comfort and energy efficiency the heating, lighting and ventilation installations in the building need to be matched to the uses within it – alternatively, the uses can sometimes be more easily matched to the environmental installations. Spaces that are only used occasionally need not be heated throughout the day and night. On the other hand, it may be possible to schedule occasional activities to share a space that is permanently heated. Where intermittent use is planned the switching and installation controls should be able to reduce temperatures, lighting levels and ventilation whenever the space is unused. For example, if night security guards are patrolling a

INLAND DESERT
Very hot and sunny in the day – cold at night. Meagre rainfall. Local winds may develop into dust storms

ISLAND/TRADE WIND COAST
Sunny – therefore warmer by day. Temperature range only 14°C throughout year. Cooling breezes and strong winds and some hurricanes. High rainfall and intense rainstorms. Much influenced by topography, seasons more marked away from Equator

UPLAND (OVER 1000 m)
Air cooler, especially at night. Warmer and cooler seasons more marked. Larger diurnal temperature range. Rainfall usually over 1 m per annum but varies with topography

MARITIME DESERT
Less extreme than inland desert, warmer and cooler seasons more marked. Humidity high – somewhat seasonal – hazy and uncomfortable when humid

EQUATORIAL LOWLAND
Near lakes or sea. Little seasonal temperature variation. Humid often still, especially at night, rain one third of days in year. Occasional strong winds in rain squalls

TROPICAL INLAND
Two thirds of year hot and dry one third warm and humid. Rain in monsoon. Wind strong and stormy, in monsoon, variable hot and dry otherwise dry. Season hotter by day and cooler by night.

Figure 4.6 Warm climate categories (based on BRE Digest 302)

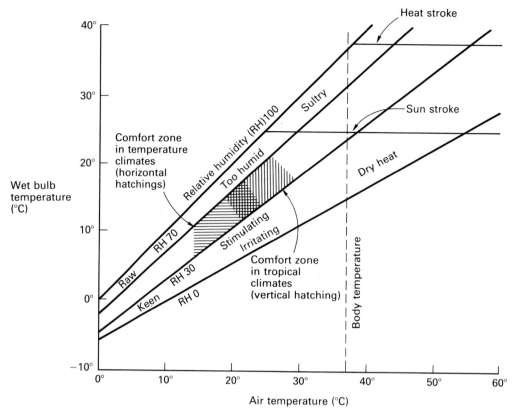

Figure 4.7 Comfort zones in different climates

building all that is required is a reduced level of lighting that can be operated for one section of the building at a time. It should not be necessary to turn on all the lights in the building each time they do their rounds. Similarly, it may be possible to schedule cleaning of a building when it is still warm after having been used rather than turning heating on early for cleaners before use starts.

4.4 Insulation

4.4.1 The extent to which the external envelope of a building should be insulated is often based on a calculation which balances the cost of the insulation against the cost of the fuel that will be saved over a certain number of years by providing that insulation. It is quite possible to go beyond this 'economic' level of insulation – for the sake of the environment it would be better not to burn any fuel at all – but the economic benefits of extra insulation and its importance in relation to controlling ventilation and the areas of windows diminishes rapidly once the economic level is exceeded. Table 4.1 shows the 'payback' periods for different energy-conservation measures affecting the fabric of domestic and non-domestic buildings. It should not be assumed that increasing insulation of one part or element of a building will always produce the same reduction in

Table 4.1 Payback periods for energy-conservation measures related to the building fabric (from evidence to the Select Committee)

Cost effectiveness	Building type	Measure
Highly cost-effective (payback less than 5 years)	Domestic	Draught stripping, loft insulation
Cost-effective in most circumstances	Domestic	Cavity fill, higher levels of insulation than Building Regulations minimum
	Non-domestic	Draught stripping, higher levels of insulation than Building Regulations minimum
Cost-effective in some circumstances	Domestic	Better loft insulation, double-glazing. Ground-floor insulation, solid wall insulation
	Non-domestic	Cavity fill, roof insulation, solid wall insulation

payback period. If 70 per cent of the heat is being lost through large areas of single-glazing doubling the thickness of roof insulation may not be an economic proposition.

4.4.2 Heat losses through roofs and windows are generally more significant than those through the same area of walls or floors. In a single-storey factory where the area of the roof is large in proportion to the wall area as much as 70–80 per cent of the heat loss can be through the roof. It pays to improve insulation where heat losses are greatest but the feasibility and cost of significantly increasing insulation must be taken into account. It may be difficult and costly to reduce the heat loss through windows without affecting their function. Even triple-glazing does not give the same level of insulation as an external wall with insulation in the cavity. Roofs and suspended floors are generally cheaper and easier to insulate than external walls. As a rough guide to priorities for improving the insulation of a building, tackle rooflights, then roofs, then suspended floors, then fixed windows and walls, then opening windows and finally solid floors. (Practical considerations and the relative areas of the different elements will often alter the order of priorities in a particular building.)

4.4.3 Besides retaining heat in cold conditions insulation can exclude unwanted heat when external air temperatures and the temperature of external surfaces is higher than the temperature required within the building. The need for mehanical cooling or additional natural ventilation is reduced in a well-insulated building. Mechanical cooling (usually provided as part of an air-conditioning installation) is not normally essential in the UK and, as it uses additional energy, it should be avoided unless either the surroundings of the building call for a sealed building envelope to exclude noise and pollutants (for example, city-centre sites, airport buildings and clean rooms) or the use of the building requires conditions that cannot be achieved by natural ventilation (computer rooms, large assembly rooms or storage of perishables).

4.4.4 Suitable insulation is one of the three measures that combine to reduce harmful condensation – the others are suitable heating and ventilation. Surface condensation occurs when warm, moist air comes into contact with a cool surface. Interstitial condensation occurs within a material as water vapour from inside a building passing through the material condenses in its outer colder layers. Surface condensation can be reduced by insulating to avoid cold surfaces. Interstitial condensation, which can saturate an insulation material and render it ineffective, can be avoided by controlling the passage of water vapour with impermeable materials and membranes. Figure 4.6 shows typical situations where surface and interstitial condensation can occur and the measures necessary to avoid condensation. Surface condensation can affect all the external wall and ceiling surfaces in a building but it is most common where there is a cold bridge in the external envelope of the building, that is, where the roof, walls, or floor is less well insulated and therefore creates a colder inside surface than elsewhere. Building Regulations recommend that lintels, jambs and sills should have a minimum U-value of $1.2 \text{W}/\text{m}^2\text{K}$ to avoid the condensation associated with cold bridges. Unevenness in insulation can also cause pattern staining without creating the damp surfaces associated with condensation. Where a supporting framework for an internal lining interrupts the continuity of insulation the pattern of the framework shows as surface staining. The colder parts of the lining pick up dirt from the air more easily than the warmer (and dryer) parts that are well insulated. The cure for pattern staining is an insulating lining that provides continuity of insulation across the supporting framework.

4.4.5 There have been instances of insulation catching fire within cavities behind the cladding of buildings. In most cases the fire has remained within the cavity but sometimes it smouldered and spread for some time before it was seen and extinguished. When fitting insulation fire risks should be considered. Use non-combustible types of insulation (such as mineral wool) where the insulation could carry the fire from one part of a building to another. Specifiers should also be aware of the damage that can be done to the performance of insulation by birds, inadequate fixing, wind redistributing loose-fill insulation, leaking pipes, rain penetration and any other source of water that can saturate insulation.

4.5 Air quality

4.5.1 Some air change (and associated heat loss) takes place by infiltration through the external envelope but there is often a need for supplementary ventilation when a building is occupied. For background ventilation trickle ventilators are recommended to avoid draughts. A minimum rate of air change is needed for background ventilation to avoid stuffiness and to disperse odours. This rate is more than double that needed to prevent too great a concentration of CO_2 for breathing comfort. In spaces that are closely packed with people or where people are physically active five or six air changes may be needed but in normal house or office use a rate of $0.5 – 1.5$ air changes per hour is adequate. There is a need for external doors or windows that can be opened for rapid ventilation to clear out damp and polluted air. In dwellings, Building Regulations call for openable windows or doors of at least 1/20 of the floor area to provide rapid ventilation with the top of the highest opening at least 1.75 m above the floor. Mechanical ventilation is required for kitchens. Where bathrooms are mechanically ventilated the fan should be capable of a rate of three air changes per hour. A proposed revision of Part F of the England and Wales building regulations extends the ventilation requirements to other building types and increases the requirement for trickle ventilation in rooms over 6 m². Combustion air is essential for open-flued gas and solid fuel appliances. When essential ventilation is blocked by occupants (or by alterations) dangerous conditions can be produced and there is a likelihood of increased condensation.

4.5.2 Ventilation controls should ideally be able to deliver different rates of air change to suit individual tastes and the requirements of different activities. They must also be arranged to change all the air within a space without leaving 'dead' pockets and without creating draughts (people are more sensitive to the velocity of the air that passes them if it is cool). In sealed, air-conditioned buildings conditions can be better controlled and monitored but it is unusual and expensive to offer individual choice. When occupants can open windows in the UK the sense of personal control makes up for the disadvantages of greater temperature fluctuations and occasional draughts.

4.5.3 In a climate where the night temperature is much lower than that during the day – for example in desert or temperate regions during most of the summer – the costs of cooling can be reduced by allowing the structure to cool at night. Simply by opening windows at night, cooling costs can sometimes be reduced or avoided.

4.6 Maintaining and improving thermal performance

4.5.1 Table 4.2 gives examples of measures that can be taken to maintain and improve the thermal performance of the building fabric. The list is divided between: measures to be taken during the normal running of the building, measures that are related routine maintenance or periodic repair and measures that can normally only be taken as part of a general refurbishment. The list needs to be considered together with the building's heating and ventilating systems. Monitoring the use of energy in the building can show up deficiencies in the fabric and ways in which the building could be used more efficiently from an energy point of view. It can also point to ways in which the comfort and convenience of the occupants can be improved with least expense. Improvements may

Table 4.2 Examples of improvements to thermal performance in existing buildings

Day-to-day operation	• Review use of building to see that rooms that heat up fast are used early in the day and those that stay warm are used late
	• Monitor opening and shutting of windows and doors. See that they are shut after rooms have been used
	• See that shading blinds are used (in empty rooms too) to avoid unnecessary solar gain where an air-conditioning or cooling system is operating
	• See that the building is allowed to cool at night (by opening windows) when daytime temperatures are high
Routine maintenance	• Repair broken windows and fastenings, replace damaged weatherstripping and door closers
	• See that insulation is replaced if moved or damaged by maintenance work (loft insulation may be rolled back)
	• See that openable windows and blinds for shading are operating freely
Periodic repair	• Remove build-up of paint from closing edges of windows and, if necessary, replace weatherstripping
	• Check fuel consumption against previous figures and investigate increases that do not appear to be directly related to the weather or the functioning of the plant
	• Replace sealants around windows and between cladding panels at the end of their service life
Improvement as part of planned maintenance or refurbishment	• Increase insulation in external envelope and floors next to the ground
	• Fit double-glazing (consider using low-emissivity glass). Stepped edge section may fit better to existing glazing rebates
	• Optimize use of solar heat gain (and daylighting) in temperate climates by replanning and enlarging or blocking windows and rooflights
	• Fit blinds (or other devices) to shade windows from unwanted solar heat gain

be made by changes to the services installations but alterations to windows and insulation can be equally effective. In most cases the best results will come from changes in both areas but the opportunities for improvement of the building fabric are the ones most likely to be forgotten or postponed.

5
Building materials and their maintenance
Alastair Gardner

5.1 Introduction

5.1.1 Many of the factors affecting the future maintenance of any buildings are determined when its purpose, location and overall budget is established. The extent to which designers can influence the cost of this maintenance will then be decided by the choice, position and detailing of materials.

5.1.2 Materials are seldom chosen on maintenance grounds alone but a study of the factors affecting the durability of materials and a full appreciation of the cost of maintenance should be established, as part of the design process.

5.1.3 When maintenance becomes a problem, rather than a routine approach to the protection or decoration of a structure, there have been earlier, avoidable errors. These may have occurred through errors in design, faulty specification or poor storage. The avoidance of unnecessarily high maintenance costs thus depends on eliminating deficiencies at the early stages, through a full understanding of the construction process and the properties of materials (see Chapter 1).

5.2 Design

5.2.1 Many maintenance problems arise where design, although satisfactory in principle, has, in practice, a low probability of success. This should not be seen as a defect in workmanship but rather as too high an expectation on the part of the designer. For example, the detail for the placing of reinforcement in concrete cladding panels may appear to provide the necessary cover on an engineer's drawing, but would need a 'watchmaking technique' to be achievable. This would be quite impractical on a building site or even in factory prefabrication.

5.2.2 Problems can be caused when legislation alters the attitude of designers or when new techniques become available. This can result in inexperienced or non-specialist labour being called upon to perform operations previously handled by separate trades. Energy-saving measures may require the accurate placing of a considerable thickness of insulating materials by a labour force more concerned with structural stability. The declining use of floor screeds in preference to self-finished structural slabs has led to problems when specialist finishing staff have not been employed, in the mistaken belief that cutting out a site operation somehow reduces the need for their skills.

5.2.3 Expectations of skill and quality control should be of a reasonable level and design should ensure that requirements are kept within the limits of materials and their exposure.

5.3 Specification

5.3.1 The specification should relate to the properly identified, measurable functions which the component is required to perform in use. It is pointless to specify that a component should be durable if there is no method of defining or measuring this before it is installed. The specification, in itself, is useless without defined quality control and quality assurance to support it. Guarantees do not eliminate failure or the need for maintenance. At best, they provide the resources for maintenance when a component has failed.

5.3.2 Specified levels of accuracy and tolerances should be necessary, realistic and achievable in practice. High levels of accuracy will be expensive and often it is better to accept that some degree of failure may occur and provide for it. Cavity-wall construction is a recognition of the difficulty of specifying waterproof performance from a single-leaf brick wall.

5.4 Storage and site conditions

5.4.1 The construction site may be compared to a factory which is both undesigned and temporary. It should provide an adequate environment both before and during construction to protect materials and components from damage.

5.4.2 The premature deterioration of materials can be attributed to a limited number of external causes. These include the effects of moisture, temperature, light, the atmosphere, impact, movement and all forms of biodegradation, such as attacks by fungi, insects and vandalism. In the details of problems affecting individual materials and systems which follow it, will be seen that the primary and consequential effects of moisture predominate as the most frequent cause of difficulty. Protecting against all these factors during storage and construction is, of course, impossible, but protection even against moisture has to be carefully planned and provided for.

5.5 Maintenance

5.5.1 Some materials such as painted timber, have a continuing maintenance requirement inherent in their use but even those such as bricks or concrete, that are regarded as durable rely on joints and junctions which become vulnerable in time. It is most important that designers fully understand the characteristics of the materials and take full account at the design stage of any factors which may lead to premature deterioration.

5.6 Clay bricks and tiles

5.6.1 Generally, clay products are extremely durable and deterioration in structural qualities or appearance is usually associated with wet conditions.

5.6.2 Frost attack occurs when saturated materials are damaged by the expansive forces of freezing. The existence of a large pore structure, high levels of saturation and repeated freezing cycles are factors in the likelihood of frost damage. Simple reference to compressive strength is not a reliable guide to the frost resistance of clay bricks.

5.6.3 Mortars are particularly vulnerable to frost attack during construction, since they will not set after being frozen and work should be abandoned if the temperature drops below 3°C and is falling. Frost resistance is enhanced only by increasing cement content (see Table 5.1).

5.6.4 Sulphate attack may occur if bricks contain soluble salts or if sulphates are present on the site (either already in the ground or through being brought onto the site with imported fill materials) and the mortar is allowed to remain wet for long periods. Chemical reactions between cement in the mortar and sulphates can cause the mortar to crumble on the outside and expand internally to cause bowing of the brickwork. This can lead to costly repairs.

5.6.5 Efflorescence and lime staining can affect the appearance of brickwork through surface deposits. Efflorescence is the product of soluble salts from the bricks and mortar left when the water in which they were dissolved evaporates. This can be removed by washing. Lime staining, however, is insoluble and is formed by the carbonation of free lime present in the mortar constituents which is brought to the surface by water movement. This problem mainly occurs if walls are left unprotected during or just after construction and become saturated.

5.6.6 The careful choice of tiles, bricks and mortar combinations will minimize the chance of unsightly blemishes and enhance their durability. The risk of deterioration will be reduced through careful detailing to avoid saturation or by promoting rapid drying out. For this reason, counter-battens should be used on roofs laid over sheet materials to allow the circulation of air to the

Table 5.1 Mortar mixes

Designation	Type		
	cement:lime:sand	masonry cement:sand (includes plasticizer)	cement:sand plus plasticizer
(i)	1:0:3 to 1:$\frac{1}{4}$:3		
(ii)	1:$\frac{1}{2}$:4 to 1:$\frac{1}{2}$:4$\frac{1}{2}$	1:2$\frac{1}{2}$ to 1:3$\frac{1}{2}$	1:3 to 1:14
(iii)	1:1:5 to 1:1:6	1:4 to 1:5	1:5 to 1:6
(iv)	1:2:8 to 1:2:9	1:5$\frac{1}{2}$ to 1:6$\frac{1}{2}$	1:7 to 1:8
(v)	1:3:10 to 1:3:12	1:6$\frac{1}{2}$ to 1:7	1:8

under-surface of roof tiles. Walls can be protected by overhangs or well-designed copings with effective drips. Care should be taken to avoid splashing at the foot of walls.

5.6.7 Calcium silicate bricks are more resistant to frost and sulphate attack and do not contain the salts which cause efflorescence, but they are attacked by heavily sulphur-polluted atmospheric conditions and sea spray.

5.6.8 Concrete and calcium silicate bricks will shrink on drying and strong mortars may cause cracking of the bricks if movement is unable to take place within the mortar (see Table 5.2 from NBS).

5.7 Concrete

5.7.1 Good concrete will require little maintenance to survive structurally for long periods, but deterioration in appearance, largely due to weathering, has restricted the use of the material as a surface finish in recent times.

5.7.2 The durability of any concrete will depend on the level of exposure to weather and chemical action, the depth of cover and type of reinforcement, as well as the quality of the mix and workmanship involved in mixing, placing, compacting and curing (see Table 5.3 from NBS).

5.7.3 Concrete which is unable to withstand exposure conditions may fail directly through spalling under chemical attack or frost action, or by corrosion of the reinforcing steel. Corrosion products occupy a much larger volume than the original steel and exert large expansion forces capable of spalling off the concrete over the reinforcement, leading to progressive failure. Increased protection in very severe exposure conditions, or if the quality of the concrete is limited, may be provided by the addition of corrosion inhibitors or the use of corrosion resistant reinforcement. Where cover to the reinforcement is limited, stainless alloys provide inherent corrosion resistance and, although not universally accepted, claims are made for the use of hot-dipped zinc-galvanized coatings.

5.7.4 To maintain both the durability and the appearance of exposed concrete, care must be taken to reduce absorptivity. Permeable concrete will accelerate the rate of carbonation and consequently reduce the alkalinity below pH 10, when it ceases to inhibit corrosion of the reinforcing steel.

5.7.5 Ordinary Portland cement (OPC) concretes will normally be used but in certain conditions, where high levels of sulphates are present, sulphate-resisting Portland cement may be necessary. Extra durability may also be gained by the use of blended cements containing the addition of pulverized fuel ash (PFA) and ground-granulated blastfurnace slag (GGBS).

5.7.6 Advances in concrete technology and the increasing use of ready mixed concrete on sites has meant that structural requirements can be met with more certainty at minimum cost. Required strength can now be achieved with reduced cement content and this has resulted in an increase in water/cement ratios and thus more permeable concrete, which is less able to protect the reinforcing steel.

5.7.7 Full compaction of concrete is required if the intended levels of strength and durability are to be achieved. Each 1 per cent of entrapped air can reduce strength by 6 per cent and the ability to protect reinforcement and resist frost action will be severely decreased if a significant volume of air is left in the concrete. Workability of the concrete as an aid to compaction should be achieved by increasing cement quantities, using rounded aggregates, adding PFA or PPFAC and employing water-reducing admixtures or superplasticizers.

5.7.8 Low permeability of the finished concrete depends on hydration of the cement to fill interstices originally water-filled, particularly those near the surface. Curing by keeping newly cast concrete moist prevents the fresh concrete from drying out too quickly and allows hydration to continue. Extending the period of curing will contribute to improved durability.

5.7.9 The presence of chloride ions in concrete can cause steel corrosion, even when the alkalinity remains high and although calcium chloride as an additive in reinforced concrete has been banned since the mid-1970s, de-icing salts from treatment of roads in winter can be a problem.

5.7.10 Air-entrained concrete should be used to improve the frost resistance of concrete structures, particularly where de-icing chemicals are present. The inclusion of entrained air in a mix of any water/cement ratio reduces strength. However, the increased workability provided by the air bubbles may allow a reduction in the water contents, so that the loss is offset by the strength gain from the water/cement ratio reduction.

5.7.11 Damage to concrete can occur when alkalis present in Portland cement react with certain

Table 5.2 Durability of brickwork in finished construction

Masonry condition or situation	Brick quality/mortar designation			
A Work below or near ground level *(if sulphate ground conditions exist)*				
A1 Low risk of saturation with or without freezing	FL (i)	(ii)	(iii)	
	FN (i)	(ii)	(iii)	
	ML (i)	(ii)	(iii)	
	MN (i)	(ii)	(iii)	
	3–7		(iii)	(iv)
A2 High risk of saturation *without* freezing	FL (i)	(ii)		
	FN (i)	(ii)[a]		
	ML (i)	(ii)		
	MN (i)	(ii)[a]		
	3–7	(ii)	(iii)	
A3 High risk of saturation *with* freezing	FL (i)	(ii)		
	FN (i)	(ii)[a]		
	3–7	(ii)		
B DPCs *(if sulphate ground conditions exist)*				
B1 In buildings	DPC1 (i)			
B2 In external works	DPC2 (i)			
C Unrendered external walls *(other than chimneys, copings, cappings, parapets and sills)*				
C1 Low risk of saturation	FL (i)	(ii)	(iii)	
	FN (i)	(ii)	(iii)	
	ML (i)	(ii)	(iii)	
	MN (i)	(ii)	(iii)	
	3–7		(iii)	(iv)
C2 High risk of saturation	FL (i)	(ii)		
	FN (i)	(ii)[a]		
	3–7		(iii)	
D Rendered external walls *(other than chimneys, copings, cappings, parapets and sills)*				
	FL (i)	(ii)	(iii)	
	FN (i)[a]	(ii)		
	ML (i)	(ii)	(iii)	
	MN (i)[a]	(ii)		
	3–7		(iii)	(iv)
E Internal walls and inner leaves of cavity walls *(where designation (iv) mortars are used)*				
	FL (i)	(ii)	(iii)	(iv)
	FN (i)	(ii)	(iii)	(iv)
	ML (i)	(ii)	(iii)	(iv)
	MN (i)	(ii)	(iii)	(iv)
	OL (i)	(ii)	(iii)	(iv)
	ON (i)	(ii)	(iii)	(iv)
	3–7		(iii)	(iv)
F Unrendered parapets *(other than copings and cappings)*				
F1 Low risk of saturation, e.g. low parapets, on some single-storey buildings	FL (i)	(ii)	(iii)	
	FN (i)	(ii)	(iii)	
	ML (i)	(ii)	(iii)	
	MN (i)	(ii)	(iii)	
	3–7		(iii)	
F2 High risk of saturation, e.g. where a capping only is provided	FL (i)	(ii)		
	FN (i)[a]	(ii)[a]		
	3–7		(iii)	

Table 5.2 cont'd

Masonry condition or situation		Brick quality/mortar designation			
G	Rendered parapets (other than cappings and copings) *(where sulphate-resisting cement is recommended)*	FL	(i)	(ii)	(iii)
		FN	(i)[a]	(ii)[a]	
		ML	(i)	(ii)	(iii)
		MN	(i)[a]	(ii)[a]	
		3–7			(iii)
H	Chimneys *(sulphate-resisting cement in mortars and renders are strongly recommended due to the possibility of sulphate attack from flue gases)*				
	H1 Unrendered with low risk of saturation	FL	(i)[a]	(ii)[a]	(iii)[a]
		FN	(i)[a]	(ii)[a]	(iii)[a]
		ML	(i)[a]	(ii)[a]	(iii)[a]
		MN	(i)[a]	(ii)[a]	(iii)[a]
		3–7			(iii)[a]
	H2 Unrendered with high risk of saturation	FL	(i)[a]	(ii)[a]	
		FN	(i)[a]	(ii)[a]	
		3–7			(iii)
	H3 Rendered *(where sulphate-resisting cement is recommended)*	FL	(i)[a]	(ii)[a]	(iii)[a]
		FN	(i)[a]	(ii)[a]	
		ML	(i)[a]	(ii)[a]	(iii)[a]
		MN	(i)[a]	(ii)[a]	
		3–7			(iii)[a]
I	Cappings, copings and sills (7)				
	Cappings copings and sills except for chimneys	FL	(i)		
		FN	(i)		
		4–7		(ii)	
	Cappings copings for chimneys	FL	(i)[a]		
		FN	(i)[a]		
		4–7		(ii)[a]	
J	Freestanding boundary and screen walls *(other than copings and cappings)*				
	J1 with coping (7)	FL	(i)	(ii)	(iii)
		FN	(i)	(ii)	
		ML	(i)	(ii)	(iii)
		MN	(i)	(ii)	
		3–7			(iii)
	J1(a) with coping exposed to severe driving rain (5,7)	FL	(i)	(ii)	(iii)
		FN	(i)[a]	(ii)[a]	
		ML	(i)	(ii)	(iii)
		MN	(i)[a]	(ii)[a]	
		3–7			(iii)
	J2 with capping (7)	FL	(i)	(ii)	
		FN	(i)	(ii)	
		3–7			(iii)[a]
K	Earth-retaining walls *(other than copings and cappings)*				
	K1 with waterproofed retaining face and coping (7)	FL	(i)	(ii)	
		FN	(i)[a]	(ii)[a]	
		ML	(i)	(ii)	
		MN	(i)[a]	(ii)[a]	
		3–7		(ii)	(iii)
	K2 with coping or capping but no waterproofing on retaining face (7)	FL	(i)		
		FN	(i)[a]		
		4–7		(ii)	

FL, FN, ML, MN refers to clay brick designations.
3–7 etc. refer to classes of calcium silicate bricks.
(i), (ii), (iii), (iv) refer to mortar designations.
[a] Sulphate-resisting cement is recommended or advisable.

Table 5.3 Durability of structural concrete

CONCRETE MIXES As given in BS 8110, Table 3.4 (20 mm nominal maximum size of aggregate)	C20	C30	C35	C40	C45	C50
Lowest equivalent grade	C20	C30	C35	C40	C45	C50
Equivalent ordinary prescribed mix	C10P	C20P	C25P	–	–	–
Deemed minimum cement content[a]	180	275	300	325	350	400
Deemed maximum free water/cement ratio	0.80	0.65	0.60	0.55	0.50	0.45
UNREINFORCED CONCRETE Suitability of the above mixes for use in various exposure conditions[b] (see BS 8110, Table 6.2)	*Mild*	*Moderate*	*Severe*	*Very Severe*	*Extreme*	*Extreme*
REINFORCED CONCRETE Suitability of the above mixes with various nominal covers to steel and links for use in various exposure conditions[b] (see BS 8110, Table 3.4) 20 mm	–	–	Mild	Mild	Mild	Moderate
25 mm	–	Mild	Mild	Mild	Moderate	Severe
30 mm	–	Mild	Mild	Moderate	Severe	Very Severe
35 mm	–	Mild	Moderate	Moderate	Severe	Very Severe
40 mm	–	Mild	Moderate	Severe	Very Severe[c]	Very Severe[c]
50 mm	–	Mild	Moderate	Very Severe[c]	Very Severe[c]	Extreme
60 mm	–	Mild	Moderate	Very Severe[c]	Extreme[c]	Extreme

[a] Minimum cement content in kg/m³, assuming use of 20 mm nominal maximum size of aggregate.
Increase by 40 kg/m³ if using 10 mm.
Reduce by 30 kg/m³ if using 40 mm.

[b] General durability exposure conditions (as BS 8110, Table 3.2):
Mild:
Concrete surfaces protected against weather or aggressive conditions.
Moderate:
Concrete surfaces sheltered from severe rain or freezing while wet
Concrete subject to condensation
Concrete surfaces continuously under water
Concrete in contact with non-aggressive soil (BS 8110, Table 6.1, Class 1)
Severe:
Concrete surfaces exposed to severe rain, alternate wetting and drying or occasional freezing or severe condensation
Very Severe:
Concrete surfaces exposed to sea water spray, deicing salts (directly or indirectly), corrosive fumes or severe freezing conditions while wet.
Extreme:
Concrete surfaces exposed to abrasive action, e.g. sea water carrying solids, or flowing water with pH less than 4.5, or machinery or vehicles.

[c] Should be air entrained if used in this condition.
Average air content of concrete by volume:
7% for 10 mm aggregate
5% for 20 mm aggregate
4% for 40 mm aggregate

forms of silica in the aggregate to form a gel, which absorbs water, swells and cracks the concrete. Although fairly rare, alkali–silica reaction may occur where concrete is subject to prolonged wetting, has a high alkali content or where reactive aggregates have been used.

5.7.12 The long-term appearance of exposed concrete structures can only be realistically approached in two ways – either to design for a graceful change with time or to restore the original appearance through periodic cleaning or painting. To defy or ignore the effects of weather and pollution in an attempt to preserve an ageless appearance in industrialized regions is not a viable option.

5.7.13 The initial colour of concrete as-cast is that of the finest particles in the mix, normally the cement or any added pigment. Dark colours are subject to the effects of lime bloom, present in all concretes, but noticeable when the light-coloured crystals mask the surface. Light-coloured, smooth concrete may display signs of crazing, conspicuous when dirt is retained in the fine cracks. Colour variation can arise if the formwork is smooth and impervious, if there are changes in its absorbency or if there is leakage. The variability in colour and porosity of the outer skin makes as-cast concrete an unreliable maintenance item on appearance grounds.

5.7.14 Removing the surface layer by chemical, abrasive or mechanical means will expose, first, the colour of the fine aggregate and second, the colour of the coarse aggregate. The finishes thus achieved are likely to be more uniform and predictable but attention has to be paid to aggregate grading, mix design and the avoidance of impurities in the aggregates. While exposed concrete finishes are capable of producing an initial high quality appearance, they can be ruined by the effects of weathering.

5.7.15 Regular cleaning has been suggested to remove weathering effects, but on as-cast finishes dirt may protect parts of the surface from acid rain attack and cleaning may only produce a 'negative' effect. Concrete surfaces have been successfully cleaned by the use of water spray, chemical treatment (usually based on hydrochloric acid) and dry or wet grit blasting.

5.7.16 Painting concrete to provide uniform colour and as a renewable finish is technically a viable option, but involves a permanent maintenance commitment.

5.7.17 Pre-cast concrete components have inherent advantages over site work, due to the potential for better control and workmanship. Brick and stone veneers can now be reliably applied to precast panels, using keyed interfaces and glued dowels. Joints between panels require specialist advice and will normally comprise two levels of sealant or loose baffles to accommodate movement.

5.7.18 Concrete block masonry construction offers great flexibility and economy, but the effects of drying shrinkage, the care required in mortar-mix selection, and the effects of weathering require careful study if it is to be used successfully. Particular attention must be paid to the location and frequency of movement joints, the provision of good cavity drainage and the avoidance of wasteful cutting by adopting modular design. Storage and protection of work during construction are essential if movement cracks are to be avoided.

5.7.19 Floor screeds have often given rise to problems and current practice suggests that direct finish to the structural slab is more reliable. If screeds are used, there should be adequate thickness with margins to allow for unevenness of the base. Nominal 70 mm unbonded screeds should be specified.

5.7.20 Renders can give long and satisfactory service but are very dependent on workmanship. The variability of backings and the effect of weather conditions during construction make detailed specification difficult and the best results will be achieved by the employment of suitable craft skills.

5.8 Timber

5.8.1 The traditional approach to timber design relied on skilled selection and inspection but procedures were seldom tested. The problems caused by movement, decay and insect attack were overcome by skilled craftsmen using large sections of hardwood and possible failure was avoided by experience, generous overdesign, the inherent durability of the timber and the use of tolerant jointing techniques.

5.8.2 The indiscriminate use of hardwoods, causing depletion of the broad-leaf forests, means that traditional species are either no longer available or uneconomical and substitutes are likely to be less satisfactory in terms of durability and dimensional stability. The development of calculated design made possible by the use of stress-graded timber and supplies of softwoods with controlled moisture content has allowed the structural use of smaller sections of cheaper materials. As a consequence, safety margins are much reduced and special care must be taken to control conditions when using a timber-trussed rafter roof or other structural elements.

5.8.3 The performance of timber (particularly large-section structural elements) is most affected by moisture content. If timber is too dry when installed, moisture will be absorbed to reach equilibrium with its surroundings and it will expand (Figure 5.1). Kiln-dried timber of too low moisture content installed without allowance for movement can warp and buckle and may need complete replacement simply through the exposure of overdry materials to normal indoor conditions. Precautions are needed when timber is installed during the construction period before heating is available and care should be taken to make the building watertight prior to the fitting of finished timber products. Exposure to water or humidity in excess of about 70 per cent RH for extended periods must be avoided.

5.8.4 Problems with timber are more likely to occur when conditions are too wet rather than too dry, and this can lead to failure through movement or attack by fungi or insects. Wood may swell by as much as 25 per cent of its volume when wet, depending on species, and cause failure in finishes and joints. Fungal attack can be largely avoided if moisture contents are kept below 20 per cent, and since most insects prefer damp or rotted wood, this problem is also reduced by dry conditions. Wood should be specified at an appropriate moisture content for its future location and use, and care taken during all manufacturing, storage and erection procedures to keep it at the specified level. In damp conditions wood will readily absorb moisture to reach contents of over 24 per cent but within buildings wood should be safe from decay if weather-tightness and adequate ventilation are maintained.

5.8.5 Proper storage on-site is essential to reduce shrinkage problems and to avoid warp and twist due to unplanned loading. New buildings involving wet trades are not recommended for storage since the atmosphere will be too damp until all concrete and finishes have dried out. Material in transit should be protected and primed off-site, although priming offers only limited protection.

5.8.6 External painted surfaces are affected by weather and it is important to shelter all wood items as much as possible, using roof overhangs and deep reveals. Where surfaces are exposed, good design should ensure run-off of water and quick drying out. The factory application of primer on manufactured joinery should not be considered adequate weather protection and covered storage is required. Problems can be caused when end grain exposed by site operations absorbs water which is prevented from evaporating by the finishes. Lack of regular maintenance of exposed painted surfaces will affect durability as well as appearance.

5.8.7 The use of vapour barriers alone to protect enclosed structural timbers should be avoided. These layers can only be regarded as vapour checks, since they are easily damaged during construction, difficult to join and are often penetrated by service runs. Adequate ventilation with clear flow passage must be provided and maintained to keep the fabric dry. Timber cladding must be fixed so that air movement is encouraged on the inner face to allow rapid drying out. If vertical boards are used, counter-battens will be required and detailing checked to ensure that vents are provided top and bottom to allow the passage of drying air.

5.8.8 The use of some chemical preservative treatments are being questioned on environmental

Figure 5.1 The expansion of a timber floor installed with low moisture content

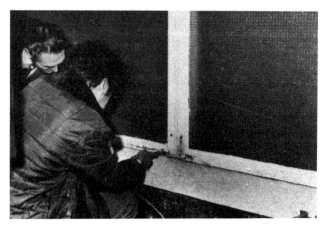

Figure 5.2 Pressure impregnation of timber *in situ*

grounds and good design and detailing practice should take care not to expose timber to unnecessary degrading hazards which make treatment essential (Figures 5.2–5.4).

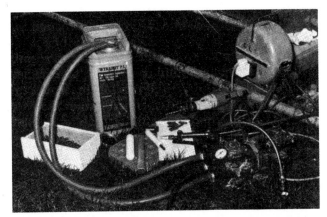

Figure 5.3 The equipment for *in-situ* impregnation of timber

Figure 5.4 A trussed rafter roof showing corrosion of the galvanized gannail aggravated by timber treatment

5.9 Metals

5.9.1 The choice of metals or their alloys for use in building will depend on the fabrication technique, function, exposure and strength required of the component.

5.9.2 Ferrous metals, which includes iron and steel, require special attention with regard to protection against corrosion (see Figures 5.5–5.8).

When oxygen and water are present with steel, corrosion will occur and be visible in the form of rust. When used below ground where oxygen is excluded corrosion rates are very low even when water is present. Inside dry heated buildings, even untreated steel will be little affected because of the

lack of moisture. Chlorides, present in the marine environment, and sulphates, a result of industrial pollution, increase steel corrosion rates. Rust layers on mild steel are porous, allowing water penetration and further corrosion, which push off the top layer as part of a continuous process which can destroy the complete component.

5.9.3 Protection is most commonly provided by coatings of paint or metal or sometimes both. They work by excluding moisture from the steel surface. Preparation is most important and determines to a large extent the durability and the type of coating to be used.

Figure 5.5 Staining of concrete by rust from Corten steel

Figure 5.6 Corrosion of steel tubes embedded in concrete

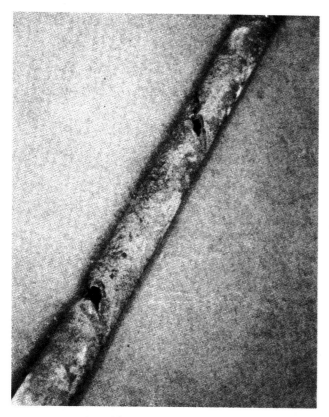

Figure 5.7 Spiral corrosion through poor pipe wrapping

5.9.4 Surfaces prepared manually by chipping, scraping and brushing are not very satisfactory, removing only about 30 per cent of rust and scale. The use of power tools will only marginally improve efficiency, and durability will depend on the effectiveness of the primer on poor surfaces. Red lead in oil primers remains the most effective treatment for poor surfaces, although it is usually ruled out on environmental grounds.

5.9.5 To prepare surfaces effectively prior to painting, blast cleaning is required. Abrasive particles are projected onto the surfaces by a centrifugal impeller wheel, compressed air or air and water jet. Recoverable systems are possible and by choice of shot or grit, rounded or angular surface profiles will be created. The durability of painting systems will be extended greatly by blast cleaning compared with manual methods.

5.9.6 Paints are usually applied one coat on another, with each layer contributing to the overall protection. Primers applied directly to the cleaned steel surface inhibit corrosion and provide adhesion for later layers. Undercoats prolong the life of the protection by building up thickness. Finish coats give immediate protection from the environment and determine the appearance in terms of gloss and colour. Paints can be classified by reference to their constituents (Table 5.4). The binder mainly determines the use of the coating and creates the film over the surface. Pigments add colour and opacity and contribute corrosion-inhibiting properties. Since most binders have high viscosity, they require to be diluted with solvents or otherwise prepared for ease of application. Application may be by brush, roller or spray, while small items may be coated by dipping. Brush coating of poor surfaces is essential to ensure the best possible adhesion.

5.9.7 Depending on the severity of exposure and the intended frequency of maintenance, various coatings may be used for steelwork. These range from conventional paints to bitumen paints, chlorinated rubbers, stoved finishes and paints containing zinc and lead salts. Plastic coatings and

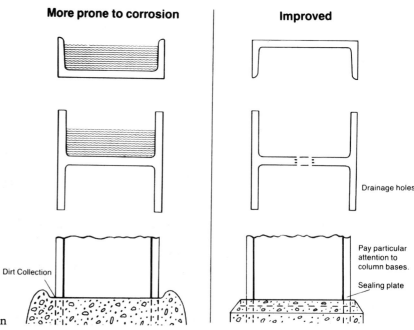

Figure 5.8 Detailing to minimize corrosion

Table 5.4 Main generic types of paint and their properties

	Cost	Tolerance of poor surface preparation	Chemical resistance	Solvent resistance	Over coatability after ageing	Other comments
Bituminous	Low	Good	Moderate	Poor	Good with coatings of same type	Limited black and dark colours Thermoplastic
Oil based	Low	Good	Poor	Poor	Good	Cannot be overcoated with paints based on sticky solvent
Alkyd epoxy-ester etc.	Low–medium	Moderate	Poor	Poor–moderate	Good	Good decorative properties
Chlor-rubber	Medium	Poor	Good	Poor	Good	High-build films remain soft and are susceptible to 'sticking'
Vinyl	High	Poor	Good	Poor	Good	
Epoxy	Medium–high	Very poor	Very good	Good	Poor	Very susceptible to chalking in UV
Urethane	High	Very poor	Very good	Good	Poor	Better decorative properties than epoxies
Inorganic silicate	High	Very poor	Moderate	Good	Moderate	May require special surface preparation

metal coatings are applied to the components off-site and provide the highest levels of protection.

5.9.8 Metal coatings on structural steel work are usually zinc or aluminium. Hot-dip zinc coating (galvanizing) or metal spraying with zinc or aluminium can be used alone or in combination with overpainting to provide durable protection.

5.9.9 The degree of exposure, proposed maintenance cycle and attention to detailed design will affect the choice of paint systems. When making a choice, specifiers should consult BS 5493 at an early stage to establish the best options for any circumstance.

5.9.10 Stainless steels containing more than 12 per cent Cr are expensive and are usually restricted to cladding for buildings. Their use for small structural fittings such as wall ties could prove effective, providing long-term security from corrosion, where inspection and replacement is difficult.

5.9.11 The main problems with other metals in building are most often the result of bimetallic corrosion. This occurs as a result of current flows generated when dissimilar metals are in contact, in the presence of an electrolyte. Electrolytes can be created by sulphates dissolving in rainwater and if contact between dissimilar metals is necessary, steps should be taken to insulate these metals from each other, and exclude the electrolyte by painting or taping. Metal fixings, in contact with treated

timber may be at risk, since the presence of copper in many of the treatments forms galvanic cells which accelerate corrosion of the metal (Figures 5.9 and 5.10).

5.9.12 Plastics are becoming increasingly competitive with copper as plumbing material for pipes but the latter retains unique appeal as a roofing material. The familiar green patina of copper carbonates and hydroxides acts as a protective barrier to further corrosion of the metal. The patina will be destroyed by sulphurous deposits from flue gases or organic acids associated with lichen growth. Designers should ensure that

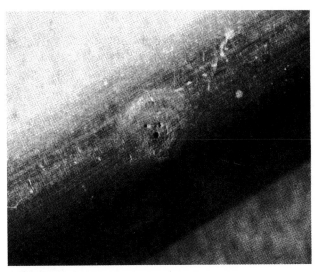

Figure 5.9 Perforated copper pipe through carbon film formed on inner surface during manufacture

Figure 5.10 Exfoliating corrosion of aluminium alloy containing 4% copper

Figure 5.11 Cracking of lead roofing sheet due to high purity

adequate falls restrict the growth of lichen and combustion gases are directed clear of the roof. Care should be taken to avoid the staining and the chemical attack of adjacent surfaces caused by run-off from copper surfaces.

5.9.13 Lead roofs may also be damaged by lichen growth and, additionally, by cement, mortar and concrete droppings (Figure 5.11).

5.9.14 Untreated timber can contain acetic acid which in moist conditions will affect steel, lead and zinc, among other metals. In exceptional unventilated conditions, acid vapour from wood can corrode metal surfaces even if they are not in contact. Within heated spaces, corrosion is unlikely to be a structural problem and galvanized steel or brass fixings are used only to maintain appearance. In unheated covered conditions the risk of corrosion exists and galvanized fixings as a minimum should be used. When preservatives or dampness are present, stainless steel or plastic-coated steel fixings should be used.

5.10 Flat roof coverings

5.10.1 Flat roofs are often regarded as problem areas, prone to failure and best avoided. However,

with attention to detailed design, workmanship and appropriate materials it is possible to construct reliable flat roofs, requiring low maintenance.

5.10.2 Design of the roof should ensure adequate falls, making due allowance for deflection in use. If external drainage is not possible, frequent roof outlets and wide gutters, away from upstands, should be provided. Upstands should be at least 18 mm and care should be taken to isolate the roof membrane from the differential movement between upstands and decks. Exposed waterproof membranes laid directly on thermal insulation in warm roof design are subject to temperature variations of over 100°C and the resultant movement puts great demands on materials and workmanship.

5.10.3 Traditional fibre-based felts, asbestos-based felts and even glass-based felts have largely been replaced except as cheap underlayers in the case of glass-based material. The high-performance polyester-based materials give improved tensile strength, flexibility and fatigue resistance, and tests have shown that a life of over 20 years can be expected. In cold weather, however, they can still become brittle and care must be exercised in the choice of insulating materials and access to the roof should be restricted.

5.10.4 Polyester mesh reinforced PVC roofing sheet with solvent- or hot air-welded lap joints mechanically fixed, using screw attached fixing plates, are being used increasingly as the waterproof layer for large flat roofs. This membrane requires no additional protection and the absence of bonding allows for structural movement. Repairs can be easily carried out with patches and 20-year durability is claimed.

5.10.5 Mastic asphalt, properly designed and laid, should last over 50 years. However, problems have arisen when laid directly over insulation in a warm roof configuration. Traditionally, asphalt was laid with sheeting felt directly on a concrete deck, which acted as a heat sink and limited temperature variations, thus reducing the possibility of failure. Inverted roof systems with the insulation laid loose above the membrane reduce the temperature range and also protect the surface from impact damage. Paving slabs used as ballast should be suspended on supports to allow water to drain and evaporate from the insulation layer.

5.10.6 Regular inspections and maintenance of all flat roofs should be carried out and re-application of solar reflective paints or chippings should take place. Gutters and outlets should be kept clear to avoid excesive loading of the structure through ponding.

5.11 Plastics

5.11.1 Due to the availability of an increasing range of plastics, these materials are being used more extensively in modern construction techniques. However, since development in the area is comparatively recent, the information with regard to durability is relatively limited.

5.11.2 Plastics are not generally expected to be maintained, and when failure occurs it is often easier to replace than to repair. However, problems may arise when fittings are not interchangeable and to avoid unnecessary complications, a record should be kept of the pattern of the original installations and the name of the manufacturer.

5.11.3 Extruded plastics materials may, in time, shrink or change shape through reversion, particularly when dark-coloured systems are exposed to the heat and radiation of direct sunlight. Changes in temperature produce a similar effect, causing expansion and contraction in pipes and gutters, allowing them to work their way out of fittings which have been poorly designed. Buckling may also take place where

there has been an insufficient allowance for movement. uPVC products and polycarbonates are particularly affected by thermal expansion and the effects of ultraviolet are usually most severe in exposed locations such as coastal areas.

5.12 Conclusions

5.12.1 In Building Research Establishment Digest 268, which studied common defects in low-rise traditional housing, attention is drawn to the fact that 'by far the largest category of faults, were attributed to performance, related to durability and maintenance. This is the area above all others where much time and thought need to be devoted to predicting future performance.' It was also shown that more than half of all faults occurred in the elements of the external envelope and concern was expressed that this information related to traditionally built houses which 'ought to be second nature to all house designers and builders' (see Figure 5.12).

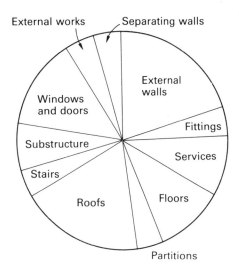

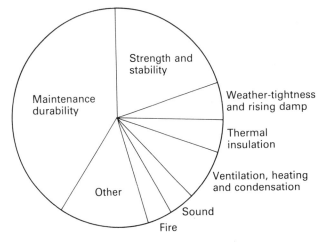

Figure 5.12 Common defects in traditional low-rise housing (from BRE Digest 268)

5.12.2 If difficulties occur in such familiar building conditions, how much greater will the problem be when new and less familiar materials are used? However, if the basic properties and characteristics of materials are understood, it should be possible to design and detail successfully, even when these materials are used in combination or in new ways.

5.12.3 The major cause of deterioration in building materials is moisture, either from the weather or by condensation. It is important to understand the movement of moisture. Vulnerable components and joints must be protected and rapid drying out encouraged if wetting does take place. Staining and other appearance defects can incur maintenance costs and detailing should address good weathering as well as protection of the fabric.

5.12.4 The deterioration of all materials is inevitable and unavoidable, but should always be predictable and never needlessly accelerated.

6
Services design and maintenance
David M. Lush

6.1 Introduction

6.1.1 The design of building services requires many skills and close cooperation between the design team and the client. The design team should draw up and agree with the client a Services Brief, in which most or all of the following objectives should be considered:

1. *Economy* The relative importance of minimizing capital, fuel and maintenance costs.
2. *Economic life* How long the equipment is expected to last before it is to be replaced, changed or upgraded for changing building use or standards.
3. *Reliability in relation to activities within the building* Provision of standby equipment, duplication, independence of public utility services.
4. *Flexibility* To meet short-term changes in activities within the building and in the weather e.g. cold spells in summer.
5. *Adaptability* To meet such longer-term effects as changes in building use, fuel prices and availability and changing climate conditions.
6. *Metering* The ability to monitor equipment performance, and fuel and water consumption by different departments or tenants within a building.
7. *Environmental issues* How the services design affects the external environment and vice versa.

6.1.2 The Services Brief must also contain as precise a definition as possible of the services and conditions required in each part of the building, in relation to the activities, equipment and populations proposed and their relationships.

6.1.3 Third, the Brief should detail the external conditions and limitations with which the services design will comply. These will include, for example, the minimum external temperatures at which the heating system will maintain full comfort conditions within the building; the storm conditions with which external drainage systems will be designed to cope; maximum demands which the public utilities can provide; noise levels limits at site boundaries and any other known site restrictions.

6.1.4 The larger the building, the more important early consideration of these matters becomes. Unless the client's objectives and requirements are clearly stated and understood, the effectiveness of the design team will be reduced, and it will be difficult for them to specify to the client the limitations he must expect within the budget limits imposed. A lack of clarity at this stage can be a source of dissatisfaction on the building owner's and occupiers' side and cause argument between them and the design team for years after the building is handed over.

6.1.5 Having established and agreed the Brief with the client, scheme and detailed designs are developed. Many factors will play their parts,[1] and the blending of these to produce satisfying buildings requires both science and art from the design team. Where services are concerned the presentation of Scheme Designs to the client is as important as agreeing the initial Brief. At this Stage the degree to which the objectives outlined in 6.1.1 above can be met by alternative solutions within the budget can be established and appropriate decisions made before detailed designs are produced.

6.1.6 Maintenance should be considered as an integral part of the project during all phases of the design and, preferably, during the preparation of the Brief. The management of building services maintenance, the reasons for such maintenance and its implementation need to be clearly defined.[2] It is no longer acceptable to relegate building services maintenance to a peripheral activity outside the main business management stream. Services do not retain their design performance unless they are maintained and business efficiency and profitability will both suffer if the performance is below par. Various definitions of

Figure 6.1 Lloyd's, London. A classic example of external services used in an architectural manner, which can be maintained without interruption to the building function and occupiers. Architects: Richard Rogers. Engineers: Ove Arup and Partners)

maintenance are given in other sections, but from the point of view of building services the following broad headings should be considered:

- *Facilities management* Total asset management including building fabric, services, office technology and all matters which affect work efficiency and staff comfort.

- *Planned maintenance* Cleaning, replacing filters, lubrication, etc.

- *Statutory maintenance and inspections*

- *Breakdown maintenance* Repair and replacement of failed components.

- *Refurbishing* Replacing, updating and changing equipment to improve standards or for changing building uses.

6.1.7 Where services are concerned, attempting to draw a line between maintenance and operation can only divert attention from the close

relationship which must exist between them. Methods and periods of operation affect both the need for maintenance and the water, energy and supplies consumed by the services. Operations may be largely automatic, as in drainage systems, or automatically controlled, e.g. domestic hot water system temperatures, but the settings such as flow rates and temperatures are operating parameters affecting both consumption and maintenance.

6.1.8 The organization of operation and maintenance and how they are carried out will depend on the size and complexity of the building and its systems, and on other factors discussed in Chapter 14 and, to a certain extent, in Chapters 3 and 18. However, some specific recommendations in relation to particular services are made here. Access for maintenance is considered in detail in Chapter 13.

6.1.9 Too often in the past there has been a concentration on minimizing capital cost in the design of services, at the expense of overall owning and operating cost. Methods of estimating these costs have been published[3] and should be well known, but their application is not encouraged by fee structures under which engineering consultants are paid a percentage of the total capital costs of the work for whose design they are responsible. This situation is now exacerbated by the percentage being based on competitive fee bidding rather than on a standard scale of charges. Enlightened clients should insist on adequate attention being paid to operation and maintenance in the designs they commission, and they should be prepared to negotiate additional fees if standard conditions of service do not provide for this work.

Figure 6.2 Embankment Place, London. An example of air-conditioning ductwork and other services above a suspended ceiling, indicating how the layout and access panels facilitate maintenance. (Architect: Terry Farrell and Company Ltd. Engineers: Ove Arup and Partners. Photograph: R. Risdill Smith)

6.1.10 Thus in summary, as far as operation and maintenance are concerned, the design of the services should take account of the objectives of the Brief (6.1.1), energy-utilization standards (Chapter 3), access, and safety requirements (Chapter 13), and the types of maintenance required (6.1.6).

6.1.11 It is incumbent on the building services designer to advise the client on the importance of maintaining the services after practical completion. The provision of a comprehensive maintenance regime will ensure that the design intent is retained in terms of the internal environment, that the occupants are satisfied and that the overall running costs are minimized. The advice should extend to offering to provide a maintenance consultancy service on the maintenance specification, the in-house staff (or contractor) to carry out the work and an evaluation of the costs proposed for the tasks. The offer may also include for short- or long-term performance monitoring of the maintenance work. The client then has the opportunity of the additional service or making a conscious decision to take direct responsibility for post-contract operational and maintenance requirements, possibly passing them on to the tenant. Before discussing particular services in detail, two other major points need to be mentioned, the provision of operating and maintenance (O&M) instructions for building services and the problems of hand-over and maintenance during defects liability periods. If these are taken too lightly, the designers' intentions may be frustrated, the equipment installed by the services contractors may be maltreated and the owners and occupiers of the building may be thoroughly (and understandably) dissatisfied with what they have to pay for.

6.2 The Building Maintenance Manual

6.2.1 The Building Maintenance Manual and Job Diary published by The Building Centre provides a framework for the contents of an owner's and occupier's manual, including sections and checklists for services systems. However, the designer's actions necessary to ensure that this owner's manual is produced effectively, that it contains all the relevant information and that it is delivered on time are worth summarizing, particularly in relation to building services. The following summary relates to the RIBA Work Stages, but should be modified as necessary, in relation to the budget and the size and complexity of the job. Some related design decisions are also mentioned.

1. *Outline proposals* Develop and present alternatives to the client, with cost estimates and summaries of operation and maintenance implications. Consider possible scope of manuals in relation to client requirements and make recommendations to client, particularly if any additional fees may be required.

2. *Scheme design* Estimate capital, energy and maintenance costs in greater detail for the chosen alternative(s), checking and recording statutory requirements. Allocate responsibility for and define the broad scope of records and manuals to be produced, and agree fees, if any.

3. *Detail design* Consider energy and water consumption, access and statutory requirements in all systems design, and durability and ease of maintenance in equipment selection. Define the contents and format of all information required for manuals. Prepare pipework, electrical and control schematics for manuals as well as for other purposes, and start collection of manufacturers' literature.

4. *Production information and bills of quantities* Complete tender drawings and specifications, including equipment performance specifications, noise level limits, etc. and commissioning requirements. Define in tender documents precisely what record information in what form all sub-contractors will be required to produce and when it must be delivered in relation to practical completion. Request separate prices for initial maintenance (and operation if required) of relevant services systems and for training of client's operation and maintenance staff during defects liability periods. Optionally, request quotations for supply of consumable items such as filters and water treatment chemicals and of spares for imported equipment.

Figure 6.3 Brooks Shopping Centre, Winchester Development. Example of an underground car park where layout, lighting and services are designed to encourage maintenance and discourage vandalism, etc. (Architect: Building Design Partnership. Engineer: Ove Arup and Partners)

5. *Tender to completion* Continue and complete collection of manufacturers' literature, editing where necessary. Produce updated, final contract drawings. Chase and check contractor's record, commissioning, operation and maintenance information. Collate, index and bind (where appropriate) all information collected and hand over to the client.

6.2.2 It should be made an absolute rule that no services installation is accepted by the designers on behalf of the client without sufficient information for it to be used safely and without damage. It may be advisable to make it clear in tender documents that receipt of such information will be a prior condition for signature of practical completion documents and start of defects liability period. It may now be a criminal offence under the Health and Safety at Work Act to operate a building without the O&M manuals.

6.2.3 The operation and maintenance of services during the defects liability period may give rise to serious disagreements in heavily serviced buildings, particularly if this is carried out by untrained staff with inadequate manuals. The temptation to alter thermostats, diffuser settings and other controls to 'see what the systems can do' is often very great. But if building users as well as maintenance staff interfere with the operation of, in particular, HVAC (heating ventilating and air conditioning) systems without recording what was done and when, it may be impossible to decide who is responsible for trouble arising before the end of the defects liability period. Placing

responsibility with the sub-contractors who installed the system, as suggested in 6.2.1(4) above, is one possible solution, which may be particularly attractive where the client or building owner is letting it to a number of tenants.

6.2.4 As noted above, the Building Maintenance Manual and Job Diary provides a framework for the services operation and maintenance manual. When completed it should provide an effective summary for day-to-day use, but the basic information from which this summary is derived is much more extensive. This chapter is not intended as a treatise on operation and maintenance manuals, but some more guidance on what basic information should be available may be useful.

6.3 Responsibility for records and maintenance instructions

6.3.1 The services designers should be responsible for producing the following information:

- The client's environmental brief – as finally agreed.

- The external design conditions assumed, summer and winter.

- The internal design conditions, including occupancy, internal heat gains allowed for, design temperatures (and relative humidities where air conditioning is provided), air change rates, illumination and maximum noise levels, occupation periods and so on.

- External services available, electricity voltage and maximum demand, gas and water pressures and maximum supply rates, gas and water analyses, maximum discharge rates into local sewers.

- Local and statutory environmental limitations, such as maximum noise and pollution emission levels, and, for example, ground conditions where waste water is disposed of on-site.

- System design data, including specifications, design drawings and schematic diagrams for pipework, ductwork and electrical systems.

- Details of intended equipment life.

- Predicted water and fuel consumptions, electrical demand curves and power factor.

- Design operating temperatures for all piped and air systems, including tolerances.

- Control system schematics.

- Details of intended operating methods and functional requirements for systems made up of a number of components.

Figure 6.4 Lloyd's, London. Standby generator indicating access arrangements and space necessary to ensure maintenance and reliable operation. (Architect: Richard Rogers. Engineers: Ove Arup and Partners)

- Advice on tuning of operating systems for changing internal requirements and energy economy.

Details of content, format procedures and updating are available.[4]

6.3.2 The main contractor should be responsible for producing the following:

- Record drawings showing locations and depths of all external services installed by all contractors and public utilities.

- Details and locations of all access manholes, valves, etc. for these services.

- Details of all tests carried out on external service mains and all statutory approvals.

6.3.3 The services sub-contractors should be responsible for producing the following record information, which should be defined appropriately and in greater detail in tender documents:

- Specially prepared drawings of all piped and wired services, generally single line, showing the services as installed, with full annotations against each component as it is labelled on site, based on the architects' final layouts, with sections, etc. for congested areas.

- Updated ductwork, wiring and control panel drawings showing the work as fitted, based on the contractors' and suppliers' production drawings, and related to the architects' drawings for identification.

- Updated schedules of all equipment installed, with full means of identification by site label numbers, component serial numbers and make, type or catalogue numbers, and ratings, duties, etc.

6.3.4 Test certificates for all equipment tested off-site should be obtained and held by the services designers. Details of which equipment is to be tested, the form of tests and test certificates and the dates for delivery of the certificates should be defined in tender documents. Particular attention should be paid to identification of the certificates in relation to the actual equipment tested and supplied to the contract concerned.

6.3.5 Dated, signed and approved commissioning sheets for all equipment and systems tested or commissioned on-site must also be supplied by all services contractors. These must contain full details of readings taken and equipment settings, environmental conditions where appropriate, noise levels where maxima are specified, and so on. Where approvals by outside bodies are required these must be appended.

Figure 6.5 (a) National Exhibition Centre, Birmingham. Service tunnel to exhibition halls, showing planned services mains to which temporary connections can be made, with ample space for working and maintenance. (Architects: Edward D Mills and Partners. Engineers: Ove Arup and Partners)

Figure 6.5 (b) National Exhibition Centre, Birmingham. Delivering equipment to the roof of one of the halls. An example of planning for awkward situations, the design of weatherproof equipment and its suitability for maintenance access. (Architect: Edward D Mills and Partners. Engineers: Ove Arup and Partners)

6.3.6 The contractors must be made responsible for obtaining and handing over by specific dates full manufacturers' literature for all equipment and systems supplied. Some manufacturers' literature may be out of date and incomplete, and the tender documents should specify as precisely as possible what is required to ensure that what is supplied is useful and complete. Summarizing very briefly, the following should be included:

- Full description, including a breakdown of parts with catalogue numbers, duties and ratings, and recommended spares lists.

- Details of the guarantee period and of all inspections or work to be carried out during that period by supplier, contractor or building owner.

- Full operating instructions, including precautions, warnings and safety measures, commissioning/balancing, starting, running, stopping, emergency stopping, shutdown and seasonal changeover procedures.

- Special attention should be paid to sequences of operation and, where appropriate, how to override controls or break into or change sequences.

- Full maintenance instructions, including precautions, warnings and safety measures, isolating methods, routine preventative maintenance (details of inspections, cleaning and servicing with recommended intervals), routine

Figure 6.6 Lloyd's, London. Boiler house indicating quality of layout, lighting and spatial arrangements to encourage proper maintenance and to provide proper plant withdrawal facilities. (Architect: Richard Rogers. Engineers: Ove Arup and Partners)

repairs and replacements, fault finding (faults, symptoms, causes and remedies), major maintenance (lifting, etc.). Details of maintenance contracts offered. Emergency contracts. Details of any statutory inspections or tests carried out or required.

6.3.7 In many types of building the majority of the services components will be concealed in ducts, trenches, false ceilings, false floors and elsewhere, and access (see Chapter 13) may become a problem in practice. The record information described above will serve to locate all equipment which has to be maintained, but will not necessarily tell one how to get at it. The architects will be responsible for the finishes, ceiling designs and so on, and they should be responsible for designing these with access in mind – based on information from the services designers and contractors – and for producing clear drawings and instructions showing how access can be gained. *In-situ* marking of ceiling tiles may be advisable, suction cup tools for prefabricated panels and various types of key may be necessary. Full information must be provided by the architect, effectively cross-referred so that maintenance staff or service engineers do not have to take down a whole ceiling in the search for a single control device, valve or drain point.

6.3.8 The record of the designers' intentions, the equipment as installed and the operating and maintenance information described above is extensive, and not all of it may be required for simpler buildings. The design team must exercise their judgement in deciding the detail necessary; failure to do this may have undesirable consequences.

6.3.9 Where existing buildings are concerned, records and operating and maintenance instructions are liable to be incomplete and inadequate for controlled building maintenance. In such cases and where buildings are being refurbished existing installations should be surveyed in detail and the necessary records produced. If the manufacturers of the equipment installed are still in business their literature should be obtained if possible. Where it cannot be obtained and the components concerned are relatively simple it may be possible to draw up the necessary basic information from simple study of similar equipment.

6.3.10 In some cases, however, where a more complicated or vital component is no longer in manufacture or information about it cannot be obtained, it may be advisable to replace the equipment with new. If literature is no longer available, spares may be equally difficult to obtain.

6.3.11 In cases where an existing installation has been gradually extended and modified over the years a number of different makes of component doing the same job may have been installed, creating great problems with spares inventories. If simplicity and economy of maintenance are important there may be a strong case for rationalization, for example replacing a variety of valves from different makers with equivalents from a single, more modern range.

6.3.12 Another aspect of the modified and extended services system is that in the course of time imbalances and overloads of systems and equipment may have arisen, not necessarily leading to frequent failures but certainly increasing the risks of this. The original loads on equipment such as boilers may have been set so that they

operated at their most efficient level, but changing loads may produce very inefficient operation.

6.3.13 A final comment on existing installations is that there has been steady improvement over the years in the performance of most items of building services equipment. Retaining the 'old, reliable' systems may be attractive, but highly undesirable from fuel consumption and energy efficiency points of view. In particular, antiquated heating controls may cause buildings to be unnecessarily heated when unoccupied, may overheat parts of buildings when they are in use – the temperature being controlled by the occupants throwing open all the windows – and may be difficult to adjust or tune for economy generally.

6.4 Standard services installations

6.4.1 There are many different types of services which may be installed in a single building, and it is not the purpose of this section to give detailed guidance on maintenance of every conceivable system. With the information given in the previous sections and by reference to standard works on maintenance of particular services it should be possible to draw up and carry out adequate maintenance schedules for most types of equipment. Chapter 14 covers policy and administration in some detail. Chapter 18 considers the problems of spare parts and statutory inspections. However, the following general and specific notes on particular aspects of some types of building service not covered elsewhere may be useful. There is also a series of maintenance guides[5] available which offer detailed guidance on the maintenance of the various elements in HVAC electrical and public health systems.

Figure 6.7 Bracken House, London. An example of pump sets installed with space for simple maintenance and dial gauges which can be easily monitored to check pump performance. (Architect: Michael Hopkins and Partners. Engineers: Ove Arup and Partners)

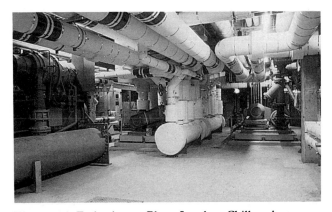

Figure 6.8 Embankment Place, London. Chiller plant room illustrating spatial and plant layout to ease access to equipment for maintenance, servicing and replacement. (Architect: Terry Farrell and Company Ltd. Engineers: Ove Arup and Partners)

6.4.2 Electrical installations

The Regulations for Electrical Installations, published by the Institution of Electrical Engineers, provide guidance to protect persons and property from electric shock, fire, burns and injury from electrically powered mechanical equipment. They are not all-encompassing and need to be read in conjunction with British Standard Specifications, Codes of Practice and relevant statutory requirements.

6.4.3 The IEE Regulations, as they are commonly known, currently published as the 16th Edition dated 1991 do *not* have statutory effect but they do list the Statutes with which installations have to comply. They can, however, be used in a court of law in evidence to claim compliance with a statutory requirement. They have now been adopted as a British Standard (BS 7671: 1992 Requirements for Electrical Installations).

6.4.4 The IEE Regulations are an invaluable guide and are intended to be cited in their entirety for contractual purposes. Electricity utilities are not compelled to commence, or in certain curcumstances, to continue to give an energy supply if the system does not comply with the fundamental requirements for safety.

6.4.5 The Electricity at Work Regulations 1989 (Statutory Instrument 1989 No. 635) which are made under the Health and Safety at Work Act 1974 are cited in the IEE Regulations. They do make non-technical as well as technical staff in buildings responsible for the safe operation and, explicitly, the proper maintenance of electrical installations to achieve such operation.

6.4.6 Both the IEE Regulations and the Electricity at Work Regulations make it clear that only personnel with the relevant technical knowledge shall be engaged on any work on electrical installations in buildings. The IEE Regulations and the cited standards, codes of practice and statutory requirements provide the basic information for safe maintenance and the equipment manufacturers' literature should be used for the detailed maintenance needs.

6.4.7 Lighting installations

A properly designed lighting installation will take account of the activities to be lit, characteristics of the building occupants, reflectivity of the surfaces within the spaces lit, any available daylight, contrast, glare, aesthetic and other considerations. Not least of these will be the gradual deterioration of the luminaires in terms of light output due to various causes. A lighting designer should consider specifically maintenance factors and the most economic policy for replacement of failed lamps. These factors are considered further in the CIBSE Code for Interior Lighting, which forms an invaluable guide to maintenance of lighting installations.

6.4.8 Not only do lamps have to be replaced but fittings have to be cleaned at intervals, depending on the luminaire design and the ambient dirt in the atmosphere. Attempts to restore lighting levels by fitting higher-power light sources in dirty fittings are singularly pointless. Illumination levels will fall if windows are not kept clean, and alterations to, or dirty, interior finishes will also affect the efficiency of a lighting installation. Thus building fabric maintenance is closely tied to maintenance of lighting installations.

6.4.9 Communication systems

For maintenance purposes public address, telephone, paging, alarm, CCTV (closed circuit television) and IT (information technology) systems may be grouped together. All of these are generally light current, pre-packaged installations wired up on site. Maintenance should be minimal, but regular and it is strongly recommended that this is dealt with by maintenance contracts with either the suppliers or installers of the systems. Only in the largest buildings or building complexes will it be economic to train and retain maintenance engineers with the special skills required for such systems. Where alarms are concerned, insurance company requirements must not be neglected.

Figure 6.9 Boots, Cramlington. Plant room with multi-level plant installation with suitable access for maintenance between separate elements of equipment and at different levels. (Architect: Faulkner Brown, Hendy, Watkinson and Stonor. Engineer: Ove Arup and Partners)

6.4.10 Lifts, escalators, hoists and cranes

All systems for lifting people and goods are subject to a variety of statutory regulations (see Chapter 18) and insurance companies may also lay down specific testing and inspection requirements. Maintenance contracts with suppliers, installers or specialist companies are strongly recommended, and these should be arranged in consultation with insurers where appropriate. The CIBSE Guide, volume D, *Transportation Systems in Buildings* (1993), provides a chapter on commissioning, inspection and maintenance.

6.4.11 Heating services

Apart from energy requirements for industrial purposes, heating systems are the main consumers of energy in most buildings. Whether lower limits of temperature are laid down by statute or not, failure of the heating prompts complaints, and calls for equally prompt action. Thus preventative maintenance is particularly important and, for all installations maintenance procedures should be set up, to be carried out by in-house or contract staff. They may require training in dealing with particular items of equipment, and suppliers may provide short courses at reasonable charges. This training may be essential where sophisticated control systems are installed for fuel economy.

6.4.12 Maintaining water quality in hot water systems is a particularly important aspect of heating system maintenance. The effects of poor-quality water are insidious, ranging from sludging up of radiators with reduced resultant heat output through internal and unnoticed corrosion of systems, to actual blockages and possible danger to plant and personnel. The problems associated with water systems of this type are covered by the CIBSE Guide Volume B, Section B7, Corrosion and Water Treatment, 1986. The necessary maintenance can be inferred from the information given in Section B7 or from BS 6700: 1987.[6]

6.4.13 Boilers and steam vessels

As for lifting devices, these are covered by both statutory regulations and strict insurance company inspection and testing requirements. Normally, in-house staff will look after such equipment on a continuous monitoring basis. For all but the smallest boilers, performance records should always be carefully made and kept up to date. Insurance companies should be consulted about the adequacy of records and maintenance routines, and they may require independent testing and inspection at specified intervals.

Figure 6.10 Bracken House, London. Illustration of the cable management necessary for equipment rooms in modern premises. This demonstrates the care which needs to be taken at the design stage to facilitate maintenance and necessary modifications after completion. (Architect: Michael Hopkins and Partners. Engineer: Ove Arup and Partners)

6.4.14 Ventilation

There are standard recommendations for air change rates for different types of building and activities within them and statutory regulations concerning ventilation of spaces for various types of activity. Obviously, ventilation must be maintained to comply with these, basically for health and comfort reasons. The increasing level of complaints about sick building syndrome and other building related illness is often blamed on poor ventilation and air quality. While there is no absolute causal connection it is good practice to always maintain the fresh air input to the design value(s). In certain circumstances duplicate systems may have to be installed, with automatic changeover in the event of failure of one system, and regular inspection to detect such failures is essential.

Figure 6.11 Bracken House, London. An example of an almost completed floor of a commercial office, illustrating that maintenance now requires access to plant at both false ceiling and false floor levels. (Architect: Michael Hopkins and Partners. Engineer: Ove Arup and Partners)

6.4.15 Extract ventilation of fumes and steam from kitchens is particularly vital, and in many cases filters will be installed in extract ducts. Grease filters in kitchen hoods are designed mainly to protect the ductwork and extract fans from build-up of dirt, and these must be regularly inspected and changed, sometimes at intervals as short as a week in such places as restaurant kitchens. Facilities for washing filters may be required. Failure to clean filters may lead to infestation with flies or maggots in the filters, and reduced air flow, which can in turn lead to condensation in duct-works and moisture leakage below. In bad cases prosecution will be a possibility under the Public Health Acts.

6.4.16 The effectiveness of well-maintained ventilation systems may be adversely affected by the behaviour of the building occupants. Doors propped open may disturb the designed pattern of air flow through and out of rooms, and the occupiers should be made aware of this possibility where it exists. Similarly, failure to maintain extract fans, for example non-replacement of worn belt drives, may lead to other fans drawing air from one space into another with resultant complaints. As an example, a restaurant and kitchen may both have extract fans, with those in the kitchen area being more powerful to produce a positive air flow from the restaurant area into the kitchen. If the kitchen filters become blocked or the kitchen fans work below normal efficiency, steam and kitchen smells may be drawn into the restaurant area.

6.4.17 In cases where quantities of dust are to be extracted build-up of dust in ducts may decrease extract efficiency. Depending on the type of dust extracted, great care may have to be exercised in the design of the system and in any maintenance work carried out on it, to avoid explosion risks.

6.4.18 Chimneys and flues

Well-insulated flues from properly adjusted combustion should rarely give trouble, but periodic inspection is advisable. If boiler plant is not well maintained there may be build-up of soot or condensation problems leading to flue corrosion, which may only become evident after damage has been done or a fire breaks out. Flue linings may be relatively fragile, and care in cleaning is necessary.

6.4.19 Air conditioning

Many of the problems and recommendations for heating and ventilating systems above apply equally or with greater force to air-conditioning systems. The variety of the latter makes it difficult to give useful general advice briefly, and makes skilled maintenance correspondingly more important. In other than large buildings where a permanent staff are employed contract maintenance is to be recommended.

6.4.20 In addition to heating and ventilating equipment, the air-conditioning system will have refrigeration plant probably involving cooling towers and, in some cases, equipment for humidity control. The testing and balancing of such systems requires special skills and equipment, and is best left to installers or their specialist commissioning sub-contractors. Full records of commissioning are particularly vital for air-conditioning systems.

6.4.21 Refrigeration plant for conventional commercial air-conditioning systems uses chlorofluorocarbons (CFCs) as refrigerants, which cause depletion of the ozone layer. The production of CFCs will soon be banned under the internationally recognized Montreal Protocol on the subject. In the past, CFCs have often been released to atmosphere during maintenance or refrigerant recharging, but this is no longer permitted. All maintenance procedures need to take this into account.

6.4.22 Cooling towers also present another problem in terms of Legionnaires' Disease. Improperly maintained towers, without suitable water treatment, can generate aerosols which can cause outbreaks of the disease. If the towers are properly maintained in accordance with the guidance from various publications[7] there should be no problems.

6.4.23 It is very important that the location of pressure and temperature tappings used for commissioning are recorded and known. This will ensure that when any item of plant is maintained, changed or replaced, or the system loads or operating conditions have to be adjusted, the system can be effectively rebalanced and recommissioned.

6.4.24 Maintenance of different parts of air conditioning systems may be carried out on different principles: routine checks and cleaning or replacements for some components, replacements or maintenance on a performance basis for others such as filters, and breakdown maintenance. A clear policy should be laid down for the maintenance of each and every part of the system.

6.4.25 A problem which can occur in heated and naturally ventilated buildings concerns the percentage of people satisfied with the environmental conditions. Everyone has varying views of what conditions are comfortable and differences in clothing exaggerate this effect so that it is impossible to satisfy everyone, however good the conditions are said to be[8]. This situation is exacerbated in air-conditioned buildings where most occupants will expect to be comfortable whatever the weather conditions, the clothing they are wearing and the work that they are doing. Thus complaints of malfunction should be carefully considered before any non-routine maintenance is carried out or changes are made to the system.

Figure 6.12 Lloyd's, London. An example of where architecture and engineering are integrated to the point where the maintenance of the building and the services need to be considered as an entity, as an essential part of facilities management. (Architect: Richard Rogers. Engineers: Ove Arup and Partners)

6.4.26 Another factor leading to complaints about the operation of such systems is the changes in occupation and heat gains which inevitably occur during the life of a building. Putting in a tea boiler or a printing machine with a high heat output, or allowing large numbers of people to congregate in a room whose air change rate has been set for a small number of occupants may produce uncomfortable conditions which the air conditioning cannot rectify. Different systems can be designed to cope with different degrees of variability of this kind, but sooner or later the extreme is liable to occur, with resultant complaints to the maintenance staff.

6.4.27 Other changes such as the insertion of fire doors or changes in glazing may have similar effects, and necessitate rebalancing or modification of HVAC systems.

6.4.28 Water services

The water services of buildings are probably subject to a wider range of Acts of Parliament,

statutory regulations and byelaws than any others (see Chapter 18). The Water Authorities have their own byelaws, based on the Water Acts, for preventing waste, undue consumption, misuse or contamination of water supplied by them, and these have detailed provisions concerning the design, maintenance and modification of water supply systems. However, the Water Authorities' responsibility for maintenance normally ceases at the point where the supply enters the site, at the stop tap, and building owners and occupiers are responsible thereafter.

6.4.29 Cold water supplies generally require little maintenance beyond replacement of tap washers and attention to ball valves, unless bursts occur due to inadequate frost protection. However, where possible, water consumption should be monitored with leak detection in mind, particularly in older installations where buried or other pipes may be subject to corrosion. Replacement of damaged pipework must be in accordance with the relevant byelaws, and in older buildings should take account of the possibility that electrical systems may rely on the pipework for earth connections. Most underground water mains are now being installed in polyethylene and metal rising mains should not be used to earth electrical systems.

6.4.30 Hot water for washing and other non-heating purposes is generally provided by boilers with some form of heat exchanger or by electrical immersion heaters. Gas- and oil-fired boilers require regular routine maintenance, best carried out on a contract basis except perhaps in larger buildings. Electrical immersion heaters rarely require maintenance, though their thermostats may be liable to drift with time, producing water hotter than necessary with consequent energy losses. Where consumption is limited to certain times of day, time controls are essential for economy, and these require periodic checking. Hot water systems, like cooling towers, can be a source of Legionnaires' Disease and temperature control at temperatures normally not lower than 55°C is recommended, together with guidance on the length of dead legs and maintenance procedures.[7]

6.4.31 Excessive water hardness may lead to gradual furring up of pipes and loss of performance in heat exchangers. Water-softening plant requiring frequent checking may be necessary in some areas in particular types of system. Where water supplies are known to be particularly hard and neither water softening nor other forms of water treatment are practised, steps should be taken to check the extent of furring up of pipework and heat exchangers from time to time.

6.4.32 Discoloration of water may occur very occasionally in water supplies, but the cause will more often lie within the building's own services. The use of unsuitable materials in pipework or cisterns leading to corrosion is a common cause of discoloration, but it may also arise from algal growth within reservoirs or cisterns. Regular checking and cleaning of on-site storage is recommended. Consideration should be given to upgrading older domestic water storage cisterns in line with the current water byelaws. The addition of lids and screens will prevent algal growth and many other potential contamination problems. Regular water analysis should be carried out to check the condition of the system.

6.4.33 Further information on maintenance procedures can be found in BS 6700: 1987.[6]

6.4.34 Drainage services

These can be broadly classified under two headings: foul and surface water drains. Foul drains take sewage and other effluents from buildings either to the Water Authority's sewers or to private disposal installations, and surface water drains carry basically rainwater to soakaways, watercourses or sewers. The Public Health Acts govern foul drainage and, to a certain extent surface water drainage.

6.4.35 Having few (if any) moving parts, drainage systems rarely require other than routine maintenance, but the importance of the latter must be stressed. Accumulations of silt, grease, solid matter and contaminants such as soil and petrol from hard-standings can lead to serious problems and sometimes to flooding. Inspection pits, silt traps, interceptors and other access points in drainage systems must be regularly inspected and cleaned out at intervals depending on usage. Manhole covers should also be inspected regularly for damage, to avoid danger to those passing or driving over them. Where trees are present, clearing of leaves may be a seasonal problem in surface water drainage systems. Tree root growth may also lead to disturbance or even blockage of underground drainpipes.

6.4.36 Where it is necessary to pump effluent up to the level of the off-site sewers daily checking is recommended, in addition to regular planned maintenance of pumps, level electrodes and control gear. Wet wells may require regular washing down, depending on design and flow rates, to avoid build-up of sewage which may turn septic and give off noxious smells. Great care must be exercised by anyone entering wet wells or sewers and breathing apparatus may be necessary (see Chapter 13).

6.4.37 Where grease traps are installed regular maintenance is essential. Traditional traps require regular removal of grease and biological traps require the regular addition of enzyme powder. Failure to do either will allow grease to pass through and congeal in drainpipes with consequential blockages. Where a pump chamber is installed the sump will act as a grease trap and grease will congeal on float switches, etc. This is a common cause of pump malfunction.

6.4.38 Petrol interceptors, where installed, also require regular inspection and maintenance. Any petrol or oil will require regular removal as an excessive build-up will allow pollution through to the public sewer and/or watercourses.

6.4.39 Further information on maintenance procedures can be found in BS 8301: 1985 Code of Practice for building drainage.

Acknowledgement

The initial work of Don Montague in producing the first draft of the chapter is gratefully acknowledged as is Gordon Puzey's current input for Water Services. The photographs in this section are reproduced by kind permission of Ove Arup & Partners.

References

1. Mills, Edward (ed.), *Planning: Architects' Technical Reference Data*, 10th edn, Chapter 7, 'Planning for services', Butterworths, London (1985)
2. TM17: *Maintenance Management for Building Services*, CIBSE (1990)
3. *CIBSE Guide*, Volume B, Section B18 (1986)
4. *Operating and Maintenance Manuals for Building Services Installations*, Parts 1–3, Building Services Research and Information Association (1990)
5. *Standard Maintenance Specification for Mechanical Services in Buildings*, Heating and Ventilating Contractors' Association. Volume 1, *Heating and Pipework Systems* (1990); Volume 2, *Ventilating and Air Conditioning* (1991); Volume 3, *Control, Energy and Building Management Systems* (1992); Volume 4, *Ancillaries, Plumbing and Sewerage* (1992); Volume 5, *Electrics in Buildings* (1992)
6. BS 6700: 1987 Specification for design, installation, testing and maintenance of services supplying water for domestic use within buildings and their curtilages
7. TM13: *Minimizing the Risk of Legionnaires' Disease*, CIBSE (1991): HS(G) 70: *The Control of Legionellosis Including Legionnaires' Disease*, The Health and Safety Executive, HMSO (1991): *The Prevention or Control of Legionellosis (Including Legionnaires' Disease)*. Approved Code of Practice, Health and Safety Commission, HMSO (1991)
8. *CIBSE Guide*, Volume A, Section A1 (1986)

Further reading

BS 8210: 1986 Guide to building maintenance management
Building Services Research and Information Association, *Maintenance Contracts for Building Engineering Services. A Guide to Management and Documentation* (September 1989)
Building Services Research and Information Association, *Building Services Maintenance*, 2 vols (1990)
Building Services Research and Information Association and Armstrong, J., *Condition Monitoring – Introduction to its Application in Building Services* (1986)
Building Services Research and Information Association and Armstrong, J., *Inspection of Building Services Plant and Equipment – A Review of Current Practice* (1986)
Chartered Institute of Building and Armstrong, J.H., *Building Services Maintenance and the Use of Computers* (1987)
Department of Education and Science, *Maintenance and Renewal in Educational Buildings, Maintenance of Mechanical Services*, DES Building Bulletin 70, HMSO (1990)
H & V News, Cooling towers. Guidance on good maintenance, 17 June, 9–20 (1989)
Heating and Ventilating Contractors' Association, *Maintenance and Disinfection of Evaporative Cooling Towers. Guidance Notes on the Method to be Adopted* (1990)
Industrial Water Society, *Keeping Your Cooling Tower Safe. Maintenance Guidelines for the Prevention of Legionnaires' Disease* (1987)
Institute of Refrigeration, *Safety Code for Refrigerating Systems Utilizing Chlorofluorocarbons. Part II. Commissioning, Inspection and Maintenance* (1989)
Institute of Refrigeration, *A Safety Code for Compression Refrigerating Systems Utilizing Ammonia. Part II. Commissioning, Inspection and Maintenance*, revised edn (1990)
James, D.B., Maintenance by contract in the public sector, *Building Trades Journal*, **166**, 16 February, 40–47 (1985)

7
The maintenance and design of security systems
David M. Lush

7.1 Introduction

7.1.1 Security systems have to be designed for all aspects of assessed risk, which may need to encompass vandalism, violence and terrorism. Within, or parallel to, these three headings the topics of burglarious entry, industrial espionage and itinerant visitors (or squatters) have to be considered. Security of services, etc. for the safety of the building occupants is also very important. What follows is intended to range as widely as possible and concerns all the design disciplines, because it is essential that the design professions overlap and communicate, particularly where their individual interests interface with one another.

7.1.2 Systems range from elements of buildings, e.g. specially strengthened doors, to very high-technology solutions, e.g. motion and acoustics sensors coupled to alarms. Their maintenance is not simply a desirable feature which needs to be considered – it is an essential and critical element which must be part of the facilities management for any building.

7.1.3 Security (and fire-defence) systems are unlike any other systems incorporated into buildings. All the other systems function regularly, either continuously or over reasonably long periods, and if they fail in their operation it is obvious to the occupants and virtually never dangerous. Security and fire systems, on the other hand, do not operate continuously or regularly, but on the rare emergency occasions when they are called upon they must be functional and operational. This is their *raison d'être*. Without proper regular maintenance this cannot be achieved.

7.1.4 The distinction between security and fire-alarm systems and other services is made to identify the fundamental differences between day-to-day use and emergency operation and does not imply that maintenance is unnecessary for other services. In modern buildings with any high-technology security and fire alarm systems the internal environment and functional operation of the building will only meet the design intent, and satisfy the occupants, if all the services are properly maintained.

7.1.5 The maintenance of security systems also differs in one other major respect from all other services, including fire defence. It can only be defined in general terms and normally will be carried out by the specialist supplier. Any more detailed instructions would be advertising precisely how to defeat the whole purpose for which such systems are installed.

7.1.6 While it seems straightforward to speak of security systems *per se* and their maintenance, it would be wrong to ignore the security requirements and risks associated with the traditional building services themselves. In modern buildings the utilities' reticulation and building services systems are essential for the production process, whether it be 'widgets' in an industrial company or mortgage statements from the head office of a building society. However, their security is guaranteed more by the design process and cooperation between the members of the design team than by adding further high-technology security systems.

7.1.7 This leads to a dilemma in terms of describing how maintenance should be carried out, but does not inhibit guidance on how suitable levels of security should be incorporated into the design process for different levels of risk.

7.1.8 It may appear simple to separate the effects of vandalism, violence and terrorism on buildings and services into separate compartments. In practice, buildings and services security is the most comprehensive general heading as both need to be covered against acts of vandalism and terrorism. The division between what vandals and terrorists can achieve in relation in some building services is a very grey area and the worst excesses of vandalism are no different from some terrorism.

7.1.9 For security against terrorism aspects, many of the points relate to very high security risk buildings, to suit the most complex situations. There is no suggestion that this is necessary for all buildings, nor would we want to live in such a society. However, these problems do arise and the level of security has to be evaluated against the risks. The necessary precautionary measures for the risk should, whenever possible, be part of the initial building design.

7.2 Vandalism

7.2.1 Vandalism is defined as 'wilful destruction', but what motivates the vandals? The physical environment of the building, its design, detailing and fittings may encourage vandalism. It cannot be condoned, but the effect of entering gloomy lift lobbies, depressing subways and walkways and unsupervised stark toilet facilities is depressing, even to the well-adjusted person. Under such circumstances light fittings, toilet accessories and lifts are magnets to the vandals who have time on their hands, no outside interests, and in such surroundings may well be encouraged in their destructive efforts by their friends. It should be remembered that access cannot be easily or properly denied to areas such as those described above, because the majority of such buildings and areas are part of the public domain.

7.2.2 What can be done? In the long term, better education should be the answer. In the shorter term there are various remedies. The first is to select fittings, lights in particular, which are classified as vandal-proof and which can also be recessed wherever possible. The term 'vandal-proof' has only a limited meaning and care has to be taken to select the most suitable items from those available. Integrated design is necessary, even for this small element, as the architect and structural engineer have to permit the fittings and conduit to be completely carcassed into the building fabric. A second remedy is to ensure that such areas in buildings are less depressing and better lit; this affects planning, architecture and cost. Both solutions carry an initial financial penalty but this will normally prove the cheaper option when applied to cost in use over the life of the building.

7.2.3 This is a very generalized statement about vandalism and specific examples are detailed elsewhere. There is no universal panacea in design terms that will stop all vandalism.

7.2.4 Maintenance, where vandalism is rife, is largely a combination of good design selection/suitable installation of components and the resources available to either prevent or repair the damage. The level of security being offered by the elements which are vandalized in this manner will vary considerably with particular circumstances.

7.3 Violence

7.3.1 If considered by its definition 'as the unlawful exercise of physical force' then violence is not related directly to the security of buildings and services. Certainly, most vandalism is based on violence but the more general aspect relates to violence external to buildings.

7.3.2 The brief here is to cover security systems for buildings and services which, by normal standards, are those within the building, or external but associated with the building complex.

7.3.3 Apart from the elements considered under Vandalism, street-mounted services equipment is outside the brief but mob violence can wreck what are normally considered to be very substantial pieces of equipment in order to use them as weapons or obstacles. The solution to this lies in the maintenance of law and order within the community as a whole.

7.4 Building access security

7.4.1 Before addressing the technology or availability of security systems, or the specific security requirements of building services, one basic element of security design concerns physical access to the building. The Security Brief for the project from the developer/owner/occupier should, at the very least, define the level of risk assessment for security purposes. It will then be the responsibility of the architect to design a building shell which meets the restrictions, whatever they may be, for access to the building, in concert with the structural engineer and the building services engineer, or security specialist, who can advise on the associated alarm arrangements.

7.4.2 The very particular aspects which may be associated with more violent forms of access are treated later. While it is necessary to be aware of the needs of very high risk buildings it is to be

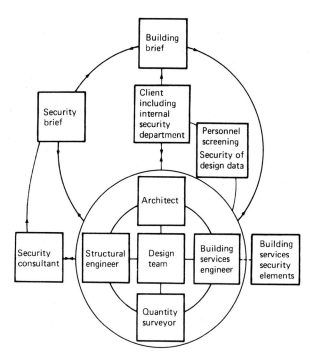

Figure 7.1 Relationship between client's requirements and design disciplines

facets of the design. Figure 7.1 illustrates how the designers should coordinate their efforts and the role of security consultant may be filled by the client, police, specialist consultant, the building services consultant or a combination of these, depending on circumstances.

7.4.6 One other point which touches on the design team, however it is made up, is the possible necessity for security screening. This is dependent on the type of building and the level of security required, but it may be essential in certain high security cases. Design data of certain elements of building structure, services equipment and their siting within the building would be very useful to organized groups illegally interested in the contents of the building. Thus design information may also need to be kept under secure conditions and this creates an additional range of problems.

7.4.7 The problems fall into two basic categories. First, an indexing and distribution system has to be applied so that data are available on a limited and recorded scale on the basis of a 'need to know' and these data must be totally recoverable. With the number of sub-contractors and suppliers on a conventional building with innumerable interface situations this is a major technical/administrative task in its own right which may be a costly extra on the project. Second, when the project is finally in operation and the data and drawings have to be used around the complex for servicing and maintenance purposes, a similar set of circumstances occur over the full life cycle of the building.

7.4.8 This is perhaps the classic situation of maintenance of security systems. The type of building or installation which falls into the category described above actually requires even the most mundane drawings, in normal circumstances, to be 'classified' and the building itself is the security system.

7.4.9 Whatever level of risk is to be countered within the building the prime objective must be to prevent access, first to the building and then to the main plant areas. In order to cover all possible aspects of access at the design stage by the shortest possible description, assume that the problem spectrum ranges from the vagrant to the terrorist with explosives.

7.4.10 Without specifying which types of building and usage are likely to be subject to terrorist attacks, one or two general design principles may be suggested. Figure 7.2 illustrates the considerations for various building services, access points, positioning and distribution.

hoped that the vast majority of building and other project designs do not need such attention.

7.4.3 The means by which the design brief for security purposes is developed is a function of the design team, in accordance with their specialized knowledge, applied to the client's building requirements. The inter-relationship between the various involved parties is shown in Figure 7.1.

7.4.4 Whether the problem is one of restricted access, damage directly or indirectly from external sources, or any other element, the subject of coordinated or integrated design should be considered. Architects, structural engineers, building services engineers and quantity surveyors working together rather than independently should, in normal circumstances, be able to produce an integrated design with improvements over a similar, conventionally designed building. In the special circumstances of vandalism and terrorism such integration would seem to be even more important. Apart from the technicalities of the design which will have innumerable interface conditions for security purposes, the cost implications of certain solutions need immediate consideration.

7.4.5 Designers from separate professional disciplines can, of course, offer acceptable solutions for the various security aspects of building design but integration will improve many

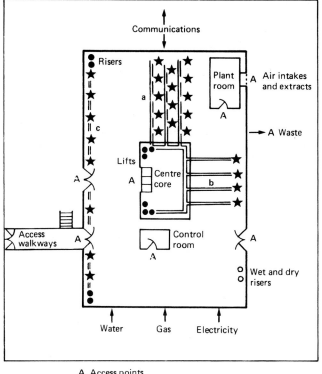

A. Access points

★ Distributed services including security systems

Figure 7.2 Building and services security elements affecting design

7.4.11 Where the building is designed for maximum security, i.e. little or no glazing, it will undoubtedly be air conditioned. In such circumstances, the services distribution should be designed so that it is secured and distributed from the core structure rather than the perimeter (a, Figure 7.2).

7.4.12 Whether the structure is then proof against explosive charges or not, the effect on the building services will be minimized. In extreme conditions completely independent plants should be used to serve separate sections of the building.

7.4.13 Other types of building which can be identified as terrorist-prone, but of more conventional construction, will presumably still require some perimeter heating (or cooling) system. Where possible, this should be designed so as to provide the minimum possible contact with the perimeter structure (b, Figure 7.2), rather than the normal distribution (c, Figure 7.2). Risers which are conventionally carried adjacent to the perimeter should be repositioned in protected areas such as the central core.

Similar principles should be applied for water, air and electrical distribution systems. The risk of explosion damage to services is one which simply emphasizes the recurrent theme of limiting access to the buildings and their services. Designing ductwork, pipework and most plant against adjacent explosions is not normally feasible in design or cost terms.

7.4.14 Access to buildings through doors or windows is reasonably easily monitored for security purposes, the level of security required governing the sophistication of the system employed. Various other means of access may also exist through certain of the building services. Figure 7.2 indicates these access points. Such access needs to be restricted, first by means of proper design and second, by suitable security measures and monitoring.

7.4.15 The building service which offers the best opportunity for unauthorized access is the ventilation or air-conditioning plant. Almost without exception, any ductwork system serving such a plant will be large enough for access. Equally, most buildings with such intake or exhaust systems have at least some of them virtually at ground level. Once past the first decorative louvred panel at the access point, the infiltrator, with the panel replaced or not, can gain entry at leisure.

7.4.16 Where such low-level access exists the first priority must be to make it secure by monitoring it as part of the general alarm system. When buildings are being specifically designed for security purposes these access points should be removed from ground level to inaccessible, intermediate or roof levels, facing inwards if the building configuration permits. To do this on a large complex may require a major change in the design philosophy and certainly dictates that all the design disciplines are involved together at the concept stage of the design. Another reason for this feature will be apparent under one of the later elements. Even with the correct siting of these intakes they should be included in the security alarm system in the same way as conventional entrances.

7.4.17 One other building service which may permit unauthorized and illicit entry is that of public health or drainage. Depending on the geographic location and methods of waste disposal, there may be means of access available from both the rainwater and foul drainage systems. Such access will be to basement or ground-floor levels and will remain there in any foreseeable design circumstances. The problem is therefore a combination of security monitoring, so that any unauthorized lifting of manhole covers or access panels can be identified early enough to prevent subsequent actions or sabotage, and of siting, so that such access only leads into otherwise empty and secured spaces. Outlandish as this suggestion may appear, it is included on the basis of 'you never know when you might need it'.

7.5 Access, monitoring and security systems

7.5.1 Following on from the subject of access via the services systems, the actual alarm, security and detection systems should also be considered. Conventionally, the building services designer is responsible for the incorporation of such systems into the services design. Security, as with fire defence, is a specialized topic which is becoming increasingly technologically complex. Some building services design practices may not have sufficient expertise in this topic to carry out all the necessary specification and design work, and it is obviously desirable that they consult the specialists. There are also practices with sufficient expertise to identify and analyse the problem, produce a solution and specify the requirements quite clearly to suitable equipment suppliers and installers. In either case, designers who are aware of security needs will readily discuss the problems with specialists, if only to keep abreast of the rapid developments in this field. A cautionary and possibly contentious note concerns the ability of the 'specialists'.

7.5.2 There are consultants who specialize in this field of activity and can produce a comprehensive design covering the complete range of security systems, wholly independent of equipment manufacturers. At the other end of the scale are large manufacturing organizations which cover the whole range of systems, frequently with separate technical and sales organizations for each system. Each systems division offers a technical service but the coordination into a comprehensive whole by the controlling organization is sometimes lacking.

7.5.3 It is certainly difficult to achieve this coordination at times, whatever promises are initially made. If the client has not, via an architect, nominated a specialist professional security consultant, then the responsibility will lie with the building services designer. He or she will then have to make the decision on the correct direction in which to proceed, dependent on the client's brief, the complexity of the system and the designer's own knowledge of the subject.

7.5.4 Whichever design philosophy is adopted it is necessary to remember the title of this chapter. The maintenance needs for the individual and combined security systems must be included as part of the design specification so that tender offers can be judged on service ability, performance, mean time between failures and call-out times, as well as apparent first cost. Life-cycle costing can be very important in terms of security systems. Despite the earlier cautionary note about how far maintenance can be defined for security systems, it is true to say that knowledgeable consultants/building services engineers will evaluate offers so that suitable maintenance is provided from the points of view of both retention of security and selection of the most appropriate tenderer.

7.5.5 What is the best solution? It must include selection of designers who have a broad-based knowledge of the possible problems with sufficient specialist skills to either provide, or obtain, the solutions. At any particular level the designer must have the ability to decide where external specialist help is required and this includes consultation with various police authorities.

7.5.6 It is not the purpose of this review to provide details of specific security systems but this is a suitable point to illustrate the range of security and associated systems for which the building services designer is normally responsible. They include:

- Door, window and access point alarms
- Perimeter fence monitoring
- Closed-circuit television, internal and external
- Acoustic, laser and similar alarm systems
- Various fire-defence systems
- Electronically controlled access cards
- Communications
- Data-control centres and software.

7.5.7 Each of these systems (and variants appear very regularly) has its own specific maintenance and servicing requirements, which should include performance checks. However reliable the systems are, proper maintenance has both value and cost associated with it.

7.5.8 At the same time, the designer must not contravene central or local government laws on safety or technical requirements, and this may create difficulties in high-security buildings.

7.6 Security of building services

7.6.1 While security systems are an identifiable area of design in their own right the security procedures necessary to ensure that building services are always available need to be outlined. In most normal situations building design teams do not have to consider the actions of third parties with nefarious reasons for disabling building services systems, but we are all aware of the problems particularly in modern highly serviced buildings, and in quite simple buildings, when there is a power failure. In high-risk and high-security buildings power outages or other means of disabling services may be created deliberately. Because these failures can totally disrupt the functional operation of buildings it is worth identifying what can be done at the design stage to minimize such disruption to services. The maintenance of the services themselves has been dealt with in Chapter 6.

7.6.2 Apart from the activities within the building, or explosive effects on its exterior, consideration must be given to possible external activities against the services feeding the building. Figure 7.2 indicates the range of these services.

7.6.3 Electricity

The most vulnerable service to the majority of buildings is almost certainly the electrical supply. Figure 7.3 is a schematic of the various forms of electrical supply which may be encountered in practice both in the UK and elsewhere. The use of the various forms for different types of buildings is illustrated in the table which is associated with the figure. Some of the more

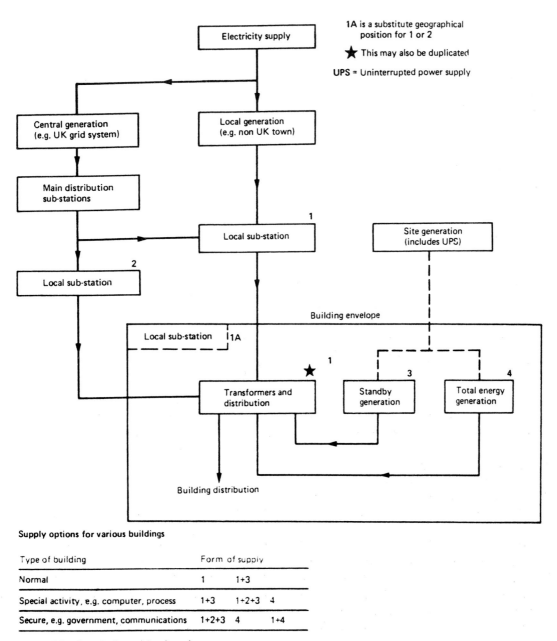

Supply options for various buildings

Type of building	Form of supply		
Normal	1	1+3	
Special activity, e.g. computer, process	1+3	1+2+3	4
Secure, e.g. government, communications	1+2+3	4	1+4

Figure 7.3 Possible electrical supply situations

complex options may be adopted independent of extremist activity, because of the non-reliability of the external supply. The schematic is only a guide to the possibilities. In many cases the transformers would be part of the sub-station.

7.6.4 The schematic and its table are virtually self-explanatory. The simpler forms of supply lend themselves to damage or destruction without any risk to the extremist or vandal who only has to obtain access to unprotected sub-stations.

7.6.5 In essence, the higher the security risk, the more independent sources of non-interruptible standby supplies become necessary. When designing the more sophisticated on-site generation and uninterrupted power supply (UPS) systems, problems of space, noise, fuel storage and suitable maintenance become additional design features.

7.6.6 Other fuels

With the exception of power stations, coal supplies are unlikely to be at risk for any of the buildings considered here. Oil and gas may be at risk. Where gas is the primary fuel its protection should be treated in much the same way as described for electricity. Like electricity, site storage is minimal but unlike electricity the site production of gas is highly improbable. Therefore, where gas is used, some form of dual-firing equipment should be employed with the secondary fuel being oil and the storage capacity being adequate for the estimated maximum gas outage period. In the UK gas tariffs for large installations do include an interruption clause which may, in theory, be for periods as long as 60 days, so that oil storage would in any case be required against this eventuality.

7.6.7 Where oil is the primary fuel it is less likely that there will be an alternative fuel. Whether oil is supplied, either by pipeline or tanker, the storage capacity must in fact be suitable for full load operation for the estimated maximum period for which supplies are not available. On highly secure buildings a strike of tanker drivers may be just as critical as any extremist activity.

7.6.8 In any case, where fuel is being stored, the positioning of large storage tanks, where security is a priority will frequently create building design problems at both the planning and detailed design stages. The tanks have to be secure from unauthorized access but accessible for checking leaks and firefighting purposes.

7.6.9 Communications

For security purposes the maintenance of communications external to the building is vital. The most accessible communications link for the terrorist is the telephone system. In the UK British Telecom no longer has a monopoly on hard-wired communication routes into a particular site or building complex and there may be independent sources of incoming communications lines.

7.6.10 It is possible to cut such incomers both literally and metaphorically and to stop normal hard-wired data transmission. This does not mean that this is a common event by reason of terrorism, vandalism, fire or flood, but it has occurred. It is therefore imperative to ensure that, where necessary, there is a duplicate incomer at a separate intake, served from a different exchange and that other forms of communications are also available which are not cable-dependent. Obviously, telephone exchanges themselves are particularly vulnerable, a point clearly understood by extremists.

7.6.11 Water

The water supply to a building or even an area can be prone to attack and dislocation, as was illustrated by the example of a pipeline explosion in Wales in the 1970s. Depending on location, multiple water inlets from separate sources are recommended. Apart from the necessity of maintaining the supply for normal security purposes, the needs of fire-defence systems are extremely important in the context of security.

7.6.12 As with oil, storage facilities have to be provided of a suitable capacity to satisfy the needs of domestic hot water services, drinking water, HVAC (heating, ventilating and air conditioning), cooling towers and fire defence for a specified period of time. These facilities should be separated for the various services and stored in different areas of the building.

7.6.13 Introduction of harmful substances

The elements so far, cover damage or dislocation caused to building services by destructive means. It is also possible that the services themselves may be utilized to create havoc within buildings, without any attempt at personnel infiltration. This can occur from the introduction of various harmful substances into air, water or fuel which is

Table 7.1 Some specific building services security problems

Source of problem	Danger related to services	Solution
Danger from within	Electrical dislocation Data passed out via waste systems	Restricted access to electrical distribution centres Filtration/monitoring of waste products (wet and dry)
Fire-defence systems	Alarms out of action Water supply cut off Water damage by vandals	Protect electrical distribution Duplicate sources and prevention of unauthorized access Prevent unauthorized access
Lifts	Lifts disabled by vandals Lifts damaged by terrorism Unrestricted access to and from lifts	Lighter and brighter aesthetic surroundings Stronger car equipment Prevent unauthorized access Individual or centralized control by card or key, programmed for security
Lighting	Cutting off electrical supply Damage to luminaires by vandals	Include maintained and emergency (self-powered) systems Prevent unauthorized access Select special luminaires Integrate luminaires into design of building fabric Remove from easy reach
Controls	Any service is vulnerable if its control system is inoperative	Secure electrical supplies Duplicate wiring by separate routes Restricted access control centre with duplicate/back-up equipment remote Security and checks on software

being supplied to the building. It is not intended to cover the detailed possibilities in this particular field of activity but certain necessary features of preventive design are outlined.

7.6.14 The suitable positioning of air intakes and extracts has already been mentioned to prevent unauthorized access of personnel. The adoption of the same design philosophy is equally necessary to prevent the introduction of substances into the air distribution system. The filtration equipment and monitoring of air quality may be extremely important on a high-security building, even with the intakes designed to be inaccessible. This may be extended to include alarms and emergency actions in the event of contamination being detected. It should be remembered that filtration equipment is very selective in terms of the large range of possible noxious substances which may be introduced and that this form of security may be very costly.

7.6.15 The introduction of dangerous substances to the water supply is also one which may be accomplished from outside the building boundary. Unlike air, it is not possible to remove the access points beyond reach. While this form of activity seems unlikely it should be referred to as one link in the security chain which may have to be considered. One basic solution is to take all water through secure primary storage tanks, which are monitored regularly or continuously for possible forms of contamination, and then pump the tested water from them to the main storage tanks.

7.6.16 The contamination of air or water will affect personnel directly. The contamination of fuel may bring the plant to a halt. The degree and form of contamination may affect certain fuel uses more than others. Here again such actions are unlikely but there is a potential security hazard now and possibly more so in the future.

7.7 Other specific problems

7.7.1 There are a number of problems which do not easily fit in the elements described so far. Table 7.1 tabulates several of these and outlines the dangers, with proposed solutions.

7.7.2 Restricting access to buildings and plant has been emphasized as being the best solution for safeguarding major systems. If, however, the staff includes personnel whose reliability is not absolute then problems can be caused from within, related to all aspects of security. There are specific areas which are relevant in this context:

1. (a) The complete disruption of the electrical system at a key point and hence the possible interruption of security and communication systems.
   (b) Sabotage of air conditioning and heating which would disrupt the building operation.
2. The less obvious but effective use of waste-disposal systems to pass out data, thus avoiding the normal security procedures. This point is clearly related to security measures

which are more likely to be related to industrial espionage or disgruntled employees.

The remedy to the first is very high security at the electrical distribution key point(s) and for the heating and ventilating plant rooms, while sophisticated filtering/monitoring arrangements will overcome the second. This emphasizes the need for security screening to a suitable level, based on the risk.

7.7.3 This group of problems is related to maintenance only in the sense of maintaining good personnel management. There may, of course, be remedial technical work if the problems do actually occur.

7.7.4 Fire defence

This covers a very wide range of systems in its own right. The systems are related to security of personnel and property and are part of the overall security system of the building. Like other security systems, the integrity of fire-defence systems needs to be assured and this is covered under normal design procedures (see Chapter 14). Where extremist activities are considered as an additional fire hazard, it is essential that alarm systems, water storage and wet and dry risers be designed into the building to minimize the possible damage from a fire so caused.

7.7.5 If we consider the mundane but more frequent occurrence of vandalism and the fact that fire-defence systems must always be operational, whether the building is occupied or not, then the access situation becomes crucial. Most buildings do not fall into the high-security category. If they are not secure they may be entered by vandals (or thieves) and the vandals can cause more damage with water than they can by breaking whatever is breakable. Fire-defence systems, once installed, are part of this risk as, to a lesser extent, are any other wet systems in the building. One can only pose a question in this respect. How far does education or discipline need to go to reduce vandalism, as an alternative to the provision of security systems and guards to prevent unauthorized access to unoccupied buildings? The answer cannot be given here as it is a complete subject in its own right.

7.7.6 Lifts

Vertical transport has many advantages to the user. It may also suffer from the effects of vandalism and be subject to the requirements of the security system.

7.7.7 Vandalism in lifts is regrettably well known and occurs in rather specific types of building, such as local authority high-rise housing. It creates inconvenience and delay rather than, of itself, being necessarily costly to correct. Solutions are probably long term. Brightening the decor and better lighting of the lobbies would help as would more sturdy control equipment in the lift cars and lobbies. This is no substitute for people respecting the environment in which they live and this must be an educational process. Guarding the lobbies and restricting access thereto has proved successful in areas where it has been tried.

7.7.8 In security terms lifts may be used to provide access to particular floors and deny access to others. Keys or electronically accepted access cards are two methods by which selective access to and from lifts, or any other areas, may be achieved. The not uncommon 'Directors' Lift' in high-rise offices is a simple example of selective access. In addition, it is not difficult to override any lift controls from a control centre such that it may be stopped, started, sent to or held at a particular floor, with doors open or closed, in accordance with security requirements. The simplest example of this is the automatic arrangement which runs all lifts to the entrance lobby when there is a fire alarm.

7.7.9 Lighting

This is mentioned for completeness in a review of building services and security. Suffice to say that buildings these days are commonly designed with normal, maintained and emergency lighting, even when there are no high-security risks. So long as this is done with special risk buildings and the electricity supply is maintained as indicated earlier, there is unlikely to be a problem.

7.7.10 Lighting which is installed in unprotected areas subject to vandalism must be carefully selected, integrated as far as possible into the building fabric and kept out of easy reach.

7.7.11 Controls

Controls have a very wide meaning, depending on the profession of the designer using the term. In normal use for building services it used to refer to the thermostatic controls and data centres relating to the HVAC system, but it does, of course, also cover lighting controls, fire defence, lifts and security systems, etc. Where there are complex systems, with or without sophisticated security requirements, the use of microprocessor-based software-driven direct digital control (DDC) data centres and outstations are now an accepted design philosophy. A cautionary note should be sounded. Do not accept any system without independent professional advice. The manufacturers may not always be totally objective when offering their own systems for this activity.

7.7.12 It is common for data centres to be located in an area or space set aside for the purpose (Figure 7.2) but access is often unrestricted. When high-risk security is involved, the area itself becomes the key point for security operations. This then becomes a segregated and secure space in its own right, normally within the building which is already protected. It is important to decide whether this area is common for security and plant surveillance, whether it is split physically with separate staff for each function or whether the monitoring equipment is centrally located with repeater and/or backup systems for specific purposes, e.g. fire and security on 24-hour working.

7.7.13 In design terms, if the control centre equipment is to be a coordinated whole, the requirements must be considered at the brief and concept stages of the project and all the design disciplines need to provide input at these stages.

7.7.14 Apart from this coordination, the duplication of wiring systems run through different routes for all security-orientated systems is a technical requirement which affects several design disciplines. Duplication of data bus systems and the needs of cable management are important issues in the design process.

7.7.15 The use of computerized data and control centres also requires an appreciation of what suitable software is available and what has to be written for a particular project. It should be made clear that the costs of any special software may frequently be very high and that the period for debugging the new programs may run well beyond the date of beneficial occupation. Security of the software packages themselves must also be taken into account.

7.7.16 Information technology (IT)

While conventional controls have always been considered part of building services, they also act as communication systems which need a degree of security. Telephones, on the other hand, have in the past been viewed less as part of the building services than as an added feature to meet the particular users' requirements, with maintenance generally provided by the supplier (the GPO, later British Telecom).

7.7.17 Telephones are now just one element of IT systems (controls data wiring is another) and the performance and prosperity of many organizations is wholly dependent on the reliability of such systems. Reliability depends on designing to the level of risk assessment specified in the brief, a knowledge of the electrical problems which can affect IT systems, the need to prevent unauthorized tapping into the systems (computer hacking being one example) and adequate maintenance systems.

7.7.18 The brief for IT systems in modern complex buildings can be a highly sophisticated specification in its own right and normally requires specialist input from professional advisers knowledgeable in the particular IT systems (projects may require more than twenty specialist suppliers in the extreme cases). The technical standards and interfaces between systems are themselves complex. Added to this, the client has to stipulate what level of outage is acceptable and what risk assessment evaluation needs to be adopted. The professional advisers can assist in formulating and developing these elements of the brief.

7.7.19 The electrical risks which may affect IT systems include high-voltage spikes, voltage harmonics and electromagnetic interference (EMI). High-voltage spikes which can be induced by switching electrical equipment, electrical faults and lightning strikes can destroy micro-electronics and corrupt data transmission. Harmonics are an increasing problem caused by solid-state power electronics, standby generation and uninterrupted power supplies. They can create problems similar to those of voltage spikes. Electromagnetic fields are created by all electrical systems, but at different strengths. Screening the more sensitive systems in a building from EMI created by other equipment in the building is reasonably straightforward, but screening from external sources can be very difficult. High-power radars, mobile radio transmitters and electric railways are sources of EMI which is frequently at high levels and intermittent. In acute cases the whole building, or those parts containing sensitive equipment, may have to be fully screened, including the windows.

7.7.20 The security from screening may need to serve two purposes: that described above and the need to protect unethical eavesdropping on electromagnetic fields created by data/IT systems within the building, which can then be translated back into useful information. This is no longer science fiction and in special circumstances needs to be taken into account.

7.7.21 The reliability and security of IT systems, like any other system, is only as good as the level of maintenance it needs and the quality of maintenance provided. Most IT systems, once fully commissioned, are remarkably reliable. None the less, comprehensive maintenance agreements should be the norm. As part of this, and depending on the importance of any system to the operation and profitability of the organization, the call-out and response times after a failure are critical.

7.8 Costs

7.8.1 One can only generalize on this subject. Almost without exception, any building, its services, facilities and security, are governed by cost. There is a conflict between absolute security and the absolute flexibility requested in many briefs for buildings. To achieve this total flexibility implies virtually infinite costs, and so it is with total security. In the design of security systems the cost of the measures taken must be related to the probablility of the risk events actually occurring.

7.8.2 The costs of maintaining security systems for buildings and building services, whatever form they take, cannot be ignored during the design phases of the project. Wrongly, normal maintenance costs tend to be ignored all too often, because first cost is paramount, but this is not allowable in security terms. While there is no regulation to ensure its inclusion, its exclusion is putting life and limb (as well as property) at risk, with all the ethical and financial implications which may result.

7.8.3 The security brief for any building, as with any other section of the brief, must be clearly expressed at the concept stage by the client and refined into specific requirements by a professional design team so that the cost implications are clear from the outset. The cost effect of some of the suggested standby arrangements may be enormous on a large complex, e.g. electrical plant and storage volume. However, with the exception of particular items such as these, good security normally costs very little in relation to overall costs.

7.8.4 On very high-security buildings it is essential that the designers coordinate their efforts to reduce the special cost element.
Non-communication between designers will be

infinitely more expensive than on the normal building and maintenance could become a nightmare for the occupants.

7.9. Conclusions

7.9.1 The conclusions which may be drawn from this review are best summarized as a list of the points that are important to the design team in terms of security planning. It should be emphasized that very few buildings are likely to require the ultimate in security precautions but the principles should be employed irrespective of the degree of sophistication. The summary of principles is listed below, and it should be noted that it includes a requirement for comprehensive testing of security systems:

1. Adequate brief for the level of risk.
2. Coordinated and integrated design.
3. Suitable specialist input.
4. Full consideration of all the available security systems for each sector of the project.
5. Specifications covering the maintenance requirements for each security system and security need.
6. Detailed comparisons of the security tender offers against every technical and maintenance requirement of the specification.
7. Cost criteria – preferably on a life-cycle cost.
8. Restricted access – fabric design and associated alarms.
9. Suitably distributed services for the risk.
10. Maintenance of essential services and standby arrangements.
11. Monitoring and centralized surveillance equipment.
12. Suitable communications – internally and externally.
13. Inclusion of full performance testing and commissioning procedures.

In the context of security perhaps the wisest dictum is that of the Scout movement: 'Be Prepared.'

Further reading

American Society for Testing and Materials and Stroik, J. (ed.), *Building Security. A Symposium on Security Systems and Equipment*, ASTM Special Publication STP 729, Gaithersberg, MD, 4–5 April 1979
BS 1722: Part 10: 1990 Specification for anti-intruder fences in chain link and welded mesh
BS 8220: 1987 Guide for security of buildings against crime. Part 2: Offices and shops

BS 8220: 1990 Guide for security of buildings against crime. Part 3: Warehouses and distribution units

Building Design supplement, Special report: doors and windows (May 1989)

Department of Education and Science, *Crime Prevention in Schools. Practical Guidance*, DES Building Bulletin 67, HMSO (1987)

Department of Education and Science, *Crime Prevention in Schools. Specification, Installation and Maintenance of Intruder Alarm Systems*, DES Building Bulletin 69, HMSO (1989)

Department of Education and Science, *Crime Prevention in Schools. Closed Circuit TV Surveillance Systems in Educational Buildings*, DES Building Bulletin 75, HMSO (1991)

Fire Protection Association, *Access Control*, Fire Safety Data Security Precautions SEC8 (April 1991)

Museums Association, *Museum Security*, MA Information Sheet 1525 (1981)

National House-Building Council, *NHBC Guidance on How the Security of New Homes Can Be Improved* (1986)

Security and Fire Protection Yearbook and Buyers Guide, Paramount Publishing (1994)

Traister, J. E., *Design and Application of Security/fire Alarm Systems*, McGraw-Hill, New York (1990)

Walker, P., *Electronic Security Systems. Better ways to crime prevention*, Butterworth-Heinemann, Oxford (1988)

8
Maintenance of the building structure and fabric
Alan Blanc

8.1 Introduction

8.1.1 A building fabric is designed to be durable but will need maintenance to keep in good condition. The points to look for are structurable stability, wind- and weather-tightness, effect of thermal movement and performance of finishes. Clearly, higher initial cost on good-quality materials and wise selection in relation to use will have an important bearing on maintenance costs.

8.1.2 Building maintenance costs are issues worldwide. Monetarism has had the effect of reducing repair budgets, witness the appalling state of London pavements, let alone the complaint of crumbling housing and school buildings from local authorities in the UK starved of adequate funds for routine repairs (Figure 8.1). No building or even a roadway looks after itself and it would be helpful if there was a greater awareness of the size and complexity of the task in safeguarding the nation's investment in buildings and infrastructure. One only has to look at our European neighbours – say, Germany or Holland – to see how much better such matters can be arranged.

8.1.3 British maintenance is, of course, defined by BS 3811, a key quotation being as follows: 'Work undertaken in order to keep or restore every facility, i.e. every part of a site, building and contents to an acceptable standard.'

It is significant that local authority tenants have successfully sued councils over poorly maintained property. However, the intent behind the Shops, Offices and Railways Act that also refers to upkeep did little to protect the public from the negligence of London Transport. At King's Cross, years of neglect in cleaning the spaces below the escalators eventually produced a catastrophic fire. Ill-advised maintenance on the treads with polyurethane coatings that dripped through to the machinery

Figure 8.1 Decay of building caused by neglect

may well have been a contributing factor.

Legal protection due to leasehold or rental agreements have more 'teeth' and have worked well in ensuring regular upkeep judging by the Crown Commissioners or the properties of the great estates in London (Bedford, Grosvenor, etc.). The licences that relate to buildings used for entertainment, manufacturing plants or places where food is prepared also appear effective in ensuring adequate repairs.

8.1.4 The traditional leasehold system where the development is built 'cheap' leaving the leaseholder to be held under a full repairing lease was a contributing factor to the lower standards of construction seen in eighteenth- and nineteenth-century domestic buildings in England. To see the difference, one has simply to look at the more solid achievement north of the border in Edinburgh or Glasgow, where the landlord has to meet full repair costs and, in consequence, spends more wisely on the original work. John Nash's shoddy inheritance has had to be continually upgraded to keep the elegance intact (Figure 8.2).

Figure 8.2 Repairing a shoddy inheritance (Nash Terraces, Regent's Park)

Figure 8.3 Minimal attitude to repairs

The landlord–tenant or user relationship (say, in local authority property) can often produce the worst of both worlds, where either side is committed to minimize expenditure to the irreducible minimum (Figure 8.3).

There are many commercial owners who see repairs as incidental and disruptive, unless the accommodation can be cleared for a total refit. Today such refits are more likely to relate to updating communication technology and services (AC and lighting) than to ageing paintwork on a 7-year cycle. The familiar 7-year breaks in lease agreements were an inheritance from the past when interior decorations were reckoned to be renewed at this frequency, hence the landlords' inspection to establish that the property was in repair before a step in rent occurred. In 1970 the HMSO prepared guidance entitled *Review of Practice Close Up in Property Maintenance Management*. It makes depressing reading. Attitudes are no better than times past and explain the number of poorly maintained buildings in the UK today.

8.1.5 There are three basic ways of maintaining buildings:

1. Day-to-day work caused by unforeseen breakdowns or damage.
2. Cyclic work undertaken to prevent failure (e.g. exterior repainting to protect joinery).
3. Planned repairs to restore elements or services to an acceptable standard.

The second and third categories should be able to be planned in a scheduled manner with least inconvenience to the building user. Wise designers present their clients with a repair manual that stipulates the time scale for cyclical inspection, and Section 8.7 describes the contractual implications.

Breakdown repairs or vandalism need special provision and urgency where safety is concerned (see paragraph 8.2.6).

Cyclic inspection should be made of building components with moving parts such as taps, ball valves, door springs, locks, etc. since replacement of a failing washer or a missing screw to a lock can save extensive repairs in the event of a flood or break-in. The aim should be to maintain each building regularly to a standard which will carry it through from one repair cycle to the next.

8.2 Causes of deterioration (see also Chapter 5)

8.2.1 Ideally, buildings should be maintenance-free, like cut diamonds. Regrettably, building materials are subject to weathering and ageing. The cumulative effect of snow and rain, wind and sun, atmospheric pollution and chemical action causes deterioration and erosion of the external fabric.

8.2.2 Many building materials are porous and a common factor leading to deterioration is the ingress of moisture. The source can be rainfall, condensation, service leaks and rising damp. Thermal or structural movement can also form paths for moisture penetration. The integrity of damp-proof courses or membranes are vitally important and often a minor shortcoming leads to a major problem. Ageing of bituminous or oil-based sealants is critical, particularly where the life is considerably less than that of the cladding. Finally, condensation is often a major cause of concern due to the way the building is occupied, especially through partial use or where heating is intermittent. Increased insulation standards coupled with airtight enclosures offering little natural ventilation have increased the risks of condensation and fungal attack. The decay of timber can generally be traced to the presence of moisture, the exception being damage due to insect attack.

8.2.3 Some defects have their origin in the building process itself, i.e. less than perfect movement joints, poor workmanship with sealants, inadequate cover to reinforcement and poorly considered fixings for cladding to framed buildings. Of current concern is the corrosion and damage caused by the ubiquitous galvanized steel wall tie that was common in post-war housing. Bimetallic corrosion is a further source of trouble in ageing buildings and where differing metals in the presence of water give rise to an electrolytic cell. This can occur within mixed metal heating systems and to metallic cladding (say, aluminium), galvanized steel and cast iron within a roofing system. More serious are foundation movements that affect the structure and ground slabs as well as fracturing services below the ground.

8.2.4 Frost damage is usually limited to the exposed parts of buildings which have been

Figure 8.4 Facing bricks performing well in protected environment and failing with full exposure

subjected to a cycle of soaking, freezing and thawing. The typical zones for frost attack are chimney stacks, parapet walls and garden structures (free-standing walls, paving and steps). It should be remembered that certain bricks which withstand frost under the protection of overhanging roofs can fail miserably when placed as copings or paviors (Figure 8.4).

8.2.5 Wear and tear on a building is a critical factor, especially when the use is changed or intensified. Excessive use means that flooring and internal wall finishes are quickly ruined and need earlier replacement.

8.2.6 Vandalism appears to be a way of life. The freedom gained in removing the Berlin Wall has given rise to a free-for-all for graffiti in Eastern Europe, a cultural perversity that has spread worldwide from the USA. Graffiti happens to be one of the easier offshoots of vandalism to remove. Overcoatings exist that can be dissolved and cleaned off together with the surplus decoration. A recent visit to downtown Chicago revealed that the city authorities had defeated the graffiti artist by chemical means. Natural materials such as brickwork and stone assume glossy textures when coated with epoxy finishes (Figure 8.5). This is perhaps the price paid for by a society that has survived the twentieth century.

8.3 Diagnosis of defects

8.3.1 Correct diagnosis is a prerequisite of a satisfactory repair. Errors in diagnosis may result in applying remedies which simply make matters worse. Lawyers have had a field day when architects have failed to correct defects. Matters may have started as latent defects (things the builders got wrong) and then after further

Figure 8.5 Glossy textures of anti-graffiti coatings on walls

instructions become patent defects (things that the designers got wrong).

8.3.2 Many defects have symptoms which are superficially similar and it is therefore important to keep an open mind, but at the same time, to recognize the signs that ensure accurate identification of the root causes of a problem. Investigators should be advised to adopt a systematic approach to assess impartially all the evidence. Drawings, specifications, contract files and catalogues of components used should be collected and studied, with subsequent checks on-site to look for differences between original design and the building as finally built. There should be careful operations with opening up and recording in detail all the relevant information. Laboratory testing of materials may be required as well as detailed interviews with those concerned on-site. An accurate and complete survey is needed and may have to be used in court.

8.3.3 The best reference sources are the BRE Digests relating to Building Defects and Maintenance. There are also useful textbooks from the recent past which set down theory with illustrated details such as *Materials for Building* by Lyall Addleson (Iliffe Books, 1972) and *Structural Failure in Residential Buildings* by Schild, Oswald, Rogier and Schweikert (Granada, 1979).

The tools of the trade are a working moisture meter and a camera to record evidence (the Polaroid type having the advantage that immediate photos are available and the evidence is recorded in a satisfactory form). Binoculars are invaluable for studying inaccessible parts of the exterior or roofscape. Photometry has advanced to the stage that survey drawings are no longer required (Figure 8.6), the photometric prints being invaluable for elevational and air views. Drawn key plans or outline elevational sketches are still

useful for reference purposes to grid lines or locations.

8.3.4 A maintenance manual is an essential tool in the professional management of property. It should be prepared by the design team and updated by the maintenance staff and building owner. The record should contain the names, addresses and telephone numbers of the building team and those of sub-contractors and suppliers. The preparation of the manual is a useful discipline for designers and aids 'feedback' as the maintenance sequence unfold. Chapter 18 refers to other aspects of this 'tool'. Clearly, the investigation of defects will be made easier by referring to a manual that has been conscientiously followed. There is protection for the designer, since lack of regard to regular inspection and repair can be placed as the building owner's responsibility.

8.4 Inspections

8.4.1 Regular inspections are a vital part of the procedure for building maintenance. A systematic approach employing a manual as guidance will help in planning the time scale and scope of inspections. It is common to divide the task into exterior and interior. There are various pro-formas that are useful sources if the building owner has not previously arranged matters on a rational basis. Hotel groups and hospital management boards have standard schedules. Another valuable reference are the studies produced by the Ecclesiastical Architects and Surveyors Association for quinquennial reviews. The Property Services Agency is another well-known authority.

8.4.2 Property managers often operate with a broad-brush approach using a grading system so that buildings can be compared regardless of age or type. Such management methods rely on a percentage assessment of condition applied to various elements – for example, foundations, structure, external envelope (roof and walls separately), internal detail, services (heating, water services, electrical, etc.), decorations and fittings. From such outline reports management decisions can be made in terms of priorities, programming or calling for detailed appraisal.

8.4.3 Access and the type of inspection are joint problems. Refer to Chapter 12 for further detail. The original design should allow for access with eye bolts for cradles or cleaning rails where elaborate facades occur. Inaccessible parts of buildings will also be at risk from neglect due to difficulty in making inspections. With modest

Figure 8.6 Photogrammetry for Covent Garden repair (GLC Architects Department)

traditional buildings useful assessments can be made with the aid of binoculars. Today, lorry-mounted hoists can be effectively used for studying the upper part of facades with a height limit of around 30 metres. The employment of such methods should be checked with the hirers *vis-à-vis* ground bearings and the local authority requirements. The costs are high and the use should be maximized by having the survey equipped with 'rising-front' cameras to avoid parallax problems with the views obtained. In some cases there is no other solution than scaffolding, as shown by the regular sequence of overhaul at the Palace of Westminster. The cause of this labour-intensive method is the continual decay of stone, not only inappropriate in choice but face-bedded in many areas. Presumably, Pugin might be blamed in retrospect but he was only paid £200 a year for details of the carved work by Barry in the 1840s. For simple inspections, scaffold towers and ladders have much to commend them.

8.4.4 The frequency of inspections used to be governed by the cycle of redecoration, 5 years

externally and 7 years internally, such were the standards of the Crown Estates. Where that pattern exists, a careful examination before and during painting work will be beneficial since ladders and access can be part of the contractor's service. Under this traditional practice, pointing, putties, flashings, gutters, pipes and roofing can be examined and routine repairs carried out.

Many buildings today no longer rely on paint finishes but use anodized aluminium, epoxy or plastic coatings, precast concrete as well as sealants that need overhaul. The time sequence is difficult to assess, perhaps every 10 years where painting is not required. That time scale also relates to advice from flat-roofing specialists. Insurers would advise examining new buildings every year for the first 5 years to look for latent defects.

There are the basic reasons of safety (governed by the Offices Railways Act and Safety at Works legislation) whereby mechanical services need regular inspections (cranes, hoists, lifts, etc.). Similar provisos exist with electrical, heating and ventilation services (see Chapter 17).

Finally, there is the need to maintain a clean building following the mandates of BS 3811.

Apart from regular cleaning, most internal finishes for flooring and walling need proper refurbishment on a time scale of 7 to 10 years.

8.4.5 Records of all inspections must be kept and obviously should form an addendum to a Building Maintenance Manual. A suggested pro-forma is given at the end of this chapter (see also Chapter 18).

8.4.6 In these days of computerized records it would be an easy matter to extract this information in chart form and to have outline plans or elevations updated with notes as to which repairs were completed.

8.4.7 Computerization has changed attitudes towards the retrieval of information. Reliance can no longer be placed upon district surveyors or local authorities to have kept the original drawings. Microfilming is common but often relates to general layouts instead of detailed construction, yet another reason why building owners should keep a comprehensive file on any new building. Refer also to Chapter 12.

8.5 Practical maintenance

8.5.1 'Penny wise, pound foolish' is an apt riposte to the owner who fails to deal with a blocked rainwater pipe while the adjacent brickwork is soaked, conveying damp for a dry-rot attack in the adjacent wooden floor. Such tales abound in the publications of EASA (see paragraph 8.4.1).

8.5.2 The same can apply to 'lash-up' repairs where the fault is due to a patent defect in the design of a building – for example, bitumenizing felt roofs that are sagging and ponding due to inadequate joists, when the slopes and decking should be put right as a first priority. See Chapter 1 for a more detailed discussion of design faults.

8.5.3 It is important to distinguish the basic forms of construction when deciding how best to maintain them and to decide whether replacement is the simplest solution. Craft-based methods common before 1939 have been largely supplanted by industrialized techniques and differing skills are needed for refurbishment and repair. Industrialization also affects the way repairs are selected, with overcladding and/or window replacement found to be a cheaper method than craft-based repairs to facades. The same economic considerations often dictate a total strip-out of interiors so that modern services can be installed (cabling, electrical and air conditioning) despite the fact that finishes and sub-structure are in sound condition but are not easily adaptable to modern services.

Buildings which have their roots in craft-based methods are more readily understood and are often tolerant of considerable neglect. It is noticeable in Cambridge or Oxford that old college buildings often appear in better condition than their contemporary additions. It is not simply that better materials were used in the past but that modern technology, with an amalgam of materials put together in a non-traditional way, needs more sophisticated repair than repointing and a painter's skill that sufficed pre-war.

The ensuing pages are dedicated to both the traditional and the non-traditional approaches. The designation is material- and trade-based.

8.5.4 The largest recurring cost in maintaining buildings is the cost of redecorations, hence the tendency to choose materials that are self-finishing, and for external paintwork finishes to joinery and plaster needing regular attention for reasons of cleanliness. In the care of painted joinery, the surface coating needs to be kept in good repair where dampness occurs by external exposure or through condensation. Softwoods can achieve 22% absorption of moisture and be at risk to decay in window and rooflight joinery. The decay is commonly wet rot which stresses joinery joints leading to failure of the glue-line (Figure 8.7). Dry rot can follow where external woodwork continues to be neglected. The critical areas are bottom rails to sashes and doors and the stub tenon joints where jambs meet wooden sills. High-quality joinery today has mastic pointing to weather-point the vulnerable bottom-rail tenons, with drainable joints on the underside as an additional precaution (Figure 8.8). Metal or plastic beads are also employed for bottom-rail glazing instead of beads or putty that fail in such positions (Figure 8.9). Overhaul of joinery is an expensive operation, and the substitution of all plastic frames is often chosen as being more economic than joinery repair. There are many historic or listed buildings where such policies are not appropriate and where the technical improvements seen in Scandinavian wood windows could be incorporated into sensitive locations. It is worth recalling that external paint films are only 90% efficient at repelling the weather when new, and that 7 to 10 years represent the average effective life of paints that are brushed into woodwork. Wood stains have a far shorter life between recoatings (say, 3 to 5 years) and will not prevent decay in frame joints unless the coating is kept in good order. (Refer to Chapter 5 for further details.)

8.5.5 The treatment of decayed joinery needs skill. The sequence involves cutting out decay, stripping adjacent paint and allowing the sound wood to

Figure 8.7 Wet rot in joinery

dry. The exposed work is brush treated with a preservative (remember that this is not as effective as pressure impregnation). Joints between old and new work need to be sealed, either by glueing and pinning or, traditionally, with priming, stopping and pinning. Priming and glueing does not work since the glue-line is deficient. Infilling timber should be impregnated softwood or durable hardwood. Another point to watch in remedying external joinery is the effective shedding of rainwater from sills and transoms. Scandinavian detail employs aluminium or plastic flashings which can often improve weathering (see Figure 8.9).

8.5.6 Loose putties or decayed glazing beads are another source of trouble with wooden windows. Putties should be rerun after the glass has been reset with the rebates fully primed, patching putty or simply infilling without back-priming will quickly cause cracking to the paintwork. External beads should be checked and solidly rebedded, and using metal components to the critical weather edge (see Figure 8.9).

8.5.7 External paintwork in new buildings needs special care so that the first repainting occurs before the original decorations have broken down. A typical latent defect from builders' work is the thinning of priming or site damage that has never been remedied. The latter is very troublesome if

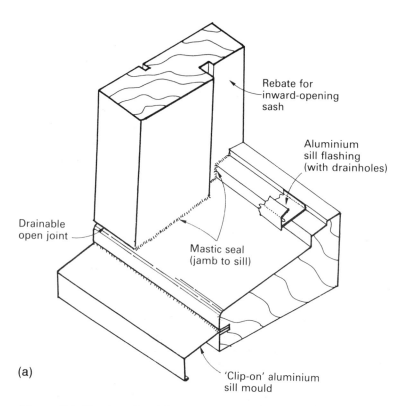

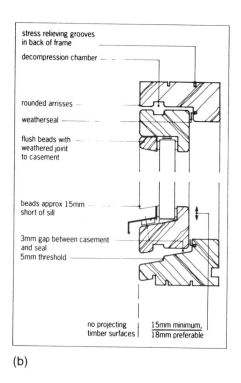

(a)

Rebate for inward-opening sash

Aluminium sill flashing (with drainholes)

Drainable open joint

Mastic seal (jamb to sill)

'Clip-on' aluminium sill mould

(b)

stress relieving grooves in back of frame

decompression chamber

rounded arrisses

weatherseal

flush beads with weathered joint to casement

beads approx 15mm short of sill

3mm gap between casement and seal

5mm threshold

no projecting timber surfaces | 15mm minimum, 18mm preferable

Figure 8.8 Mastic seal to joinery joints. (a) View of mastic seal. (b) Open drainage

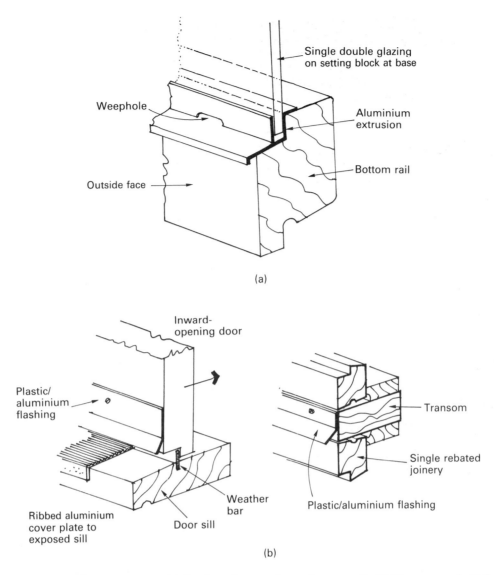

Figure 8.9 Aluminium and plastic flashings. (a) Beads and protection to bottom rail. (b) Transom and sill flashings

factory coatings to steel cladding, railings or windows have been touched up without preparation. Undamaged epoxy or powder coatings to aluminium or steel components eventually suffer 'chalking' and fading with external exposure after 15–20 years and need repainting by brush, roller or spray coat. Prolongation beyond 20 years will give rise to difficult repainting work at edges and vulnerable corners. Severe rusting of exposed steel surfaces will need expensive preparation by flame cleaning or grit blasting and the careful application of a complete protective system. It may be cheaper to replace steel elements in flashings, gutters and RWPs. Zinc coating should always be considered for steel elements like railings, windows and wrought iron. Workshop preparation with stripping and grit blasting means that zinc coating (hot-dip or electrodeposition) will prevent further corrosion and be cheaper to repaint in future.

Leaving the zinc coating to weather for 3 to 5 years will improve the subsequent bond between the coating and paintwork. Working components (hinges, locks, bolts) on external joinery, metal gates and windows should be overhauled and lubricated before decorations are commenced.

8.5.8 Painting for protection

Site and 'off-site' painting are the dual treatments employed where external painted components need weather protection. Paragraph 8.5.7 indicates that today's factory-applied coatings have a life limited to between 15 and 20 years. It follows that future maintenance of external paintwork will still rely upon skilled painters applying their skills, whether by brush, roller or spraycoat.

The traditional lead-based paints dating from the eighteenth and nineteenth centuries are no longer permitted as they are a health hazard. Old painted buildings can still be found in excellent

Figure 8.10 Build-up of old lead paint

external condition where successive coatings may have built up to layers over 4 mm in thickness (Figure 8.10). It is not possible to leave such work. Chemical stripping has to be employed and precautions taken to protect the workforce and the environment. Contemporary paints, whether cement, oil, epoxy, polymer or water-based, still depend upon bonding one coat to the next. It is not advisable to mix differing makes of material, or to overlay one form of paint upon another. Manufacturers will supply general specifications and will prepare detailed advice that is specific to a project. Natural materials like soft-facing brick or pervious stonework need to be able to 'breathe' through paint films and the specifier should listen carefully to the advice offered. Highly effective sealants over soft brick or stonework can lead to rising salts breaking down the surface with exfoliation instead of leaving the materials to breathe and weather naturally in the open air.

8.5.9 Wood stains have been mentioned in paragraph 8.5.4. Ongoing maintenance, although cheaper than painting, involves a more frequent cycle. The situation of maintaining the system within one group of materials is important. It is worth studying the options as the technology improves, as many treatments today give an opacity almost equal to traditional paintwork. The key to successful specification is the match that must be made between initial preservatives, original coatings and subsequent finishings. Overpainting with traditional oil paint is seldom successful.

8.5.10 Joints and junctions

Many defects in external cladding can be traced back to the joints between components. The vexing topic can be categorized into the following 'systems'. Advice regarding upkeep needs to be addressed to the specialists, installing 'systems' such as precast/composite panels.

- *Filled joints* These rely wholly upon seal and or gasket. Typical examples are door/window units set within an opening in masonry or concrete.

- *Drained joints* Such joints rely upon a sophisticated system of baffles and rebates to provide an airtight seal in defence against the weather. Skilled tooling is needed for assembly and for taking apart. Typical examples are the panel-to-panel joints in precast concrete cladding or between composite panels (e.g. metal and plastic or metal framed with glass/stone cladding).

- *Lapped joints* Usually applied to formed sheet materials but also to metal/plastic sills and copings. Undersleeves are needed for weather surfaces (Figure 8.11).

- *Tongued and grooved joints* These can be employed in vertical or horizontal cladding joints for thin components, such as plastic/timber. Vertical forms need drainage.

- *Cover batten joints* These rely on 'clip-on' or mechanical fixed mouldings to weather-seal

(a)

(b)

Figure 8.11 Undersleeving to copings. (a) Failure of coping joints. (b) Improved detailing

external panels. Typical examples are vertical timber battens for ply/cement fibre sheets.

- *Coupling joints* Specialized joints engaged for joining window elements in metal, plastic or timber and for coupling sill components.

- *Mortar joints* Traditional practice for artificial stone/stone sills and copings. Requires underbedding with a damp-proof course to deal with failure at mortar joints between elements.

8.5.11 Apart from specialist advice from installers it is possible to add the following general guidance on repairing cladding joints and junctions:

- *20- to 30-year life* Most oil-based sealants have a limited solar life under full exposure to the sun. Filled joints with face mastics will therefore need replacement. Cover battens or baffles will give greater protection and extend life to 40 to 50 years but involve more elaborate overhaul.

- *Metal cladding* The greater range of thermal movement in contact with metal means that oil-based sealants become brittle, and details that depend upon flashings and gaskets will be less troublesome.

- *Mortar joints* Repointing will be needed at exposed situations for traditional brick/precast/stone copings and sills at 30- to 40-year intervals (Figure 8.12). The same problem will occur with south-west exposure to traditional ashlar facades or brickwork after 50 to 60 years. Silicone treatment is offered as an alternative but to date it is difficult to assess its effectiveness.

- *Cleaning* Regular cleaning of facade elements will keep glass/panel and framing components in a better condition. Clearing weepholes in cladding should be an essential part of contemporary cladding systems. Incidental washing of the facade does not forestall a thorough maintenance programme for drainable joints in panel and rainscreen facades. Specialists advise 5- to 10-year intervals and that tracked cradles should be provided for inspection. Many local authorities now require that the arrangements for maintenance access are according to Building Regulation Consents. The fitting of cradle facilities should be seen as the essential complement to modern systems incorporating sealants and drainable joints.

8.5.12 Roofs

The 'for and against' arguments for pitched or watershed roofs versus flat waterproof roofing are familiar. Suffice it to say that both need maintenance and overhaul on differing time scales. Table 8.1 is based upon material collected for technology lectures and drawn from many technical sources.

8.5.13 Flat roofing, by virtue of having to be waterproof, requires greater levels of maintenance than pitched roofs employing slates and tiles. Regular inspection at yearly intervals should be made of flat roofs to clear gutters, cesspools and rainwater pipes. Wise designers will have installed overflows to parapet and valley gutters, a point of detail that should be borne in mind with overhaul (Figure 8.13). Leaks are usually associated with poor workmanship and can be traced by electronic means, and locations around outlets, flues, gutters or roof vents are the most likely culprits. Water travels considerable distances by capillary attraction between felt/asphalt layers and the roofdeck, so that internal signs of dampness do not necessarily indicate failure at that position. Design defects occur such as inadequate falls or lack of height with upstands (150 mm minimum under Codes of Practice but 200 mm with exposed sites). Lack of solar protection is found, particularly at kerbs and upstands and where paint may be inadequate. Cover flashings in white-coated metal or plastic can be shaped to provide sun shading (Figure 8.14). Thermal movement is another common problem and bonding failures occur between roof coverings and deck and more particularly between decks and the surrounding structure, hence cracks and leaking at roof edges.

Publications such as *Flat Roofing – A Guide to Good Practice* published by Tarmac describe the recognized detailing for movement joints in surface planes and at all kerbs. Unsuitable decking, incorrect falls or condensation problems within the substrate will not be cured by overlaying with higher-grade roofing felts or additional coatings. The basic problems outlined above should be cured before renewal of roof

Figure 8.12 Repointing needed to parapets after 40 years

Table 8.1

Roofing	Average life	Overhaul
Three-layer bitumen felt	20–30 years	Additional felt or renew
Three-layer bitumen felt with 75 mm shingle	30–40 years	Additional felt or renew
Two-layer mastic asphalt and solar paint	50– 60 years	10-year cycle of overhaul after 20 years
32 g copper	30–40 years	Renewal
24 g copper	100 years or more	Renewal
Lead roofing		
Code 4	40–60 years	10-year cycle of overhaul after 40 years
Code 8	100 years or more	10-year cycle of overhaul after 80 years
Cast lead	200 years or more	St Paul's dome lead work survived for 250 years
Zinc roofing		
14 gauge	10–15 years	Renewal
12 gauge	20–25 years	Renewal, but in France where heavier gauges are used, patching is normally employed for repairs
Slating		
New Welsh	80–90 years	Renewal
Westmorland slate	200 years or more	Patching and renewal
Cement fibre	30–40 years	Renewal
Cedar shingles (with preservative treatment)	40–60 years	Renewal
Tiling		
Well-burnt fully vitrified clay tiles	Indefinite	Patch and renew where damage occurs
Gas-fired clay tiles	Not known, probably 60 years	Patch and renew where damage occurs
Concrete tiles	Oldest examples now 70 years	Patch and renew where damage occurs
Corrugated and formed sheet		
Stainless steel	Probably indefinite	Oldest roof, Chrysler Building, New York (1930) cleaned after 50 years, found perfect
Glass fibre-reinforced polyester	30–40 years	Renew
Galvanized iron	30 years	Overpaint after 20 years to prolong life
Coated sheet steel	20 years for paint film	Overpaint to prolong life
Coated aluminium sheet	20 years for paint film	Overpaint to prolong life

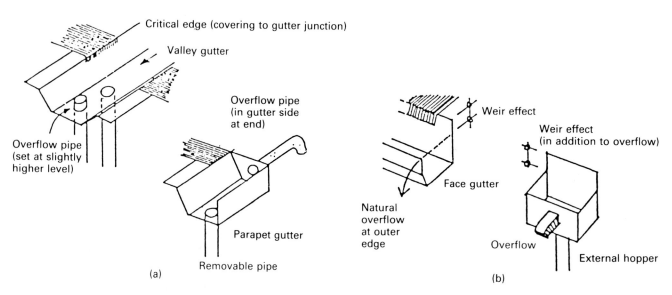

Figure 8.13 Overflows to gutters. (a) Face gutters and hoppers. (b) Parapet gutters

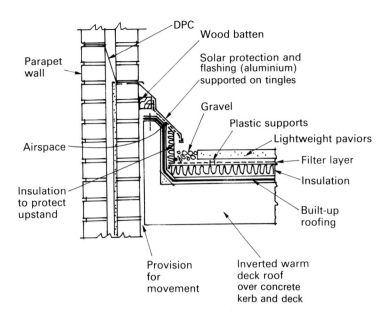

Figure 8.14 Solar protection to upstands

coverings. Many clients will insist on guarantees and these can be obtained if the roofing firm is also made responsible for the decking. Asphalt specialists will ask for insurance premiums to be added to their estimates and this aspect could be explored more generally when a guarantee is needed.

Overlaying with high-performance single-layer bitumen polymer roofing is often employed in the upgrading of existing flat roofs. Such materials can also be protected from solar radiation and today ecological or green coverings are feasible that employ a pre-seeded fibre carpet that protects the coverings (Figure 8.15).

Traditionally, felt roofing was stripped of chippings and additional bitumen provided plus new chippings or an additional layer after 20 years or so. Old roofs exist from before 1939 where the original felt is still protected by 75 mm of shingle. However, those seen have face-mounted iron gutters and not internal felt gutters. The latter are

the most suspect detail with felt since the felt rolls that run down roofs are found in the wrong laying pattern, with the felt laps within the main waterway.

Asphalt is the ideal material for inward-draining roofs but requires skilful patching where cracks develop. Solar protection with paint is simple to maintain but 75 mm shingle will give greater protection. Vertical asphalt work is usually single-coat and will be in need of overhaul at 5-year intervals unless shielded (see Figure 8.14). The new generation of high-performance polymer roofing will no doubt present fresh problems for upkeep, and observation of the new technology in operation reveals that patching is easier to undertake.

8.5.14 The Roman architect Vitruvius said that sound second-hand roofing tiles should be saved for copings or damp-proofing work. His theory was based upon the sound concept that pitched

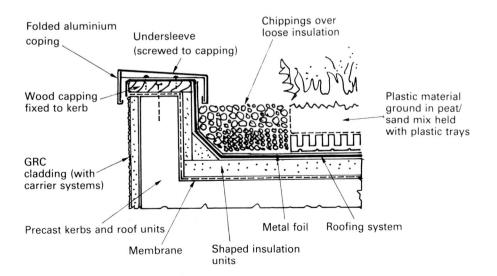

Figure 8.15 Ecological roofing

roofing materials are subject to the most severe exposure. It is true that well-fired clay products have an infinite life, but the same does not apply to the majority of so-called 'traditional' coverings employed today, unless the budget permits buying the best available.

Most nineteenth-century roofs in London have been totally renewed. It is unwise to re-use old materials unless the roofers have a Vitruvian skill in sorting out sound old slates and tiles. Matching sizes represents a problem, let alone the colour and texture of the modern equivalents. Gas-fired tiles have a different quality from their coal-fired predecessors. 'Nail sickness' is the prime cause of failure with slating, and total stripping will be needed and replacement using non-ferrous nails for fixings. Current regulations require improved insulation, underfelt ventilation and establishment that the roof structure complies with a stated wind resistance. Counter-battening augments ventilation under close-fitting coverings such as slate (natural or artificial) (Figure 8.16) and shingles (natural or felt).

Concrete slates and tiles weigh considerably more than natural materials and strengthening of the substrate will be needed under the revised Building Regulations. Short-term overcladding is offered by specialists with bitumen-coated hessian or glue-bonded underlays to stick slipping slates to existing battens. Both forms can provide dry-rot nursery beds.

8.5.15 Flashings, gutters and rainwater pipes are essential parts of water shedding, whether flat or pitched and their effectiveness warrants constant vigilance. Removal of debris from gutters (face, parapet and valley) is an annual chore. Face gutter-joints need resealing on a 5- to 7-year cycle (when repainted). Short-life flashings (14-gauge zinc, coated steel) will need renewal long before the main coverings. Long-life flashings (copper and lead) have a high retrieval value for metal thieves. Tern-coated stainless steel or lead-coated steel may be better value to the building owner and are a less attractive proposition for thieves (Figure 8.17). This policy could also apply when reconsidering reroofing in copper or lead.

Valley or parapet gutters may give longer service in one-piece construction, either moulded in GRP and lined with polymer or cold-formed in stainless steel with welded outlets and end-pieces. Short-term gutter repairs can be effected with bitumen tapes or by metal-faced polymetric sheet (e.g. Evode's Flashband) but will not defer long-term repairs.

8.5.16 A common cause of damage is access without adequate protection. Coverboards are essential for work over flat roofing and crawling

Figure 8.16 Counter-battening for reslating at Durham Cathedral (Architect: George Pace)

boards when access is needed across pitched roofs. Some accidents are not foreseeable as, for example, the seagulls that demolished T-bone scraps from the local steakhouse on a nearby single-layer flat roof. A colander-like covering quickly developed. The solution lay in a replacement white polymetric roof but decorated by black cutouts (to enlarged size) of gull-hawks, who, of course, feed on gulls.

8.5.17 Walls

Most brick walls will be almost maintenance-free for the life of the building. Where the bedding mortar is sound, repointing will be required only in limited areas where frost action has resulted in spalling. Leaking rainwater pipes or overflows will also cause local damage (Figure 8.18). Other vulnerable areas are chimneys, gables, parapets and garden walls. Joints of sills and copings will

Figure 8.17 Lead-coated steel roofing at Lloyd's offices, Chatham (Architects: Arup Associates)

need raking out and repointing. Undercloak damp-proof courses should be provided where non-existing or failed.

Capping walls should be considered where copings have failed, and plastic or metal profiles which provide overhangs will give greater protection than flush-fitting designs (Figure 8.19). Concrete blockwork, silicate brickwork and rendered walls need more protection in exposed situations, lack of projecting sills or copings giving rise to erosion and poor weathering. Movement cracks in brickwork (Figure 8.20) due to expansion with clay products need to be relieved, such as rebuilding parapet walls off a slip joint, and greater complications are caused by panel-wall performance in framed buildings. Consult the Brick Development Association for advice. Shrinkage cracks with blocks and silicate bricks are not so serious and can be remedied by repointing or by limited rebuilding employing movement joints.

Differential settlement movement represents serious problems and demands structural advice on underpinning, etc.

8.5.18 Limestones and sandstones will decay in exposed situations, and the rate of erosion or chemical breakdown of masonry depends on the type of stone, exposure and nature of pollution. Acid rainfall is deleterious to limestones, and vigorous cleaning can often aggravate decay by removing the protective crust. The same problem can occur after grit-blasting sandstones (Figure 8.21). Conservationists will point out that cleaning the surface of masonry will remove salts that

(a)

(b)

Figure 8.18 Leaking rainwater pipes and gutters. (a) Defective hopper and rainwater pipe. (b) Broken gutter end

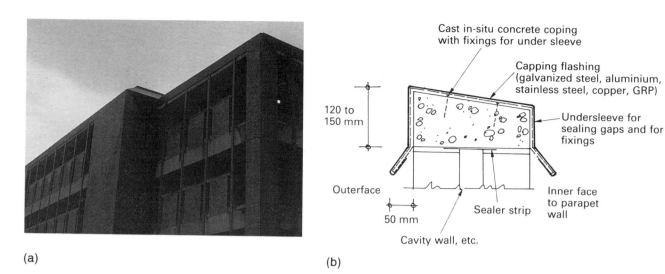

(a)   (b)

Figure 8.19 Capping to walls. (a) Walls wetted by lack of adequate capping. (b) Typical capping detail to improve weathering

Figure 8.20 Thermal movement of parapet brickwork (top right-hand side and through quoin)

Figure 8.21 Grit blasting sandstone

would otherwise cause injury. The advice is coupled with the need to use fine watersprays and brushes without chemicals or grit blasting. Repointing and stonework repairs will follow, with the employment of the same scaffolding. Silicone treatment may help with water-proofing but epoxy coating will not only change the texture but will also increase risks with exfoliation from decaying stones. It is a useful technique for protection of weather surfaces that are out of sight, i.e. tops of copings and sculpture (Figure 8.22), or weather surfaces on mouldings and sills.

8.5.19 Rising damp is a critical problem with ageing buildings and is due increasingly to failing damp-proof courses in nineteenth-century work. The reasons are the hardening and cracking of bituminous material or settlements that crack slate. Cutting out and insertion of new damp-proof courses may be the only long-term solution.

Electro-osmosis and chemical injection are processes developed for damp-proofing old masonry structures and where reasonable success has been achieved. It is difficult to believe that either method will solve rising damp in nineteenth-century brickwork, often of one-brick thickness built in London Stocks with totally perished lime mortar. In such cases, bitumen coating, waterproof rendering and cavity construction to the inner side is probably the only worthwhile remedial work other than demolition and rebuilding. The basic masonry construction needs to be studied – first-class work was achieved in the past without damp-proof courses, Fowler's Market at Covent Garden is a case in point. There, waterproof mortars (to be exact, eminently hydraulic lime, exemplified in Blue Lias limes) were used in conjunction with bricks of engineering quality, the subsoil is gravel and no tanking was employed. That level of perfection did

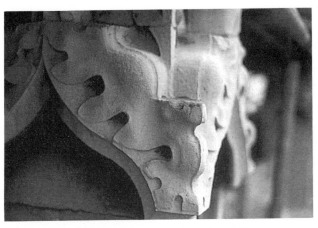

Figure 8.22 Epoxy 'skull caps' at Trinity College, Cambridge. (Architects: Donald Insall & Partners)

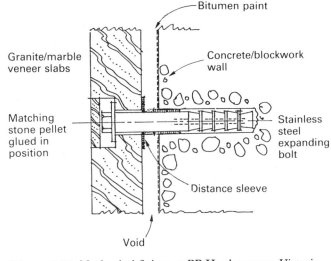

Figure 8.24 Mechanical fixings at BP Headquarters, Victoria Street, London. (Architects: EPR)

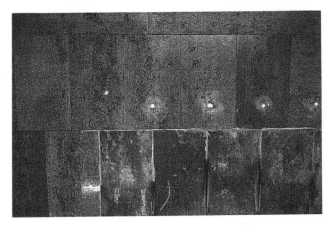

Figure 8.23 Loosening of veneer slabs at the Royal Festival Hall (GLC Architects Department)

not exist with 'spec builders' be they Nash or Thomas Cubitt.

8.5.21 Brick or stone-clad framed buildings represent different problems since the cladding needs movement joints to counter shrinkage stresses with concrete frames. On the other hand, there are expansion stresses caused by the interaction of mortar with the majority of brick and sedimentary building stones. The failure of brick and stone-veneer facing to concrete frames (Figure 8.24) has been well publicized, and guidance is available from the BDA and the Stone Federation. The corrosion of galvanized ties and their replacement is also dealt with in BDA publications. Less familiar is the fatigue failure of bronze cramps. Stainless steel cramps and fixings overcome corrosion and fatigue problems. The other solution adopted is to mechanically fix every veneer slab, with testing as to security of metal and stone (Figure 8.24). Current ideas also call for

a drainable cavity behind the stone and for the employment of sealants to all masonry joints to cope with differential movement between the cladding and the frame (thermal and structural). Few marbles as chosen for veneer slabs seem able to withstand the extremes of frost and high summer temperatures (Figure 8.25). It is noticeable in London that few marble-veneered facades exist from before the Second World War. Shop fronts do exist but these are found to be protected by polish or other coatings.

8.5.22 Reinforced concrete either in 'self-faced work' (as in the National Theatre) or in precast of artificial stone form (like the Hayward Gallery) weathers like limestone. There are highlights where rain erodes the surface and darker patches where algae takes over (Figure 8.26). Water washing will restore the natural colour. 'Concrete cancer' is the phrase that covers decay due to carbonation. The problem is related to acid rainfall where the outer 25–40 mm of concrete assumes acidic instead of alkaline qualities. This creates an electrolytic cell at the outer reinforcement zone with enhanced corrosion to the steel. Another effect is surface cracking (Figure 8.27). Stainless steel or galvanized steel reinforcement will counter the corrosion risks, hence the high performance of the National Theatre's concrete. Buildings constructed in plain bar reinforced concrete that is self-faced need elaborate repair to be protected from acid rain. Overcladding or rendering are the popular options. Initial paint decoration can save repair bills. The best example is the work by Sir Owen Williams on the Boots mushroom frame building at Beeston, built 1930–32 and where a minimal repair bill has been incurred.

Figure 8.25 Failing marble veneer at the Finlandia Concert Hall, Helsinki. (Architect: Alvar Aalto)

Figure 8.26 Weathered concrete at the Hayward Gallery, London (GLC Architects Department)

8.5.23 Rendering is commonly used in Britain as remedial work on older buildings or as facing material where sound bricks or building stone is not available. Matching repairs to natural or self-finished rendering is difficult, and the usual remedy is to patch and paint. The critical areas for weather penetration are copings and sills and where overhanging metal or plastic patterns will perform as well as stone and precast forms. Sulphate attack from the backing materials are common causes for the failure of bond (Figure 8.28). This can be caused by expansion of rich Portland cement mortars and common bricks with a higher than normal sulphate content (such as the common fletton) or by the use of clinker-based aggregates in concrete blocks. Overcladding with rendering supported on an independent mesh may be the best solution. The other options are slate or tile hanging over failed rendering.

Figure 8.27 Carbonation of concrete British Rail Offices, Lisson Grove, London

8.6 Feedback

8.6.1 Feedback is normally regarded as an important procedure in other technical fields. The building industry appears backward in this respect. A retrospective view on the behaviour of

Figure 8.28 Failure of bond to rendering over fletton bricks at Liverpool's Roman Catholic Cathedral

Figure 8.29 Mason's scaffold at St John's College, Cambridge, for carrying out traditional work

materials and detailing would be of great benefit to designers and could reduce maintenance costs. Refer to Chapter 12 for a more detailed view.

It is of considerable advantage if designers can have client representatives who will be responsible for repair as part of the initial team (see Chapter 1).

The Building Research Establishment has been at the forefront in technical reviews on building repair. The present trend that research should be market-led has diminished the impact of original research into building failures. The problems are, however, cyclical and reference to papers published by the BRE, the former GLC and the PSA reveals that similar queries arise repeatedly.

8.6.2 Continuing professional development

It does appear that the learning curve for designers is lengthy. One can only conclude with the hope that Continuing Professional Development is not just a catchphrase to appease insurers. Let us trust that it will be a real tool to ensure that building maintenance will not be such a crippling burden for the taxpayer and user as the slipshod inheritance of the 1950s and 1960s.

8.7 Execution of maintenance work

8.7.1 Erection of scaffolding for access to the external fabric constitutes a significant percentage of the total refurbishment cost (Figure 8.29). It is prudent to plan full use and to specify carefully the type of inspections to be made in a predetermined cycle. The cycle and disruptions should be agreed with the owner and user. Provisional quantities and the judicious use of spot items will cushion unexpected costs. The use of two contracts, one to open up and another to

repair, may also reduce expenditure, although there is the continuing cost of scaffolding. Additional works should, as far as possible, be priced by the contractor, including 'claims arising' to avoid the more expensive fare of dayworks and their excessive profits. A fully priced specification is superior to a bill of quantities for works up to £250 000. If Bills are insisted upon, then use provisional quantities throughout and prepare the contract in such a way that contractors claims for extension with variation orders are struck out.

8.7.2 Weather conditions are important for many trades. Spring and summer working will be cheaper than winter, and be more convenient for painting and roofing work. The insurance policies of the client and contractor should be carefully inspected before the work starts and steps taken to make sure that both parties have taken the necessary measures in case of fire or flood. The retention of fire extinguishers in roof spaces and on the scaffolding may be of great help to quench fires caused by painters, plumbers or welders.

8.8 Conclusions

The 'survival of the fittest' could be one way to describe the fate of a building fabric. The preparation of this chapter has not changed the writer's views. A recommended walk is along the South Bank, observing the relative condition of County Hall (pre-1914 and post-1950), the Festival Hall, the Hayward Gallery, the National Theatre, Waterloo and Hungerford bridges, and finally the Shell Centre. Waterloo bridge is the clear winner and it has never been cleaned.

INSPECTION REPORT	DATE BUILDING/ BLOCK		
	CONDITION SOUND +	SUSPECT O	DEFECTIVE −

EXTERIOR DECORATION
Timber surfaces Paint/Stain
Wall coating
Other surfaces

INTERNAL DECORATION
Ceilings
Walls
Other surfaces

ROOFING Flat/Pitched
Finish
Insulation
Structure
Roof lights/glazing
Parapets
Gutters
RWP's
Roof interior (pitched)

FLOORS & STAIRCASES
Ground Floor
Finish
Screed
Structure
DPM
Ceiling
Under floor space (suspended floors)

UPPER FLOORS
Finish
Screed
Structure
Ceiling
Suspended ceiling

STAIRCASES
Structures
Treads
Finishes
Balustrade
Soffits

EXTERIOR WALLS
Masonry/Brickwork
Cladding
Rendering
Structure
Jointing

	CONDITION SOUND +	SUSPECT O	DEFECTIVE −

WINDOWS & DOORS (EXTERNAL)
Glazing
Construction
Ironmongery
Floor springs/closers
Finish
Jointing

PARTITIONS ETC.
Structure
Finish
Doors (internal)
   Glazing
   Construction
   Ironmongery
   Floor springs/closers
   Finish
Fixed furniture
   Construction
   Finish

MECHANICAL INSTALLATION
Heating and Ventilation
Heat source
Controls
Distribution pipework/ducts
Radiators
Insulation

ELECTRICAL INSTALLATION
Switchgear
Distribution
Fittings and Equipment
Burglar Alarm
Lightning Conductor
Lift: date of last inspection

GAS INSTALLATION
Carcassing
Fittings
Equipment

SANITARY
Drainage
Plumbing
Fittings: Sinks
      Basins
      Baths
      Urinals
Sewage disposal
Manholes

EXTERNAL WORKS
Boundary walls
Fencing
Paths and paving
Landscape
Roads and curbs
Drainage
Manholes
Gullies

Structural movement

Telltales

Failure of materials

Design of construction defects

Remarks

9

The spaces between and around buildings
Hugh Clamp

9.1 Introduction

9.1.1 Landscape is defined in the 1973 Town and Country Planning General Development Order, as the treatment of land for the purpose of enhancing or protecting the amenities of the site and includes

> the screening by fences, walls or other means, planting trees, hedges, shrubs or grass, the formation of banks, terraces or other earthwork, the laying out of gardens, courts or other amenity features. All require regular maintenance to a greater or lesser degree.

9.1.2 Landscape maintenance concerns the routine care of land, vegetation and hard surfaces in the manner prescribed for their satisfactory establishment and continued future performance.

9.1.3 The true cost of a design is the price of the materials and their planting, plus their annual maintenance over a period of, say, 10 years. Because of the living nature of plants, as distinct from bricks and mortar, constant attention is required throughout the seasons of the year in the contemporary environment. If a natural mixed woodland is left to itself with no maintenance of any kind it will clearly survive, but compared with woodland sympathetically managed it will not be particularly attractive to people.

9.1.4 The usual function of maintenance is by establishing and preserving a landscape composed of a mosaic of different layers to create an ecosystem which will be interdependent and self-sufficient. Maintenance is thus a composite and complex business which may consist of arresting the natural progression of vegetation and holding the different types of plants at that stage of their development which suits our own particular purpose at the time. Most maintenance today is by machines but even bare ground (except under mature forest trees) needs much more labour than grass, mown grass more than rough grass, rough grass more than shrubs and shrubs more than trees.

9.1.5 Low-level ecosystems require more effort to preserve than high-level systems since herbaceous plants and shrubs at the intermediate level have to be eliminated. This is easily exemplified, as every gardening enthusiast knows there is no such thing as a labour-saving flower bed nor an easy herbaceous border. The routine weeding and yearly top-dressing of 'easy' annual and perennial flower beds is a back-breaking and costly service, notwithstanding the fact that to many people the delight to the eye and the scent of flowers is well worth the time and effort involved.

9.1.6 It is therefore essential to determine at the outset of the design of any project which of the following three types of treatment is to be chosen for the particular site in question:

1. A natural and semi-wild low-maintenance solution.
2. An intermediate-maintenance, suburban garden type of solution.
3. A formal inner-city high-maintenance labour-intensive solution.

9.2 Hard landscape

9.2.1 If no funds can be found for frequent maintenance then, provided sufficient capital expenditure is initially available, it is possible to devise a scheme, to be virtually free of the cost of any subsequent regular maintenance, by designing a totally 'hard' landscape for the spaces between and surrounding the building. If well designed, and the necessary finance is available, such areas may require virtually no expenditure on annual maintenance other than that arising from vandalism and other such malicious damage.

Figure 9.1 Informal arcadia: a natural and semi-wild low maintenance site

9.2.2 Even if finance is not available for such an extreme option most schemes incorporate some elements of hard landscape, each of which requires varying expenditure on regular maintenance. These can be considered under the following categories.

Vertical enclosure – walls and fences

9.2.3 Natural stone

Walls of natural stone have not earned their reputation as one of the finest building materials over the centuries without good reason. Their weather resistance, mellowing and timeless characteristics have made them a natural choice; for the farmer building the dry stone walls to fields, and the stonemason for the cathedral cloister or nineteenth-century city bank. If properly detailed to throw off the rain and soot,

they require virtually no maintenance whatsoever over a long period of time.

9.2.4 Brick

Free-standing brick walls are more prone to deteriorate than when used in buildings, being exposed on both sides. It is essential therefore, if subsequent maintenance is to be minimized, that no porous bricks are used in exposed situations, the mortar is of a consistent strength with an appropriate pointing and the proper damp-proof courses are provided both at ground and coping levels.

9.2.5 Concrete

Free-standing concrete block walls require the same care and attention in the choice of materials

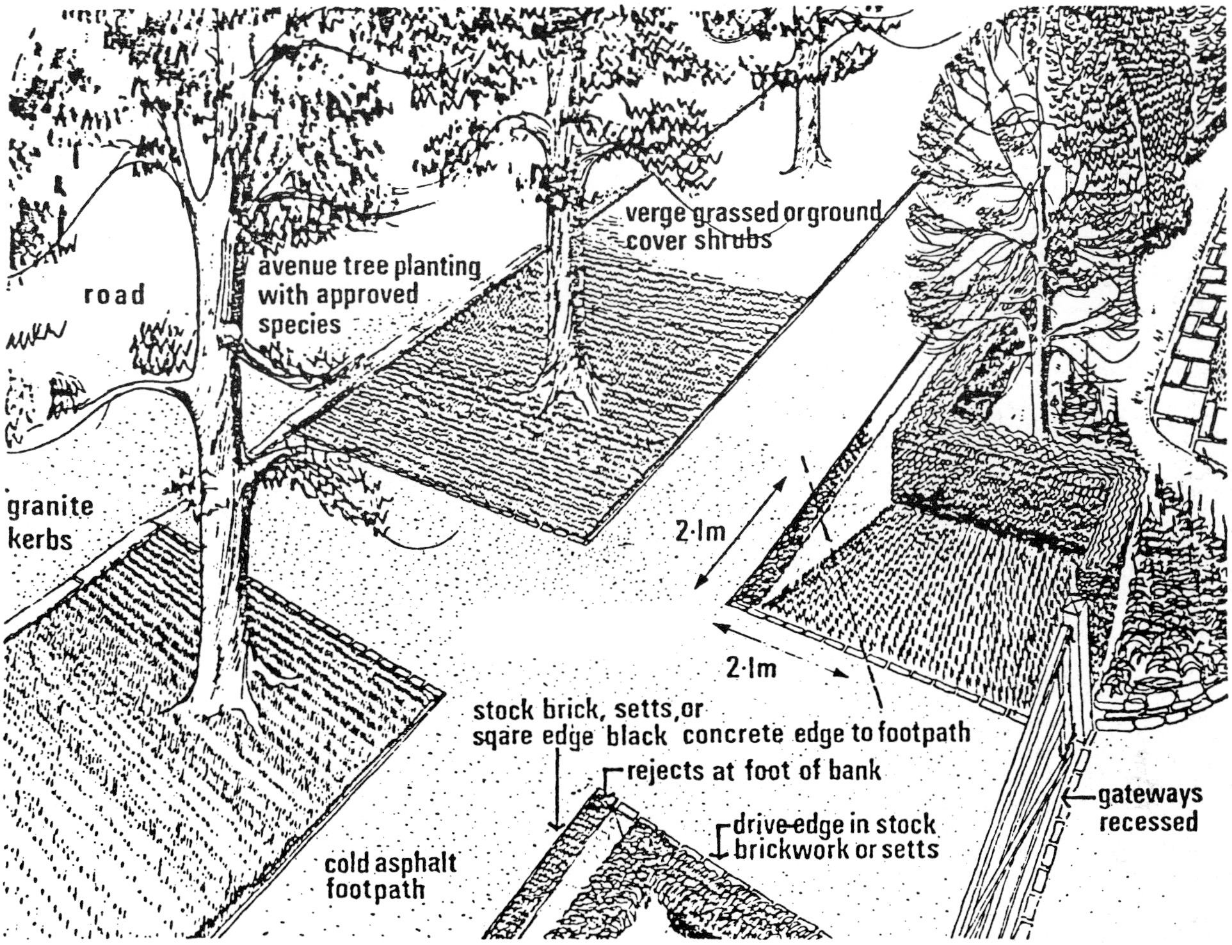
verge grassed or ground cover shrubs
avenue tree planting with approved species
road
granite kerbs
2·1m
2·1m
stock brick, setts, or sqare edge black concrete edge to footpath
rejects at foot of bank
gateways recessed
drive-edge in stock brickwork or setts
cold asphalt footpath

Figure 9.2 A typical formal layout of paths, trees and shrubs

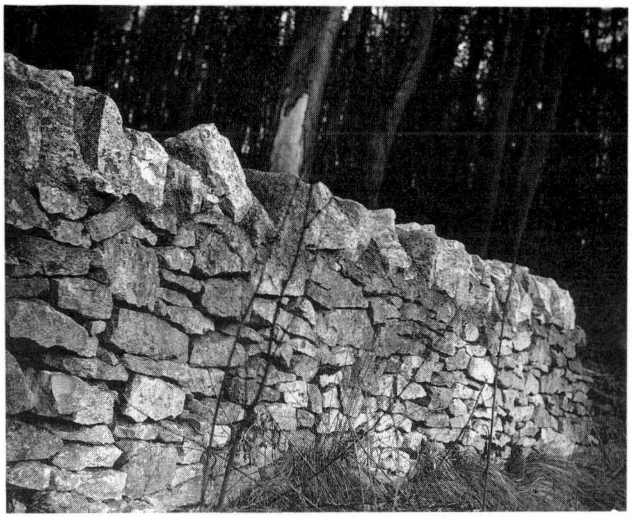

Figure 9.3 Dry stone walling – Cotswolds

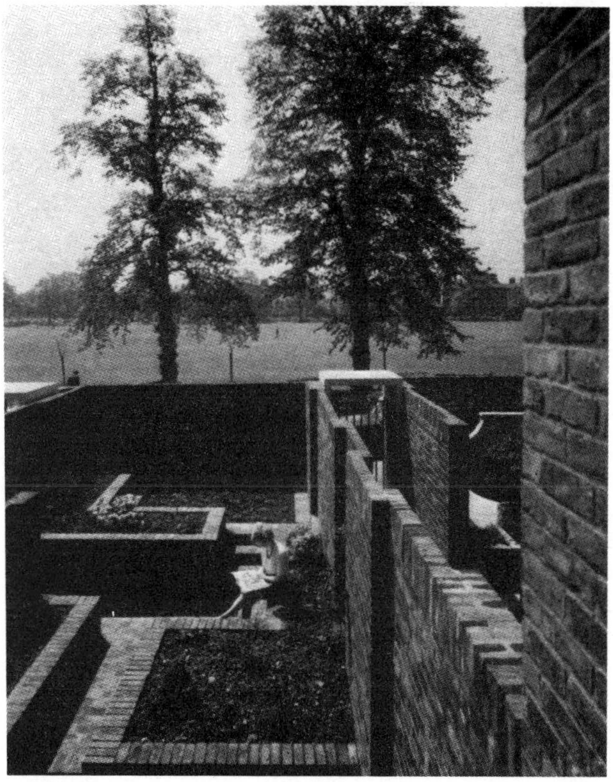

Figure 9.4 Brick walls – Richmond Green

Figure 9.5 Timber chestnut pale – Infant's school, Hampton, Middlesex

and workmanship as brick. The proviso is that being a cast material, and due to the uniformity of colour and texture, meticulous care in detailing and specification are essential if future unsightly staining is to be avoided and continuous and costly maintenance eliminated.

9.2.6 Timber

Where timber fences are oak, sweet chestnut or cedar no preservatives at all above ground are required. Posts into the ground and other softwoods require pressure-creosoting or impregnation with other similar preservatives if the effects of rot are to avoided. Surrounding the posts in concrete is advisable only in certain circumstances.

9.2.7 Metal

Wrought iron and mild-steel fencing should always be to BS 1722: Part 809 with posts set in concrete and protected ex-works with red oxide, zinc chromate primers or hot-dip galvanizing after fabrication. With the advent of PVC and plastic coating facilities subsequent maintenance has been virtually eliminated from chain-link and mild-steel vertical fencing protected in this way.

Horizontal surfaces

9.2.8 Vehicular and pedestrian paving

All paving, vehicular or pedestrian, precast or *in situ*, requires a hardcore sub-grade of an adequate thickness commensurate with the bearing capacity of the subsoil over which it is laid and the anticipated superimposed load and wear, if subsequent maintenance problems are to be eliminated. The choice has also to be made as to whether the surface is to be permeable or impermeable. If the former, precast units have to be laid with sanded joints and if the latter, adequate falls and rainwater drainage gullies are essential.

Figure 9.6 Loose gravel – Richmond Hill

Figure 9.7 Bridge Place, Croydon

9.2.9 Gravel

Subject to proper attention being paid to the requirements in paragraph 9.2.8, the choice of a suitable blend of aggregate and binder, adequate compaction and the absence of traffic for which it is not designed, gravel paths and drives can provide a relatively maintenance-free finish. Regular treatment with herbicide at the appropriate times of the year is advisable.

9.2.10 Bitumen macadam

It is easy to forget that asphalt, bitumen and tarmacadam roads and footpath finishes have been in existence for little more than 150 years. The continued existence of Victorian roads and playgrounds is adequate evidence of the low-maintenance characteristics of these materials together with the added advantage of flexibility

and ease of subsequent access to underground services.

9.2.11 In situ *concrete*

The rigid nature of *in situ* concrete involves the added cost of suitable expansion joints at centres of 60 m. Its extensive use on motorways provides irrefutable proof of its cost effectiveness; the additional initial capital cost is outweighed by its greater load bearing, wear and low-maintenance performance.

9.2.12 Precast units

Precast paving units are frequently chosen for their attractive appearance, scale and, like bitumen macadam, the ease with which they can be taken up for access to underlying services when required. The units can take the form of the traditional and precast concrete paving slab, granite setts, interlocking concrete units or paving bricks. The growing habit of allowing vehicles to park on the footpath to ease congestion on the roads, however, has recently tended to outweigh the advantages, since it is now increasingly becoming advisable in pedestrian areas vulnerable to vehicular traffic to lay 600 × 600 × 75 mm thick units or on a rigid *in situ* concrete base.

9.2.13 Drainage

Irrespective of the materials used – salt glazed ware, cast iron or PVC – soil and rainwater drainage disposal systems need a negligible amount of regular maintenance. With the advent of main drainage systems the Victorian preoccupation with the public health Acts would seem less relevant and the 'testing' of drains before the acquisition of a residential property is now rarely a priority of the purchaser. At present, drains are usually only infrequently cleaned out except when a blockage arises, but the regular rodding of drains and hosing down inspection chambers on an annual basis is still essential in the interest of public safety. New materials such as PVC have been introduced but the advantages of these have yet to be proved over a period of years. Where the economic 'safticurb' or glazed half-round channels with cast iron gratings are used for rainwater drainage a high degree of regular maintenance is essential to prevent silting up.

9.2.14 Statutory services

It has been sometimes suggested that public footpaths are provided solely as a convenient route for excavation at regular intervals for the laying of the statutory service mains. Regrettably, this is not true since the authorities in question would appear consistently to refuse to coordinate the route of their mains so that they coincide at least on a common line, if not in a common trench. Nevertheless, to ensure that the cost of subsequent maintenance and upgrading statutory service mains is kept to a minimum it is considered essential that all designers attempt to follow the recommendations of Ministry of Public Works in their publication *The co-ordination of underground services on building sites; the common trench* (HMSO), still relevant today.

9.2.15 Street furniture

Street lamps, seats, litter bins and other items of street furniture are frequently overlooked both from the capital cost of their initial provision and the cost of their subsequent maintenance. Precast concrete street lamps are frequently specified on the grounds that they require no subsequent maintenance and a wide range of attractive designs is now available. The comparative cost of a 5.0 m high concrete lamp standard is an average at 1992 prices £85, in mild steel £95 and in aluminium £105. The cost of painting once every three years is £5 per year. Of far more relevance is the type of lamp used. A blue mercury vapour (MBFU) lamp costs £7.50 and up to £25 per year to run compared with the yellow sodium (SON) costing £32.50 and £6 a year running cost for the same lumen output.

9.2.16 Seats

The initial capital cost of a hardwood local authority seat is in the region of £150 with an anticipated life of 15 to 20 years. This, however, is subject to it escaping the attention of vandals,

Figure 9.8 Bitumen macadam – Parade Ground, Royal Artillery, Woolwich

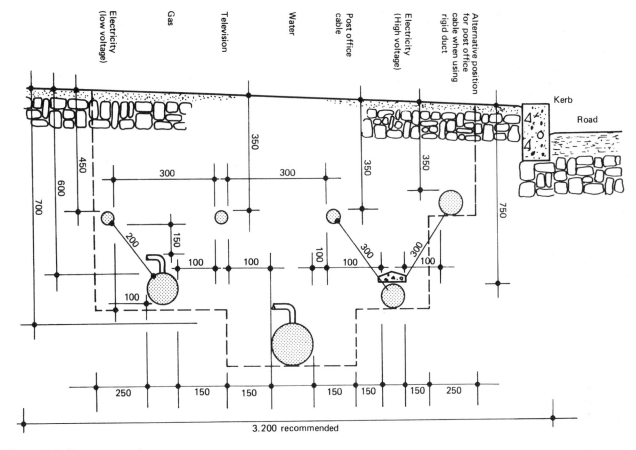

Figure 9.9 Statutory services

where in certain inner-city areas it can be expected to have an anticipated life of only a few weeks. It is essential therefore that the most important criterion for the choice of a seat in a public area is its resistance to malicious damage. Litter bins must be equally robust for the same reason and the size should be no larger than that necessary for the collection of normal litter (as distinct from domestic refuse which will otherwise be deposited in it). Full consultation with the local authority to arrange for regular clearance is obviously essential.

9.2.17 Playground equipment

Notwithstanding the provisions of BS 5696: 1979 covering the minimum standards of construction, installation and maintenance of playground equipment from the point of view of safety, regular inspection to reduce the number of accidents and their severity is equally essential. Those responsible for the provision of such equipment must ensure by regular recorded inspections every 1 to 3 months and certified inspections every 6 to 12 months that the following features are visually checked:

- *Structure* No bending, breaking, warping, cracking or loose parts.

- *Surface finish* Protective coating remains, no corrosion, cracks or splinters.

- *Edges* Still no sharp points, edges or protrusions.

- *Crush points* Mechanism, junctions and moving components still covered.

- *Moving parts* Bearings, etc. properly lubricated and operating.

- *Guard rails* None missing, bent, broken or loose.

- *Access* No missing, broken or loose steps, treads or rungs.

- *Seats* None missing, damaged, loosened or insecure.

- *Foundations* All stable, firm and free from cracks.

- *Horizontal surfaces* Free from defects, hollows, irregularities and projections.

- *Safety surfacing* An approved impact absorbing surface is essential to avoid the risk of injury.

9.3 Soft landscape

9.3.1 The regular maintenance of external areas must be capable of being executed with the

Figure 9.10 Playground equipment – Windmill Road, Teddington, Middlesex

minimum of inconvenience and obstruction of access to the users and occupants of adjacent buildings. Maintenance work must therefore be able to be carried out as discreetly and inconspicuously as possible. No work can be tolerated which prohibits the normal use of the site, except for very short periods. Detailed recommendations for the provision of landscape maintenance are given in the various British Standard Specifications.

9.3.2 Trees

Trees require the least regular maintenance of all soft planting and, provided they are attended to twice a year, they should remain in a satisfactory

Figure 9.11 *In-situ* concrete, paving – Carnatic Halls of Residence, Liverpool University

and healthy condition at a negligible cost. In May and October roots of newly planted trees should be watered with an approved liquid manure, such as 10 per cent nitrogen, 15 per cent phosphoric acid and 10 per cent potash (10:15:10 NPK), applied at a rate of 60 g/m² measure. Weeds and grass should be kept clear of the stem for a diameter of at least 1.0 m until the tree is properly established. Stakes should be checked periodically, tightened or replaced, ensuring no chafing of the bark takes place. All dead, diseased and branches broken by wind or malicious damage should be cleanly removed. After severe frost, soil around roots should be firmly trodden back. In order to maintain the natural shape and habit of the species no pruning should be allowed other than that carried out by an experienced horticulturist. Trees destroyed by vandalism should always be replaced immediately, weather conditions permitting.

9.3.3 Shrubs

Shrubs should also be maintained and fertilized in spring and autumn in the same way as trees. In addition, at least once per month during the growing season, shrub borders should be lightly forked over to remove any weeds and to keep the soil open and pliable, taking care not to disturb the roots of the plants, and the soil refirmed around them. Weeds should be cleared away or buried and perennial weeds removed with their roots and destroyed. Informal flowering hedges should be lightly pruned after flowering to remove dead heads, straggling side or upright shoots but not clipped under any circumstances or pruned back to a rigid line unless specifically so instructed. Shrubs planted at comparatively close centres ensure weed-free borders reasonably quickly with little subsequent maintenance and the consequent minimization of labour costs.

9.3.4 Grass

Short mown grass requires the most frequent attendance and maintenance of all grassed areas, needing to be cut at least once in April, three times each in May, June, July and August, twice in September and once in October, a total of a minimum of 16 cuts, together with any necessary trimming of edges and mowing margins which should always be provided to all grass areas. In addition, grass should be rolled in April, June and August. Where bulbs or meadow flowers are growing the grass should not be cut until the late summer and grass under heavy shade never closer than 75 mm. Fallen leaves should be swept up in October and then periodically as and when

Figure 9.12 Low-maintenance landscape – Fountain Court,
London

Figure 9.13 Low-maintenance landscape – Hampstead
Heath, London

Figure 9.14 Medium-maintenance landscape – Parliament Square, London

necessary during the winter. Edges adjoining pavings and kerbs, etc. should be cut back to their original line in spring and autumn. All grass areas also require dressing in spring and autumn with an approved turf fertilizer and selective weedkiller. Between May and August grass cuttings may be collected and spread as mulch over shrub borders, worked into the soil and under the foliage of groundcover, as may dead leaf mulch in the winter.

9.3.5 Watering

It is essential to arrange for the regular application of sufficient water to all planted areas to maintain the healthy growth of trees, shrubs, herbaceous plants and grassed areas. This should be carried out using a fine rose or sprinkler until the full depth of topsoil is saturated and will usually be necessary once a fortnight in June and September

and weekly in July and August, i.e. a minimum of 12 per annum.

If a hosepipe ban is brought into force by the Water Authority it should be noted that this only applies to washing cars and private gardens, not to car parks, commercial business parks and local authority open spaces, particularly those where the landscape has been provided as a condition of an earlier planning consent.

9.4 Maintenance plans

9.4.1 Planning authorities in granting consent to development or change of use often include a condition of their consent specifying a time limit during which an agreed landscape scheme must be planted and maintained. A few authorities specify the number of years during which plants must be replaced but most merely ask for the scheme to be maintained 'to the satisfaction of the local planning

Figure 9.15 Medium-maintenance landscape – Exterior of the Hayward Gallery, South Bank, London

Figure 9.16 High-maintenance landscape – Courtyard of the Shell Building, South Bank, London

authority'. This implies maintenance in perpetuity but the Department of the Environment has suggested that this is unenforceable since it implies a continuing obligation on future unknown owners subsequent to the original applicant.

9.4.2 A frequent condition is one requiring that: 'All planting, seeding or turfing shall when approved, be carried out in the first planting and/or seeding season following occupation or completion of the development whichever is the sooner. It shall subsequently be maintained to the satisfaction of the Local Planning Authority for a period of X years, such maintenance to include the replacement of any plants which die within the period specified.'

This is best covered by the requirement for provision of a bond to ensure that the necessary

Figure 9.17 High-maintenance landscape – Peabody Estate, Cumberland Market, London

Figure 9.18 Cavendish Square, London

finance is made available by the bondsman in the event of the owner or tenant not complying with their obligations to maintain the planting and to replace any that are dead or dying.

9.4.3 It is frequently forgotten by both the architect and the building owner that, in addition to the daily cost of emptying waste bins, cleaning floors and the monthly cost of cleaning windows from the moment of practical completion or the commencement of the occupation of the building, they must also make the necessary arrangements and pay for the following:

1. The regular removal of litter from the areas surrounding the building;
2. The pruning of trees and shrubs in spring and autumn;
3. The cost of regularly weeding and watering planted areas during the summer months;
4. Cutting grass once or twice a week during the same period. This is often the most costly of all maintenance.

9.4.4 Storage

In the same way that all buildings should contain cleaners' stores in convenient positions so should they include a gardeners' store in a central location. This means that equipment and materials do not have always to be specially brought each time they are needed on the site.
Landscaped areas must also be designed to permit the maximum use of labour-saving mechanical equipment, with ramps provided to enable all parts of the site to be accessible to mechanical equipment, irrespective of the existence of steps in other locations.

9.4.5 It is essential at the completion of each maintenance visit that all machinery plant and

hoses are returned to the store, all rubbish, weeds and superfluous materials removed and the site left clean and tidy on completion.

9.4.6 Care must be taken to ensure that no plants on-site are damaged during the maintenance operations. Damaged plants must be reported so that remedial measures can be taken immediately.

9.4.7 All maintenance work must be performed at suitable intervals for the total number of times specified and in suitable weather conditions. Allowance must also be made for any necessary measures such as dusting and spraying for the control of weeds, pests and disease. In addition to the central store, hose points must be provided in convenient locations at approximately 50 m centres.

9.4.8 Most planting conditions, including the JCLI *Standard Form of Contract for Landscape Works*, include provision for the option of regular maintenance of planted areas by the firm who has provided them, for the first year after practical completion. At the same time, the contractor accepts responsibility for the replacement of any trees or shrubs which may die during this period, for any reason other than theft or vandalism. Only in the most rare instances is it advisable for a building owner to use staff or pay another firm to carry out this work during the first twelve months after handover; the risks of argument arising from such a divided responsibility are self-evident.

9.5 Maintenance costs

9.5.1 The cost of maintaining the hard landscape is negligible provided that it is not subjected to traffic greater than that for which it was designed, e.g. juggernaut lorries parked on the footpath.

Table 9.1 Typical maintenance budget

Item	Percentage of overhead	Percentage of total budget
Wages (maintenance crews)		45
Equipment operation		2
Supplies and equipment, etc.		5
Subtotal		52
Overhead (itemized)		
Supervisory salaries	27	13
Office operation	6	3
Payroll tax and insurances	6	3
Unemployment compensation	4	2
Accounting costs	13	6
Equipment depreciation	14	7
Interest on loans	2	1
Property insurance	5	2
Rent (property)	2	1
Utilities	2	1
Public relations, including dues, subscriptions, etc.	2	1
Vehicle operation and maintenance	4	2
Equipment repairs and maintenance	4	2
Equipment rental	2	1
Miscellaneous	2	1
Lost time (vacation, sick leave, etc.)	5	2
Subtotal	100	48
Total		100

Table 9.2 Time required to complete some landscape maintenance operations

Operation	Minutes
Turf	
Mowing (and catching clippings)	
Small area (100 m²) with hand-manoeuvred 20-inch mower	15
Large area (0.5 ha) with rider-operated 60-inch rotary mower	30
Edging (30 m)	
By hand	45
By mechanical edger	
Curbs	20
Other edges	20
Fertilizing	
Small area (100 m²) with rotary spreader	5
Large area (0.5 ha) with rotary spreader	30
Spraying	
Small area (100 m²) with a back-carried tank	10
Large area (0.5 ha) with a 300-gallon tank Spray gun operated by two men for one hour	120
Spray hawk with boom on sprayer, in open area	105
Trees	
Spraying (not included are times for filling, mixing and travel)	
Small (3-inch caliper)	25
Large (8-inch caliper)	10
Pruning (heavy)	
Small (3-inch caliper)	15
Large (8-inch caliper)	60
Watering	(Variable)
Shrubs	
Spraying	
Small (910–1200 mm)	1
Large (215–250 mm)	2
Pruning	
Small (910–1200 mm)	5
Large (215–250 mm, for rejuvenation)	30
Watering	(Variable)
Ground cover	
Spraying 100 m²	10

Table 9.1 and 9.2 are reproduced by permission of W. H. Freeman & Co.

Replacing a park bench once a week as a result of malicious damage likewise would not be covered.

9.5.2 Planted areas are, however, a different matter and, as can be seen from the preceding paragraphs, may often incur not inconsiderable sums of money if they are to be properly and regularly looked after. Hence there is the need to classify planted areas into the high, medium and low categories previously mentioned.

9.5.3 It is obviously essential therefore to cost the regular maintenance plan for the building owner, to put in hand and implement as soon as the building is finished. Having scheduled the various tasks required for a particular site throughout the year, it is then possible to determine the time needed to carry out each of them. From this information the total maintenance costs for the site in question can be estimated.

9.5.4 The varying seasonal requirements of different plants throughout the year make the efficient deployment of labour and machinery throughout the twelve months of the year difficult. Trees require sucker growth and formative pruning only once a year between June and October and shrub borders forking over and spraying with weed killer only once a year in March or April. The same border will, however, require weeding and hoeing six times between May and September and grass cutting no less than sixteen times.

9.5.5 A site which may require 10 working days per month to maintain during the summer will probably require only 3 days in the winter. Alternative tasks for the 7 working days surplus to requirements during the winter months have to be found since plants are far too vulnerable to be entrusted to casual labour taken on solely for the summer months even under the close supervision of permanent staff. Activities such as the repair and maintenance of the hard landscape are therefore better left to autumn and winter months, weather permitting, allowing the replacement of dead and diseased plants in the spring, and leaving the summer free for weeding, watering and grass cutting.

9.6 Maintenance contracts

9.6.1 It must be evident from the foregoing that, notwithstanding the regular nature of landscape maintenance, it is rarely economic, except on the largest sites, for a building owner, tenant or authority to attempt to carry out landscape maintenance work using his or her own staff. It is therefore necessary to select either by competitive tender or by negotiation a landscape contractor capable of carrying out the work with the necessary expertise, skill and care; and especially in view of the labour-intensive nature of the works during the summer months coinciding with staff holiday periods. Whether on the basis of a lump sum or basic rates for the valuation of the individual hours of work as and when they are required, an accurate set of contract documents, comprising drawings specification and contract conditions, is clearly essential if subsequent disputes are not to arise.

9.6.2 Maintenance contract drawings

Although drawings of landscape components, their location and assembly are rarely required, as they are for contracts for new work, plans and maps indicating, the location, rights of way access, site facilities and the areas of different planting (e.g. existing trees, shrubs, borders, close-mown and long meadow grass) should always be provided for those firms invited to tender to ensure comparability of tenders submitted. The provision of such drawings and maps should in no way relieve a tenderer from his or her obligation to visit the site(s) to ascertain the nature, scope and site conditions prevailing and to make provision accordingly in pricing. It is, of course, always as well to make specific reference to this in the accompanying specification.

9.6.3 Maintenance contract specifications

By far the most difficult task in any landscape work is the preparation of a specification accurately describing the quality of materials and workmanship to be provided. Although compliance with the relevant latest British Standards and Codes of Practice, such as BS 1722: Parts 1–13, BS 3882 (Classification for Topsoil) and BS 3936: Parts 1–11 (Specification for plants) in addition to those previously mentioned should always be included; ensuring that strict compliance on-site is always easier said than done; and for this reason, only nursery owners and landscape contractors of proven reputation and inspection of current and recently completed previous projects should be approached. Landscape work, like building, can be specified either by finished effect, British Standard or method, and this is particularly relevant in maintenance contracts where, for example, grass cutting can be specified either by the number of cuts per year to be included and to be provided and paid for 'as and when instructed by the employer' or, alternatively, 'as often as necessary to maintain the minimum standards laid down' (e.g. 'to cut the grass as often as necessary, to ensure that the grass is at no time longer than 75 mm or shorter than 25 mm') throughout the growing season.

9.6.4 Maintenance contract conditions

In addition to drawings and specification conditions of contract are also necessary which, while having for obvious reasons, no defects liability clauses, extensions to the contract period, damages for non-completion, partial possession or provisions for plant failures or vandalism other than for the replacement of any damaged work at the employer's expense. While there is a liability for employer's liability insurance, insurance against damage to the works is, of course, the employer's liability.

9.6.5 JCLI grounds maintenance model form of tender and contract documents

The Joint Council for Landscape Industries (JCLI) produced in 1987, a model set of contract documents comprising formal tender specification and contract conditions suitable for both public and private landscape maintenance contracts,* the specification clauses covering both playing fields and amenity planting incorporating the choice of standard clauses to give effect either to the management option of maintenance by specific instructions or by performance 'as and when required'. Although primarily designed for public and private sector lump sum or remeasurement contracts, with drawings and specification with or without quantities for new works since it includes in the Recitals for a schedule of Rates it is easily adapted for maintenance work by deleting those clauses previously indicated as being inapplicable.

9.6.6 Compulsory competitive tendering

Chapter 9 of the Local Government Act 1988 obliges all local authorities to invite competitive tenders for a range of services subsequently defined in DoE Circulars 8/88 and 19/88 including 'cutting and tending grass, planting and tending trees, hedges, shrubs and flowers, and controlling weeds'. If they have their own horticultural Direct Works Departments they could only carry out the work if they had submitted the lowest tender in accordance with clearly defined competition rules. Detailed guidance to ensure compliance with the Act is published by The Institute of Leisure and

Amenity Management (ILAM) in their 1988 booklet *Competitive Tendering – Maintenance of Grounds and Open Spaces.*†

9.7 Conclusions

9.7.1 The high cost of regular maintenance makes it essential to consider at the outset of a project whether the high-, medium- or low-maintenance solution is appropriate to the particular circumstances, so that the external areas can be designed accordingly. As a general rule, it would seem that inner-city areas require a combination of maintenance-free hard landscape and high-maintenance planting, the suburban areas – the garden style intermediate type, and the more rural areas the natural, semi-wild low-maintenance approach; each site, however, having to be considered on its merits, taking all relevant factors into account.

* Obtainable from The British Association of Landscape Industries (BALI), Landscape House, 9 Henry Street, Keighley, West Yorkshire BD21 3DR.
† Obtainable from The Institute of Leisure and Amenity Management (ILAM), Lower Basildon, Reading, Berkshire RG8 9NE.

10

Conservation: the maintenance of older buildings

John Earl

10.1 Introduction

10.1.1 Public and professional interest in conservation is now so intense that there is, paradoxically, a risk of misunderstanding arising from the mere use of the word. It has acquired a new meaning – or, rather, it has gathered new, emotionally charged overtones – in recent years. It must be explained, therefore, at the outset, that this chapter will be of little assistance in campaigns to preserve historic buildings and neighbourhoods from the threat of deliberate destruction. It is concerned with the day-to-day care of old buildings (which may or may not be buildings of special architectural or historic interest) for the benefit of their owners and users.

10.1.2 'Conservation' is used in this chapter in a fairly narrow sense to identify a particular aspect of building maintenance. The care of older buildings involves all the ordinary processes of slow renewal and decay prevention comprehended by the term 'maintenance', but it makes special demands and may call for special skills. There may even be special motives involved, beyond those of maintaining utility and economic return. The term 'conservation' is helpful in this context in that it is, by usage, associated with specialized repair and protective measures. It has no other implication here.

10.2 The life and death of buildings

10.2.1 It is a common misconception that the desire to conserve old fabrics and adapt them to modern uses is the reflection of a relatively new enthusiasm which, like all fashions, will eventually fade away. The swing of opinion which has taken place in this direction in the last two decades has certainly been impressive and it has produced occasional excesses, but there are no grounds for dismissing it as a temporary aberration. In historical terms it can be argued that it represents a reasonable questioning of the wasteful results of comprehensive renewal policies and a return to long-ingrained habits of good husbandry.

10.2.2 In most periods and places a soundly constructed building has not been regarded as having a predetermined 'life', at the end of which, by the mere passage of years, demolition and rebuilding would be imperative. The motives for terminating the life of a building have commonly been provided by social, political and economic pressures, changing fashions and architectural ambitions or the desire for greater profit. More subtly, the same forces have often weakened the will for continued care, setting in train processes of decline which eventually make demolition inevitable. In such circumstances it is often possible to make calculations which show that it is economically unsound to continue to spend beyond a certain point on the old structure, which thus has a 'life' determined for it. Such calculations are valid only for the particular time, place and circumstances.

10.2.3 The most extreme examples of this process of decline are seen when city areas – and particularly industrial areas – are robbed of the functions which gave them life. For example, large complexes of warehouses in abandoned dock areas suffered catastrophic dereliction in the 1970s and 1980s, the best and most adaptable being dragged down with the rest. At the opposite extreme, the survival prospects of buildings have been radically improved and expenditure on future maintenance given new incentives by planning decisions and environmental and social improvements outside the control of the building owners.

10.2.4 Even in an area which retains its original function and appears to be economically resilient,

(a)

(b)

Figure 10.1 Monuments to consistent maintenance: (a) A
gentlemen's club in Pall Mall, built in 1829–1832. (b)
Almshouses of 1600–1636

some building types may suffer untypical obsolescence. Windmills in areas still producing cereal crops provide one obvious example, but ecclesiastical buildings are among the most vulnerable in this respect. The redundancy of large numbers of established churches has not been solely due to a decline in faith but also to population migrations. Where the population has not been reduced numerically but has changed in ethnic character, a redundancy of churches may be accompanied by a shortage of mosques. With no pressure towards preservation in the campaigning sense, buildings which have reached the end of one 'life' may, after a pause, embark on another. Interesting and apparently opposed trends can be observed. In Britain a number of redundant churches and chapels have been converted to theatres. In the United States some architecturally spectacular redundant theatres have been successfully adapted as churches for charismatics.

10.2.5 In fact, given that a building was prudently sited and soundly made at the outset, its date of erection and particular structural characteristics rarely, of themselves, determine how long it is capable of surviving. In the absence of external pressures of the kinds cited, there has historically been a tendency for buildings to be maintained in some kind of working order for so long as they could be made to serve a useful purpose. The separate components and materials have always been recognized as having widely varying and more or less predictable rates of decay, but by renewing each part as it became outworn, the life of the whole could be progressively extended. Ignorance, neglect, misuse or misfortune might bring about the demise of an otherwise healthy structure, but the number and variety of buildings which have survived from distant times to become 'preservable' objects today provides impressive evidence of the effectiveness of regular routine maintenance.

10.3 Maintaining utility: minor adaptations

10.3.1 The motive for thus continuously extending the life of buildings has overwhelmingly been one of utility rather than sentiment or antiquarian interest. The capacity for survival of old buildings (national monuments aside) is firmly tied up with the value of the accommodation they provide. An ancient timber barn will be maintained principally for its usefulness as a farm building rather than as an example of medieval carpentry. Owners of Georgian houses may derive great pleasure from their architectural character but they too, must find in them the accommodation they need. The owners of both

the medieval barn and the old house may be under exceptional statutory constraints as to the kinds of alterations they may carry out, but some small measure of controlled change to meet practical needs is quite likely to occur even in these cases.

10.3.2 This is not the place to consider the kinds of major adaptations by which buildings are converted to completely new uses but no consideration of old buildings can ignore the fact that they are commonly subjected to a continuous slow process of minor adaptation to meet changing patterns of life and work. The kinds of small alterations which are usually carried out in conjunction with recurring maintenance work will be kept in view in this chapter.

10.4 The older building: a question of kind

10.4.1 Having said that we are not concerned solely with statutorily protected historic buildings it is necessary to consider what is here meant by 'older' building. The meaning is more easily understood than defined. It is as much concerned with kind as with years. Many soundly constructed buildings which are in beneficial use today are the products of architectural, structural and craft traditions which, if not now completely extinguished, are becoming attenuated.

Metrication, standardization and mass production of components are all speeding up the rate of change so that some not obviously exceptional buildings erected as recently as the second half of the twentieth century present special maintenance problems today. In some respects, at least, they must be regarded as belonging to an older tradition.

10.4.2 Whether built 50, 100 or 200 years ago, and however varied their forms or the details of their construction, the mass of old buildings share a common evolution and they exhibit many common patterns of decay. Buildings, for example, which are constructed of load-bearing masonry, bedded in flexible mortars, having short-span openings, pitched roofs, framed timber floors, joinery doors and windows and comparatively few factory-made components can, whatever their date, be regarded for many purposes as members of one family group. Steel and reinforced concrete framed buildings, rigidly constructed to withstand calculated stresses with long spans, flat roofs, non-load-bearing external walls, demountable partitions and many prefabricated components, cannot be regarded simply as younger members of the same family. They have a different ancestry, they are of a completely different physical nature

(a)

(b)

(c)

Figure 10.2 The economic life of a building is only partly decided by its structural character. External circumstances and disincentives to consistent maintenance are important determinants. The early eighteenth-century houses (a) are in a desirable residential area and have been continuously cared for. Those in (b) have fallen into low-grade industrial use. The life of these robustly built warehouses in London Docks (c) was abruptly shortened when dock activity ceased

(a)

(b)

Figure 10.3 In carrying out repairs to old buildings it is generally sound policy to conform as closely as possible to the materials and craft characteristics of the surrounding work. Some materials, like this flint walling, (a) or fine-gauged brickwork offer no scope whatever for the use of modern substitutes. Repairs, if needed, can be carried out only in the original materials, using highly skilled craftsmen. It follows that work of this kind should never be unnecessarily disturbed. Where the fabric also contains important architectural evidence (b) no work should be undertaken without specialized professional advice

(a)

Figure 10.4 Unnecessary disturbance in the name of maintenance can damage rather than preserve the fabric. The soft lime mortar joints in this eighteenth-century facade ((a) and (b)) could be raked out with ease, but the work is sound (a) and the wall shows no internal sign of damp penetration. Some selective repointing has been done at lower levels (b), but general repointing would offer no advantage whatever. Any attempt to rake out and repoint the gauged work (again, completely unnecessary) would irreversibly disfigure a perfectly preserved facade.

(c) This is a London stock brick wall where some repointing of decayed joints was clearly advisable. Deep raking out has been achieved without damage to the brick arrises. The new mortar might have been more closely matched to the old in colour and texture, but the joints are reasonably sightly and they have weathered well.

(d) This is a wall, originally very similar to (c), whose appearance has been radically changed by grit blasting and unsympathetic repointing with 'straps' of cement mortar. Only prolonged exposure will show whether this degree of disturbance was prudent in terms of weather resistance. Visually it is an affront.

The decay of the old, probably late seventeenth century, wall in (e) has been accelerated rather than arrested by ill-advised repointing at some time in the past. The rock-hard scab of cement applied after shallow raking out of the joints was clearly harmful treatment for a wall composed entirely of soft, more or less flexible materials. Saturation and frost damage have been worsened, leading to the destruction of sound bricks. Already decayed bricks which should have been replaced have continued to crumble, leaving the cement joints hanging free of the wall face in several places.

Figure 10.4 *(continued)*

(f) This is the result of careful repairs carried out to a wall of similar age and conditions of exposure as (e). The soft, porous character of the old materials has not been regarded as a defect. Badly decayed bricks have been replaced with well-matched new bricks, bedded and pointed in a fully compatible lime mortar, marginally weaker than the bricks themselves. Seen here after nearly 20 years of weathering the repair has proved itself to be effective.

and their maintenance problems are quite different in kind.

10.4.3 Although the most significant and readily described distinctions (for the present purpose) between modern and older buildings relate to physical characteristics, longevity alone can, of course, accentuate and complicate maintenance problems. It is quite likely, for example, that the present occupants are not the ones for whom the buildings were first erected. Even owners who have been in occupation for some time and who have carried out extensive adaptation works may, at best, have an imperfect knowledge of the fabric and be quite unaware of the concealed effects of earlier alterations and past periods of neglect.

10.4.4 Where the present use of the building is not the one for which it was originally designed, the effects of change (on, for example, floor loading) may not have been continuously monitored. Even where there has been continuity of occupation and where the building is thoroughly documented and its designed use unchanged, modern practices and statutory requirements may impose new demands on the old structure with unforeseen consequences. Changing craft practices and the extinction of traditional building materials may make the repair or replacement of some elements expensive or impossible.

10.5 The special problems

10.5.1 The special considerations affecting the maintenance of old buildings can now be summarized as follows:

1. *Physical character* The materials and skills required may differ in significant ways from those called for in more recent buildings and this may on occasions have repercussions on both the cost of the work and its programming. Modern substitutes and modern techniques may safely be used in some cases but choices are likely to be limited by practical and

(a)

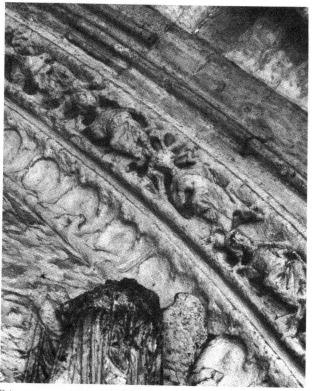

(b)

Figure 10.5 (a) The conservation of decaying stonework requires special care. Repairs should normally be carried out by skilled masons using identical or fully compatible natural stone. In some cases, plastic repairs carried out by a specialist firm may provide an acceptble solution, but inexpert patching with cement or similar hard filler may cause more trouble than it cures.

(b) Irreversible damage can be caused by the application of preservative solutions which, especially if they fail to penetrate to a considerable depth, can hasten the disintegration of the surface. Independent scientific advice should always be taken before applying any such treatment to precious carved stonework. The almost total loss of the stone carvings illustrated here resulted from the disastrous application of a shellac-based 'preservative' in the nineteenth century

aesthetic considerations. The approach to maintenance and minor adaptations and the choice of materials may be subject in some cases to the further constraint of statutory control over historic buildings (see paragraphs 10.9.1 to 10.9.10).

2. There has often been no continuity of occupation and documentation. The present user may lack essential information about the structure. Ignorance may lead to wrong judgements, unintentional misuse and neglect of necessary maintenance operations. Changing patterns of use and new legal obligations may lead to imprudent modifications being made to an old fabric.

10.5.2 These considerations call for a special approach

1. The owner and occupier must be able to call on the services of professionals experienced in the care of such buildings, knowledgeable in obsolete and obsolescent forms of construction and traditional crafts and materials; skilled and resourceful in the diagnosis and treatment of defects (see paragraphs 10.12.1 and 10.12.2).
2. The building must be investigated to establish and record, so far as is practicable, the facts regarding its site, structural character and the properties of its components and materials.
3. Maintenance and minor adaptation must have regard to the facts so established.

10.5.3 Given the professional skills referred to in 10.5.2(1) above, the requirements set out in (2) and (3) should be met without undue difficulty. The detailed investigation of the building provides an ideal opportunity for the preparation of a maintenance manual and the existence of a well-designed manual should produce major benefits in both the effectiveness and cost of future care (see Chapter 18).

10.5.4 It should not be supposed that the preparation of a manual offers no advantages where a building has been satisfactorily maintained by the same body for a number of years. Large, hierarchical, maintenance organizations caring for extensive estates have a particular need to collect and share knowledge through the medium of departmental guides and manuals. Seemingly efficient agencies with sophisticated filing and retrieval systems sometimes rely to a surprising, even dangerous, extent on oral tradition for bedrock knowledge of the simple facts and routine procedures on which the effectiveness of the organization depends. The loss of a single veteran in the middle ranges of such an organization can be seriously disruptive if essential keys to the working of a system or important items of knowledge about particular buildings are contained in that person's memory and nowhere else. The wisdom and experience of the veteran are unavoidably lost on departure. Nothing else need be.

10.6 Compiling the maintenance manual

10.6.1 While the *Building Centre Manual* (see Chapter 18) provides a reliable framework, careful consideration must go into designing the detailed contents of a maintenance manual to suit the building. The manual for an old building will need to be kept under constant review. Information which would be readily available in the case of a modern building may be completely lacking at the outset. Amplification and amendment must take place as experience is gained.

10.6.2 The description Section (Part 1(i)) is the one which is likely to need the most careful attention in this respect and it will almost always be useful to augment it with a sub-section devoted to those structural and other peculiarities of the building which influence the manner in which maintenance work is specified and executed. The information contained here will need to be signalled at all relevant points elsewhere.

10.6.3 Constraints of this kind are sometimes far from obvious. Maintenance surveyors should be able to share fully in the experience of their predecessors. References should be given wherever practicable to correspondence files and reports of investigations which bear on past discoveries and mishaps.

10.6.4 The starting point in the preparation of a manual must necessarily be an accurate and full record of the building as it exists. In most cases this will involve at least a measure of site investigation and the production of some new drawings, but existing sources of information should be thoroughly inquired into before embarking upon expensive survey work.

10.6.5 Where, exceptionally, there has been an unbroken line of management since the building was designed, full records, including the original architect's or engineer's drawings may be on file. Where the author of the building is known and the practice still exists, it is, in any case, worth enquiring whether plans and construction details have been preserved in any form (microfilm is better than nothing). Deposited plans in local authority offices sometimes provide useful information. What appears at first sight to be a perfunctorily drawn plan attached to an old drainage application may reveal facts which could otherwise be established only by lengthy site investigation.

10.6.6 All old records need to be checked for both dimensional accuracy and completeness. Details of

past alterations should be scrutinized with special care, since they may represent unexecuted intentions or proposals which were varied during execution. Apparently obsolete drawings, showing earlier states of a building prior to alteration and refurbishment should, however, never be discarded. A wall of partition which has been resited or a part of a building which has been removed, leaving no visible trace, may nevertheless have left substantial concealed relics. The existence of a redundant beam in a floor thickness, old flues in a wall or foundations and cellars under surrounding open ground can affect the behaviour of a building in a variety of ways, while the unexpected discovery of such impediments during minor adaptation works or the installation of new services can be a cause of embarrassment and expense. Risks of this kind should, wherever possible, be signalled on the modern survey plans, by annotations such as 'See plan 486K of 1922 for details of old engine beds below this floor slab'.

Figure 10.6 Quinquennial detailed inspections are mandatory for Church of England churches and a similar system could produce marked advantages for the care of other kinds of buildings. A theatre, for example, is a vulnerable organism whose effective maintenance – and continued licensing for its designed use – depends on far more than the simple upkeep of bricks and mortar, plaster and paint. Backstage, it is a complex working machine which must be kept in perfect order. The public areas must comply with stringent safety requirements as well as meet the obvious practical demands of utility and comfort. The maintenance of such a building demands the utmost care in programming and monitoring.

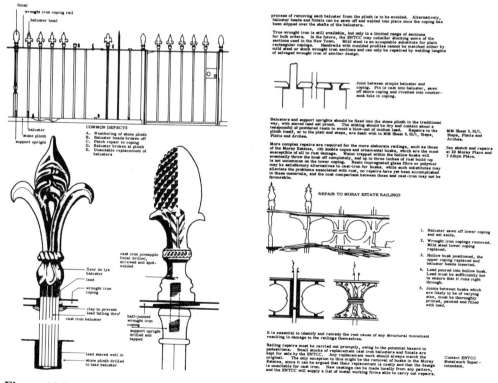

Figure 10.7 An opening from the maintenance manual for Edinburgh New Town: *The Care and Conservation of Georgian Houses* (Davy, Heath and others)

10.6.7 Where it is necessary to produce new record drawings the traditional measured survey can still be a cost-efficient method, especially if it is combined with an investigative structural survey. This procedure has the considerable advantage that it demands a space-by-space, surface-by-surface examination of the fabric, raising questions which promote further investigation (arising, for example, from the checking of apparently inconsistent dimensions) and thereby leading to the discovery of facts which would otherwise have escaped detection. The survey which is carried out 'in-house', using traditional measuring methods (tape and rod rather than electronic devices) is, for this reason, likely to be more valuable than one which is farmed out to a contractor.

10.6.8 The source of, or authority for, all new drawings should be clearly stated in some such form as:

'Measured by: Drawn by: Date:
or 'Based on a drawing by   Dated:
    Checked on site, revised and redrawn to this
    scale by: Date:

10.6.9 A mixture of Imperial and metric scales sometimes cannot be avoided. Where very full and accurate old plans exist it is usually preferable to go on using them rather than incur the risks inherent in reproducing them to a scale so slightly different (e.g. 1:50 from $\frac{1}{4}$ in to one foot) that errors may creep in unnoticed.

10.6.10 Notwithstanding the advantages of uncomplicated forms of physical investigation (see paragraph 10.6.7 above), technical aids can be used with advantage to record additional information which would either by unobtainable or obtainable only with unreasonable expense or unjustified disturbance. Modern non-destructive surveying techniques have the capacity to reveal important facts about hidden structural features and concealed defects. Rectified photography and photogrammetry can be useful in reducing the labour involved in recording elaborately modelled facades and deformed structures.

The explanation of some complex features may, in any case, require photographs in addition to drawings. Full use should be made of photography to record normally concealed parts of the structure which are revealed during works. A clearly marked rod may be used to give scale. A convenient way of filing photographs is to mount 10 in × 8 in prints into ring-back folders, each print being captioned and related to a key plan. The prints should bear a negative number and a progressive list of the numbers should be fixed in the front of the folder,

so that the loss of a print can be detected and a replacement obtained. Occasional losses are to be expected in a working collection which is subjected to frequent handling and it follows from this that negatives should be kept in a separate, secure filing system. Details of photographic and photogrammetric records should be noted in the manual.

10.6.11 The maintenance log section of the manual assumes particular importance with older structures where every building operation is, to some extent, to be regarded as an opportunity for furthering investigations into the nature of the building. Knowledge gained in the course of works should be recorded and, if necessary, the general maintenance instructions should be modified in the light of new knowledge. The information to be provided under other headings is self-evident. It is particularly important with old buildings to ensure that the occupiers are fully aware of the need to observe limitations imposed by the structural and other peculiarities of the building and are also alerted to report signs of cracking, rot or insect activity. The completion of the 'User Guide' manual pages needs special care.

10.7 Maintaining the fabric: general observations

10.7.1 The wide variety of materials and structural details and the range of maintenance problems likely to be encountered in old buildings precludes any really detailed consideration here. It is more sensible in the compass of a short section to attempt to take a view of the whole field and suggest some principles of general validity.

10.7.2 If a single aphorism is to be offered it must be to the effect that sound and stable conditions should not be subjected to unnecessarily radical disturbance. This applies as much to routine maintenance works as to major alterations. The more closely the specification adheres to the characteristics of the original work, the less likelihood there will be of unfavourable reaction or failure.

10.7.3 This principle of minimum disturbance scarcely needs stating where the fabric is a fragile one of great archaeological value and the works proposed – perhaps the removal of important structural divisions – would not only harm its interest but also upset a delicately poised equilibrium. The need to minimize disturbance where all that is proposed is the repointing of the brickwork of a terrace of plain Victorian cottages is less obvious. Even here, however, departures from good traditional practice can cause serious damage. Brick joints may be raked out with ignorant enthusiasm, damaging the brick arrises and disfiguring the appearance beyond recall. At the other extreme, a too shallow raking will provide insufficient key and the new pointing, being little more than a skin over the old joint, will be neither effective nor permanent.

The commonest and worst error, the use of a mortar mix stronger than the old bricks and denser than the original bedding mortar, can aggravate water penetration, delay subsequent drying and lead to severe physical damage to the brick face.

10.7.4 Perhaps the first question to be asked before embarking on works like repointing is: 'Is this disturbance necessary at all?'. The answer to even so simple a question calls for some knowledge and experience of old brickwork. The fact that the mortar is soft and can easily be raked out with a blunt blade is not necessarily evidence of a need for repair. The mortar may be quite sound enough for its condition of exposure and provided it is not visibly deteriorating and there is no supporting evidence of water penetration or other damage *resulting from the nature and condition of the mortar* the assumption should generally be in favour of leaving well alone.

10.7.5 To urge a minimum of disturbance is not, of course, to argue for minimal maintenance. One of the objectives of regular, planned maintenance is avoidance of the necessity for the kind of radical and expensive renovation campaigns which inevitably follow periods of parsimony. What is being commended is an attitude to the repair of old buildings, a strategy which it is believed tends to the extension of life. Whether the works being executed are of ordinary recurrent maintenance nature or major works of infrequently recurring repair and adaptation, it suggested that a guiding principle should be to specify materials and techniques (and design any new works) so as to be closely compatible with and to involve the least disturbance to the sound parts of the old structure.

10.7.6 Even when dealing with historically sensitive fabrics, it is rarely possible to adopt a strict 'museum' approach to conservation, doing nothing which cannot later be undone. But, nevertheless, we must be extremely cautious in applying irreversible treatments (especially newly developed treatments) to old materials. Spectacular failures have occurred following unwise attempts to arrest the decay of stonework by surface applications of preservative solutions. Grit-blast cleaning has permanently robbed some brick and

stone buildings of their surface textures and crisply moulded details, while the application of 'protective' renderings has all too frequently aggravated rather than improved condition of old walls, while simultaneously destroying the pleasing appearance of the weathered brickwork.

10.7.7 All work to an old building should be designed and specified with a careful eye to future maintainability. Major repairs and adaptations are sometimes carried out with a kind of finality, as if the building were at last being brought to the desired, static state which it had failed to achieve in its previous life. A 'once for all' approach is justified in cases where limited works are being carried out to prolong the life of a building for a few years, after which demolition is to follow as a certainty. In such circumstances, the irreversible treatment, which writes off the long-term life of an element for the sake of immediate advantage or which strengthens the building in such a way as to limit its future adaptability, may be the right one. In all other circumstances, such a short-sighted approach can only lead to future expense and frustration and may accelerate the slide into obsolescence.

The sound maintenance of all kinds of buildings must be greatly assisted by the adoption of timetabled inspection programmes. The older and more complex the building, the more important it is to monitor condition and performance. Preventive care is invariably less costly than crisis action. The physical character of a large public building like, say, a theatre, with an extensive system of roofs and gutters covering large undivided voids and richly ornamented, suspended ceilings, represents only one aspect of its maintenance demands. The survival of the building in its designed use also depends on the constant upkeep of complex permanent technical apparatus and a variety of public safety provisions. For such a building the Church of England practice of making major quinquennial surveys should be taken as a minimum standard of monitoring, to be backed up by frequent spot checks.

10.7.8 The detailed technical aspects of the conservation of old fabrics are dealt with at differing levels in such works as Melville and Gordon's *Repair and Maintenance of Houses*, Brereton's *The Repair of Historic Buildings* and the Ashursts' *Practical Building Conservation*. As modern building practices increasingly diverge from those of the past, the need for specialized textbooks and guides will be accentuated. These will need to deal not only with the more recondite aspects of conservation but also with simple, workaday matters which, until the mid-twentieth century, were so much a part of the common experience of the professions and crafts as to call for no particular notice or explanation.

10.7.9 The resurgence of interest in husbanding the resource which old buildings represent, coupled with growing experience in satisfying their needs with the limited means now available, has led and will continue to lead to a great expansion in the literature of conservation (see the *Bibliography* for a significant selection). It is to be hoped that this expansion (coupled with the realization that work on old buildings accounts for a very large proportion of *all* building activity) will be accompanied by action on the part of industry – notably the materials' manufacturers and supppliers – to meet the needs of conservation at reasonable cost.

10.8 Unobtainable materials: use of salvage

10.8.1 There is some cause for concern in this area. Some very commonly encountered materials and components have become dimensionally incompatible due to metrication or have changed radically in appearance to meet the demands of large-scale, non-labour-intensive production. Others are now completely extinct or are obtainable only as expensive luxury goods. 'Recognized' substitutes for some traditional materials are obtainable and may occasionally be regarded as appropriate, but in many cases they have to be rejected on grounds of unacceptable appearance, weight, dimensions, performance, permanence, weathering characteristics or physical or chemical compatibility.

In all these respects, the claims of manufacturers of substitute materials (and preservative treatments for old materials) need to be treated with caution. The possibility that an apparently useful new product may have a 'time bomb' effect has to be kept in view.

Until the industry has geared itself more effectively to the demands of conservation as suggested in paragraph 10.7.9 above (and the prospects are, it must be said, not entirely hopeless) there will continue to be difficulties in this area. In the historic buildings field grants have occasionally been made in the UK to achieve limited 'runs' of particular materials or components to permit fully conforming repairs to be made to buildings in particularly sensitive areas, but this kind of special rescue action is not appropriate to meet the needs of the millions of ordinary, traditionally constructed non-metric buildings whose continuing care must be a matter of national concern.

10.8.2 In some circumstances materials and components salvaged from demolished buildings provide a solution but the use of salvage is not desirable in all circumstances. It requires no emotional commitment to preservation to see that the increased use of salvage may encourage the undesirable practice of 'quarrying' old buildings for the sake of a quick profit, but this is not the only question to be weighed. For the greatest economic advantage, salvaged material needs to be removed from demolition site to repair contract in one operation. Double handling and storage of comparatively fragile materials such as roof tiles and facing bricks often causes wastage and invariably increases costs. Even where apparently well-matched material is readily and cheaply available it may be unacceptable in use. Walls built of salvaged bricks, for example, tend to present a very different appearance from undisturbed walls which were built with the same bricks when they were new, and the most skilful pointing is unlikely to disguise the difference. 'Domestic' salvage can, of course, be specially useful in small repairs, where a few selected bricks from an abandoned structure can be used to patch a wall of a neighbouring building.

10.8.3 Large timbers and complete, sturdy components such as iron columns, railings and paving slabs with exact counterparts in the building to be maintained, should of course, always be salvaged. The expense of having such items made anew and their ability to resist casual damage may justify long-term storage on-site awaiting future use.

10.9 Historic buildings

10.9.1 In most developed countries historic monuments are subject to some kind of statutory protection. Where privately owned buildings in beneficial use are so protected, the statutory controls may be an important influence on the manner in which maintenance and minor adaptation works are executed.

10.9.2 The legal provisions vary widely from country to country, but most systems depend on the existence of a national or provincial inventory of monuments, sites, 'landmarks' or buildings of special interest. The effect of inclusion in such an inventory is usually one of delaying destruction or mutilation. Some form of notice of intended demolition or alteration is usually required to provide an opportunity for the claims of the building for preservation to be reviewed. The more developed systems make works to such buildings subject to formal consents and there may be heavy penalties for carrying out unauthorised works. Consents can usually carry enforceable conditions.

10.9.3 Although the basic apparatus of control almost always has to have the essential features described above, there are wide variations in style and procedure between one nation and another. The most effective systems now in operation tend to be of two kinds with contrasting, if not opposed, philosophical outlooks. One type is highly centralized, rigidly supervised by an official elite professional corps and well funded, with grants, loans and tax exemptions available to owners of a relatively limited number of buildings and sites. France is the international exemplar of this type. The other is reactive rather than directional, dependent mainly on local administrative provisions and presupposing the willing cooperation of the owners of a relatively large number of economically active buildings which can look for little assistance from the public purse. The UK, with nearly half a million protected buildings and some thousands of conservation areas in England and Wales, is of the second type.

10.9.4 The British system may be taken as an example of the effects of such controls on the building owner. First, a distinction must be drawn between the 'ancient monuments' and 'listed buildings' provisions. The centrally administered laws protecting ancient monuments (nearly all objects of great antiquity, hardly any of which are in modern beneficial occupation) will rarely concern users of this guide.

10.9.5 Many readers, however, may be involved with listed buildings – that is, buildings listed by central government under the relevant planning legislation (currently, in England and Wales, the Planning (Listed Buildings and Conservation Areas) Act 1990) – as being of 'special architectural or historic interest'. The demolition or alteration

(a)

(b)

Figure 10.8 (a) The fire protection of old buildings may present problems to which there are no ready-made solutions. Cast iron members, for example, are prone to failure if they are cooled suddenly, as may happen when water is hosed into a burning building. This column in a listed historic building has been protected with an intumescent coating. In this instance, the protection thus afforded was acceptable to the regulating authority. The special interest of the building has not been improved by the slight blurring of the profiles, but the process is reversible. There is no permanent loss of character, as there would be with concrete encasing and a Listed Building consent was, therefore, granted.

(b) This iron staircase is unacceptable for escape purposes and incapable of modification. Adequate provision for escape having been made elsewhere, it will be retained for its ornamental character and daily utility.

(including internal alteration) of such buildings requires 'listed building consent' which must normally be obtained from the local planning authority. Early consultation is always advisable to ascertain the authority's views on such matters as the degree of disturbance likely to be permitted to the old structure, the materials to be used and the detailing of the works. Works of simple facsimile repair can usually be executed without consent but where the building is a particularly sensitive one, even operations of this kind should be entrusted only to professionals and craftsmen experienced in conservation work.

10.9.6 Consultation is also advisable over the treatment of unlisted buildings where they occur in designated conservation areas. With few exceptions, consent is required for demolitions in conservation areas.

10.9.7 If it is essential to be aware of the statutory obligations imposed by central and provincial governments in respect of monuments and historic buildings, it is also sensible to take advantage of whatever advice is available. Most central governments with developed control systems publish explanations of their preservation laws. A controlling authority (often the province or the municipality) may issue more particular guidance notes as to the operation of the controls in its own area and these may occasionally take the form of detailed statutory conservation plans which, as well as setting constraints, may offer positive inducements to building owners.

10.9.8 Publications of this kind, however slight some of them may appear to be at first sight, provide valuable guidance and can help to avert wasteful conflicts. In England, the central government Circular DoE 8/87 (to be replaced in 1994–5 by a Planning Policy Guidance note) is essential reading for anyone dealing with a listed building. English Heritage issues advisory material ranging from fully illustrated technical treatises to brief laymen's guides, while many controlling authorities publish advisory leaflets on matters of particular concern in their areas, such as the repair of stucco mouldings, stone-slated or thatched roof coverings, or the correct detailing of replacement sash windows. The Edinburgh New Town manual on *The Care and Conservation of Georgian Houses* (see Bibliography) is a model of its kind, containing a wealth of detailed information on every aspect of the physical care of houses in the New Town.

10.9.9 Financial assistance available for the repair and restoration of protected buildings, as indicated in paragraph 10.9.3 above, varies considerably

from country to country, as do the terms attached to its provision. In France, for example, the statutory grants, loans and tax reliefs are coupled with strict control of works and sanctions against neglectful owners. Serious neglect carries a threat of expropriation. In the UK virtually all grants from both central and local government are discretionary. Central government grants are available for a minority of buildings or areas of outstanding importance. Many local authorities make no provision at all for grant aid. Tax advantages (in the form of VAT relief on some kinds of works to listed buildings) are ill-designed to achieve better care. Sanctions against neglect exist but their effective use depends almost entirely on the will and skill (extremely variable) of local planning authorities. In the United States legal controls and related provisions vary widely from state to state (federal laws are of minimal significance in this connection) and are generally open to more legal challenges than in most European countries, but tax reliefs have been important there in securing the preservation of many buildings, as have the financial advantages accruing from officially sanctioned transfers of development rights from one site to another.

Figure 10.9 Insufficiently supervised alterations to wiring, pipework and other services, internally or externally, can damage the structure and hamper future maintenance. Characteristic of uncontrolled installations is the way in which this conduit compounds its unsightliness by failing to make use of the cut already existing in the stone moulding

10.9.10 The special requirements of historic buildings and monuments must be reflected in the maintenance manual. Where the estate is large and varied, the fact that a particular building is subject to additional statutory controls ought to be brought to the reader's attention immediately on picking up the manual. Part I contains references to applications and procedures. Copies of consents included here will ensure that conditions which are still operating continue to be observed. They will also give some guidance as to the kinds of limitations likely to be imposed on future alterations.

10.10 Fire precautions in old buildings

10.10.1 The achievement and maintenance of modern standards of fire protection and provision for means of escape in old buildings are subjects for study in their own right. Statutory requirements (see Chapter 17) especially in respect of hotels, offices and flats, can be onerous and may necessitate major works to buildings in use. The problems of compliance are most acute in buildings where safety requirements have to be reconciled with the need to preserve the character of architecturally or historically interesting interiors.

10.10.2 Controlling authorities are normally prepared to be helpful in such cases, but they will always err on the side of safety. It is no part of their responsibilties (at least in the UK) to solve architectural and ergonomic problems. The onus rests on the building owner to satisfy what may at first sight appear to be irreconcilably opposed requirements both of which may have statutory force if the building is a protected historic building) while at the same time maintaining the utility of the accommodation for the occupants.

10.10.3 Resolution of such conflicts often calls for considerable ingenuity. An experienced and resourceful designer may be able to propose ways in which the accommodation can be replanned, the uses re-allocated and the protective works detailed so as to produce a safe, workable building with a minimum of structural and architectural disturbance. Discussion with the various statutory authorities is always advisable before alternative proposals are completed and presented.

10.10.4 Despite the experience which has been built up in the offices of controlling bodies and private practices specializing in conversion, little has been published in the way of practical exemplars for the architect and maintenance surveyor. Some authorities distribute guidance

leaflets showing how some kinds of old panelled linings and doors can be made fire-resistant without serious damage to their appearance, but specific formal research into ways of protecting valuable interiors has been far from systematic.

10.10.5 More information on the actual behaviour in fire of various kinds of traditionally constructed buildings would be valuable. Experience has been gained in recent years on the use of thin intumescent coatings to protect exposed cast iron structural members, but the acceptability or otherwise of this treatment and its long-term protection from mechanical damage may still raise difficult questions for the practitioner in particular cases.

10.11 Services

10.11.1 The installation and renewal of services in old buildings – electric wiring, gas, hot and cold water pipes, radiators, waste and soil plumbing – may require more than ordinary care. While it is obvious that ill-considered installations can be unsightly, physical damage is also to be guarded against. The running of pipes and cables in floor spaces must be particularly closely supervised to prevent the weakening of joints and beams by notching. Unlagged hot water pipes in confined spaces may also produce a build-up of heat, causing shrinkage of structural timbers or panelled finishes. Unobserved leaks in inaccessible water pipes can be a cause of extensive damage to plaster and set up conditions conducive to dry rot in woodwork.

10.11.2 The routing of new services is worth careful forethought. Advantage should be taken of any pre-existing 'natural' ducts and access traps should be formed at strategic points to facilitate future inspection and maintenance. All service runs, access points and controls should be recorded on plans in the manual. The Occupier's Handbook should also contain information on the operation of services, whereabouts of intakes, meters and controls and action to be taken in case of failure of any service or on observing signs of water leakage.

10.12 The Conservation Specialist

10.12.1 The most important of the special requirements set out in paragraph 10.5.2 above is the need for professionals competent in the care of old buildings. The tide in favour of conservation is now running so strongly that architects, surveyors and engineers who specialize in maintenance and restoration work find that their expertise is more highly valued than it was in the comparatively recent past. Greater esteem has not, however, as yet, led to any marked improvement in the ways in which professionals are trained for this specialized activity.

10.12.2 The training of architects gives little weight to the study and care of old buildings. Some architectural schools offer a conservation option at one stage or another but there is no single qualification or post-graduate diploma which has established itself as an unchallengeable mark of competence in this field. The training of building surveyors is more closely concerned with the upkeep of existing buildings but pays limited attention to the older forms of construction and offers scant instruction in design. All the professions concerned with the maintenance and architectural conservation of old buildings tend to acquire their skills less from their formal professional education than by what amounts to post-graduate apprenticeship to expert practitioners. While apprenticeship is no bad thing in itself, there is a clear need for their formal education to be designed to prepare them for what must now be accepted as one of the more important activities of the building professions. Post-graduate courses in conservation are available for all the building professions, but no such course should be expected to make good gaps in basic professional education.

The steady decline in training for the traditional building crafts, amounting in practical terms to the deskilling of the industry, should also be a matter for concern. In this connection the SPAB's William Morris Craft Fellowships represent a most important step forward, but the Society's

pioneering project remains a shining light in a generally gloomy scene.

10.12.3 While the present situation persists, students who, in their future professional careers, do not intend to avoid all responsibility toward old buildings, must attend to their own education as best they can, studying traditional techniques and materials and grasping every opportunity to observe conservation work in progress.

10.12.4 Practitioners who are already committed to the care of older buildings should not neglect the building literature of the past. Building construction textbooks, craftsmen's guides, architect's and engineer's pocket books, manufacturers' catalogues and similar works contemporary with the buildings being maintained will be found to be valuable repositories of information not readily available in modern publications.

Further reading

Alan Baxter & Associates with others, CIRIA Report No. 111: *Structural Renovation of Traditional Buildings* CIRIA (1986)

Ashurst, J. and Ashurst, N., *Practical Building Conservation*, Vols 1 to 5, Gower and English Heritage (1988)

Brereton, C., *The Repair of Historic Buildings: Advice on Principles and Methods*, English Heritage (1991)

Michell, E., *Emergency Repairs for Historic Buildings*, Butterworth/English Heritage (1988)

Powys, A. R., *Repair of Ancient Buildings* (1929; reprinted by Society for Protection of Ancient Buildings 1981)

11

The conservation of modern buildings

John Allan

11.1 Introduction

11.1.1 As the twentieth century draws to a close questions as to whether and how its most significant architectural achievements are to be preserved and restored are attracting an increasing amount of public and professional interest.

11.1.2 This may be attributable to a combination of factors – economic or social pressure to deal with buildings whose original purposes or technology have been superseded by new requirements; the widening media attention and popular debate surrounding architecture and the environment generally, or, simply the 'distancing effect' of the passage of time that enables people to rediscover and more objectively reassess once-familiar artefacts as products of history.

11.1.3 Rather as the once-neglected monuments of the Victorian and Edwardian eras regained popular and professional acceptance, so it seems the ever-moving cursor of architectural taste may now be reaching the Modern Movement, at least selectively and in certain circles. Indeed, the very fact that in the dozen years since its first publication this book now requires a special chapter on modern buildings is indicative of the relatively recent emergence of this architectural period as a subject of systematic study in relation to matters of conservation and repair.

11.1.4 The conservation, adaptation, re-use and subsequent maintenance of Modern Movement buildings involve a broad range of issues, including aesthetic and philosophical questions, commercial judgement, historical research and statutory protection, repair technology and estates management, all of which confront the public authorities and private owners who have responsibility for their stewardship. At the same time, the building professions, more familiar with

the rehabilitation of traditional architectural styles and building types, are being challenged to expand their competence and knowledge base to deal with the special problems and responsibilities inherent in working with Modernism.

11.1.5 This chapter is intended to reflect this diversity by ranging from the historical, theoretical and statutory background, through the principles and strategic issues raised by modern conservation in practice, and on to some of the detailed technical problems and remedies for the material most closely associated with the development of modern architecture, namely reinforced concrete.

11.1.6 The discussion does not purport to provide a definitive treatment of any single one of these loosely defined spheres – each of which could occupy at least a chapter, if not a book on their own. Specifically, the description of concrete repair techniques in the final section should not in any sense be interpreted as a manual for site application. The aim has rather been to offer an overview of the current debate, the developing methodologies and the generic technical procedures, illustrated, wherever possible, by examples from practice.

11.1.7 The hope is that this general and holistic approach will help to narrow the gap that has so far tended to characterize the development of this field as between, so to speak, the artists and the scientists; that is, between the theoretical conservationists, including the architectural historians and academics on the one hand (who have little practical experience of the tasks involved), and on the other, the actual practitioners – for the most part, specialist contractors and product manufacturers, many (though by no means all) of whom have little appreciation of the historical and cultural significance of the Modern Movement in architecture, or the 'ethics' of its preservation. In the latter category (regrettably) must also be included many owners of modern buildings, not to mention the public at large.

11.1.8 Of course, the above observation is obviously a generalization, yet it is this author's view that much stronger links must be cultivated between the various parties concerned with this emergent field if the conservation of modern buildings is to be *both* architecturally authentic *and* technically informed. By engaging the expertise and concern that already exists in these somewhat disparate camps, the development of new initiatives and education programmes for modern conservation, which has only just begun, surely provides the ideal opportunity to work towards such an integration.

11.2 Historical, theoretical and technical background

11.2.1 The European dimension

Although the focus of this study is necessarily on the British scene, it must not be forgotten that a defining characteristic of the Modern Movement in architecture was its international scope, influence, discourse and personnel. Indeed, in our own case the new direction in architecture dating from the early 1930s is clearly rooted in the formative developments in Europe that immediately preceded and followed the First World War. Many of the most accomplished pioneers that practised here were emigrés or refugees from European countries or elsewhere abroad, while the modernists of native origin openly acknowledged their dependence on Continental exemplars.

11.2.2 This international dimension should constantly remind us that the questions and challenges posed by the Modern Movement inheritance and its selective conservation are being addressed in a wider context than merely our own.

11.2.3 Europe institutions are now well established in the general field of conservation. In 1983 the Parliamentary Assembly of the Council of Europe declared its concern with the protection of twentieth-century architecture. The Cultural Heritage Committee of the Council of Europe has promoted the Convention for the Protection of the Architectural Heritage of Europe which came into force in December 1987. International legal instruments have been prepared for the guidance of member states.

11.2.4 ICOMOS, the International Council on Monuments and Sites, is a non-governmental body composed of specialists professionally concerned with conservation drawn from over 50 nations. ICOMOS has pioneered conservation principles in an international context through a series of charters of which the Venice Charter (signed in 1964), on the philosophy of restoration, is probably the best known.

11.2.5 Although much of this work has been concerned with historic monuments and archaeological sites, the body of theory and protective controls thereby established provides a framework upon which specifically modern initiatives can be developed. Modern conservationists seeking official backing for their cause can utilize these achievements by claiming equal rights of access to the benefits of listed status and grants in aid. Indeed there is agreement among many modern conservationists that although different technical problems may be involved in dealing with twentieth-century buildings the fundamental principles already developed for traditional conservation are of equal validity in a modern context.

11.2.6 A list of contemporary buildings considered worthy of inclusion by the UN World Heritage Council in its world heritage list was prepared by an international gathering of experts convened in 1985 by ICOMOS in Paris. In 1988 under the auspices of the Council of Europe Steering Committee for the Integrated Conservation of Historic Heritage the first meeting of a specialists group on twentieth-century architecture took place in Strasbourg. Three areas of work were identified: the preparation of inventories and selection criteria, problems of legal protection and physical conservation, and the dissemination of information for education of political decision makers and the public.

11.2.7 At a meeting in Barcelona in October 1990 a committee of the Council of Europe agreed an outline for a policy on the protection of the twentieth-century architectural inheritance in Europe. In September 1991 their proposal was adopted by the Committee of Ministers to member states of the Council, resulting in a recommendation for the *identification* of the heritage, *protection* of significant items, *management and training* in modern conservation, the *promotion of awareness* and *European cooperation*.

11.2.8 Meanwhile, in 1979 the Australian branch of ICOMOS promulgated the Burra Charter, which attempted to relate the precepts enshrined in the Venice Charter to Australian conditions, while at the same time broadening its relevance to modern architectural applications and whole sites. The Burra Charter contains a series of definitions and principles which should be of value to any practising conservationist. Further reference is made to this useful document elsewhere in this study. It was successfully applied to the restoration of the Rose Seidler House in Australia in 1988.

11.2.9 Despite these positive developments, it would be incorrect to suppose that even in continental Europe, where the 'International Style' originated, the idea of Modernist conservation was not still a cause that requires to be fought. DoCoMoMo – the international working party for *do*cumentation and *co*nservation of buildings, sites and neighbourhoods of the *mo*dern *mo*vement – was formed in the Netherlands in 1988 and has now established a network of national groups and representative contacts in over 20 countries in Europe (Eastern and Western) and the Americas. Having held two major conferences (Eindhoven (1990) and Dessau (1992)) and published regular editions of its newsletter, DoCoMoMo has become the leading international voluntary organization concerned exclusively with the issues and problems of modernist conservation.

11.2.10 At its founding conference in 1990 DoCoMoMo issued its manifesto, the so-called Eindhoven Statement which is worth quoting in full here as it effectively outlines the key elements of the field under consideration and indicates the tasks to be undertaken as they are perceived by the consensus of participants involved.

11.2.11   The Eindhoven Statement

- Bring the significance of the Modern Movement to the attention of the public, the authorities, the professions and the educational community concerned with the built environment.

- Identify and promote the recording of the works of the Modern Movement, which will include a register, drawings, photographs, archives and other documents.

- Foster the development of appropriate techniques and methods of conservation and disseminate knowledge of these throughout the professions.

- Oppose the destruction and disfigurement of significant works.

- Identify and attract funding for documentation and conservation.

- Explore and develop the knowledge of the Modern Movement.

11.2.12 In the two years between its first and its second conferences DoCoMoMo has made useful (if also uneven) progress in many of these objectives. Through its network of local groups and central secretariat it has achieved a definite national and international profile that has begun to raise the awareness of modern conservation as a live issue with the public, politicians and educational community as well as the building professions and relevant official agencies.

11.2.13 Several of the more active national groups have also embarked on the task of establishing registers of the significant buildings and sites in their countries. This exercise has already revealed the regional variations that may affect the criteria for selection, and the ensuing differences in content and character that the respective registers are likely to illustrate. This may be regarded as a reflection of the richness and diversity of the field and should not be suppressed in the name of uniformity – provided some consistency of formatting ensures ease of access across national frontiers.

11.2.14 DoCoMoMo's transactions have also brought together a considerable number of case studies featuring buildings at risk, buildings where conservation strategies have been formulated but await implementation, and buildings where restoration work has been successfully completed.

11.2.15 In the last category are already such notable works as Aalto's Sanitorium at Paimio in Finland, Le Corbusier's Maisons Jaoul, Paris, and Cité Fruges at Pessac, Duiker's Gooiland Hotel, Hilversum, Terragni's Casa del Fascio in Como, Italy, and the Bauhaus in Dessau. Other model reconstructions have been undertaken – Mies van der Rohe's Barcelona Pavilion and the celebrated Pavillon de l'Esprit Nouveau by Le Corbusier, whose Maison La Roche in Paris has also been restored and adapted to become the Fondation Le Corbusier.

11.2.16 It is of the utmost importance for the development of expert knowledge and popular and official support that the experience and insights gained in these projects are shared and evaluated in an international context. The benefits to be gained should be not only technical and professional but educational and cultural. The antipathy to modern architecture that characterized much of its early history is quite as likely to reappear in the debate over its survival, and those of the conservationist persuasion –

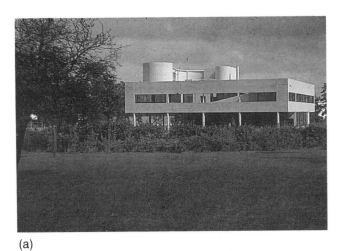

(a)

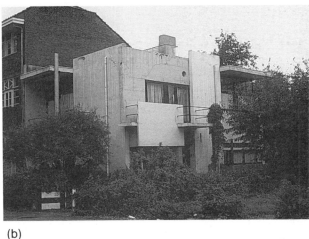

(b)

Figure 11.1 (a) Villa Savoie, Poissy by Le Corbusier (1930); (b) The Schroder House, Utrecht by Gerrit Reitveld (1924). Milestones in the European Modern Movement and now restored as national monuments. (Photographs: John Allan)

usually dependent on the financial support of others – must ensure that their cause does not remain just another minority interest. Perhaps in no country is this qualification more applicable than in Britain.

11.3 The British scene

11.3.1 In so far as such generalizations are meaningful it may be supposed that the peculiar mixture of inventiveness and nostalgia associated with the British 'national character' accounts for our continuing ambivalence towards modern culture generally and modern architecture in particular. The country that became a safe haven for the progressive European intellectual and artistic refugees in the 1930s continues to be uncertain whether Modernism is desirable or even whether we are really part of Europe. The much-publicized royal aversion to modernity has not helped this state of affairs.

11.3.2 The Modern Movement arrived later in Britain than in Europe, where the major milestones in principles and practice were all achieved either before the First World War or by 1930. In the UK nothing of significance in the so-called International Style appeared before the early 1930s, and the volume and range of modernist buildings completed before the outbreak of the Second World War was extremely modest by Continental standards.

11.3.3 Such work as was accomplished was largely the product of private patronage, much of it in the form of individual houses, again in contrast to the situation in many European countries, where modernism was adopted by public authorities and

corporate clients in large-scale municipal or industrial programmes.

11.3.4 After the Second World War the changed social and economic priorities of national reconstruction effectively converted the experimental architectural precepts of the 1930s into the orthodoxy of the Welfare State, and between 1945 and approximately the mid-1970s modernist architecture – albeit in many interpretations – was the prevailing tradition of building design used in Britain to meet the vast requirements of new national infrastructure in housing, education and leisure, health and welfare, transport, industry and commerce.

11.3.5 The well-deserved criticism of much of this output for its social and technical failures has led – at least in some quarters – to indiscriminate and unjust condemnation of modern architecture in its entirety, regardless of the great benefits it has also brought to post-war society in the many fields indicated above.

11.3.6 Meanwhile, from the conservationist perspective, the sheer volume of production and the fact that much of this work was undertaken at great speed, often with untried materials and techniques and less than generous budgets (to say nothing of the uneven standards of after-care and maintenance), has now made for commensurate problems of evaluation and classification.

11.3.7 The two leading voluntary organizations involved in this field in Britain are The Twentieth Century Society (founded in 1979 as the Thirties Society) and DoCoMoMo-UK, the British arm of the international network centred in Eindhoven. The activities of these bodies (both registered

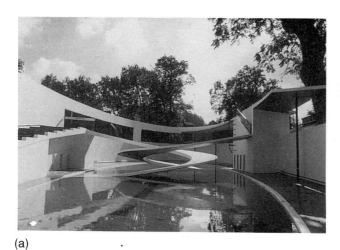

(a)

(b)

Figure 11.2 Examples of modern conservation in Britain. (a) The Penguin Pool, London Zoo (1987).

(b) New Farm, Haslemere (1993). (Photographs: John Allan)

charities) in organizing seminars, exhibitions, lectures and visits to significant works as well as engaging in campaigns to prevent the loss or disfigurement of buildings at risk have done much to bring the subject of modern conservation to the attention of the media and a wider public.

11.3.8 In addition to these advances in 'consciousness raising' there have been a number of restoration projects carried out recently on well-known Modern Movement buildings in Britain, including Lubetkin's Penguin Pool at London Zoo and his Highpoint apartment blocks in Highgate, Mendlesohn's De La Warr Pavilion at Bexhill-on-sea, and Amyas Connell's New Farm near Haslemere in Surrey. The fact that all of these and several other such projects have been funded, at least in part, from the public purse may be an indication that the best of modern architecture may slowly be beginning to find official favour. Indeed, the desire on the part of that most conservative of bodies, the National Trust, to acquire the celebrated modern house of the late Erno Goldfinger in Hampstead, London, suggests that the historic interest and cultural value of such works can be expected to claim an increasingly appreciative audience.

11.3.9 These success stories are, however, but the tip of the considerably more problematic post-war iceberg. While a recent exhibition to stimulate new interest in some of the more worthy buildings of the 1950s and 1960s was mounted by English Heritage under the hopeful title 'A Change of Heart', the controversy surrounding campaigns for the retention of such landmarks as Goldfinger's Alexander Fleming House, the South Bank Arts Complex or Sir Basil Spence's tower block housing estate at Hutchestown, Glasgow, is

probably a more realistic reflection of the resistance to modern architecture that will continue to characterize the British scene for the foreseeable future.

(a)

Figure 11.3 The conservation dilemma of 'unpopular' post-war buildings. (a) Alexander Fleming House by Erno Goldfinger, empty and languishing (1993). (Photograph: John Allan)

(b)

(b) Sir Basil Spence's flats at The Gorbals, Glasgow,
demolished September 1993. (Press Association, Photograph:
Chris Bacon)

1.4 Philosophical aspects

11.4.1 Quite apart from the residue of popular
hostility towards, or alienation from, modern
architecture that continues to colour the
conservation debate, there is another problem
perhaps unique to Modernism that has
preoccupied those who in the last few years have
attempted to develop an informed discourse on the
subject.

11.4.2 This concerns the seeming paradox of
seeking to prolong the life of buildings whose
design intentions and physical fabric were
purportedly determined solely by their *operational*
programme. To one school of thought at least,
modern architecture's defining *raison d'être* in
contradistinction to all preceding traditions was its
commitment to the idea that buildings should not
be conceived as monuments, rarefied artefacts
defying time and change, but as functioning tools
valuable only for their capacity to serve the social
requirements or economic processes that caused
them to be built in the first instance.

11.4.3 To this philosophy was added a
constructional aesthetic that, ironically, by seeking
to defy the passage of history by inhabiting a
constant present actually became uniquely
vulnerable to the ravages of time. To the sceptic at
least, in attempting to supersede the cycle of styles
with a permanent concept of 'style', the modern
movement became identified with an architectural
vocabulary of extreme ephemerality.

11.4.4 To attempt to preserve an authentic early
modernist building, so it has been argued, is thus
to misunderstand the most fundamental premise of
Modernism, to subvert and dishonour the
intentions of its authors, and to fall victim to the
same 'museum culture' that has produced the
acres of themed experience and ersatz history that
increasingly characterize our towns and cities in
the name of heritage, and that the self-professed
successors of the original moderns so vociferously
deplore.

11.4.5 By this logic, the only consistent response
to the question of what to do with modern
architecture in decay is to salute its destruction
(albeit after conscientious documentation), or
consign it to terminal neglect. A prime example at
the very centre of the DoCoMoMo initiative is the
famous Zonnestraal Sanitorium in Hilversum
(1928–1931) whose original mission to alleviate
tuberculosis in the local community has long since

(a)

(b)

Figure 11.4 (a) Zonnestraal Sanitorium, Hilversum, opened in 1931 to combat tuberculosis, (Photograph: E. J. Jelles). (b) Now derelict, its original purpose accomplished. (Photograph: DoCoMoMo International, Eindhoven University of Technology)

been fulfilled. The fact that the achievement of this goal was apparently also in the mind of Duiker the original architect, who therefore deliberately designed the centre with a relatively short life expectancy in mind, only adds to the conundrum.

11.4.6 Despite a certain intellectual appeal, this argument to exclude modern architecture from the tradition of conservation is not without its own difficulties. For one thing, it does not really address the most common predicament confronting those actually involved with ailing modern buildings, which is more often not a simple question of demolition and redevelopment but how best to modify, improve and repair. Moreover, most of the advocates of modern conservation can agree that it should only be applied selectively, and that even among those buildings for which there is a consensus for retention there are likely to be few so special, so valued for their own sake, that they are able to 'pay their own way' simply by being monuments – open to an admiring public for admission charges.

11.4.7 A related argument to that in favour of architectural euthanasia is also often advanced to support the unconstrained alteration of a modern building to suit new requirements that may have superseded the original programme. Thus the modernists' commitment to progress, with its implication that the value of a modern building lies only in its continuing 'tool value', is supposed to justify the licence to adapt and amend surviving works as radically as necessary.

11.4.8 Most buildings need to work for their living, to be sure, and as a consequence must assimilate the changing functions that economic or social circumstances demand. It is this pressure for change, rather than the sentence of death, that usually constitutes the challenge for conservationists, though there will always be the 'campaign cases' of buildings to rescue from oblivion before the evidence is destroyed. However, the acceptance of this general proposition does not license any intervention in a particular case, regardless of the architectural consequences. Conservation is a branch of architecture and, accordingly, no less than any other sort of design, entails judgement.

11.4.9 The most difficult (and therefore most interesting) cases where this judgement is exercised will be where the 'iconic status' of the most significant works renders them particularly resistant to change. Just how difficult it may prove to establish a consensus in such cases is illustrated by the Mendlesohn-Chermayeff house in Chelsea, London, where alteration proposals from a modern architect of unimpeachable international standing have not satisfied a substantial section of the modern conservation lobby.

11.4.10 Another flaw in the anti-conservation argument is that it attributes to all so-called modernists a universal monolithic belief system based on the 'expendability principle' outlined above which, historians have already discovered, is very far from an accurate account of the widely divergent ideas and ideological struggles that actually characterized the modernist period. Research and study has shown that many pioneers supposedly committed to a technocratic future were in fact deeply influenced by the traditional or classical culture of their own training and, while certainly impatient to jettison the excess baggage of historicist stylistic rules, were by no means simple iconoclasts in their attitude towards history. Indeed, many of the inter-war *avant-garde* played an active part in the conservation battles of their own day.

11.4.11 But over and above these empirical considerations, it may be contended that even an architectural movement predicated on the idea of transience *en route* to a better future cannot uniquely escape history but is also a cultural expression of its own period, and as such deserves to be safeguarded – at least in its most significant manifestations – for later generations, just like the greatest achievements of previous periods.

11.4.12 Even this brief rehearsal of the current debate suggests that when it comes to dealing with the modern period, just as with any other, the central problem of all conservation – the difficulty of sifting the past and deciding what is important – cannot be avoided by either the advocates or the sceptics. Even after the many attempts to articulate a permanent and universal set of principles, the underlying objectives must always be argued out again and refocused on the particulars of the individual case.

11.5 Preservation through legislation

11.5.1 The process of 'listing' is one of the most widely adopted measures for providing protection to buildings or sites of special interest, and is well established in legislation of one form or another in most of the countries where modern conservation is being pursued. Listing may have a particularly significant role to perform in contexts where the architecture of the period concerned does not enjoy an automatic popular appreciation. In other words, the statutory authorities involved must accept a responsibility to lead, as well as reflect, public opinion.

11.5.2 In this country listing (until 1984 undertaken by the Department of Environment, who still retain formal jurisdiction on certain casework) is now the responsibility of the Department of National Heritage under the Planning (Listed Buildings and Conservation Areas) Act 1990. The Secretary of State may list buildings for their architectural or historic interest – the latter criterion encompassing the interest attaching to a building or group by virtue of some association with a famous individual or event in the past. (An example of this is the Grade I house in Ebury Street, London, where the 8-year-old Mozart composed his 6th Symphony.)

11.5.3 It is, of course, normally for their architectural (rather than historic) interest that buildings of the Modern Movement may be considered eligible for listing. Entries to the list are ranked according to their perceived architectural value – Grade I being the highest grade for buildings 'of exceptional importance' and very sparingly awarded, then Grade II* for those of 'special interest', and the lowest Grade II for those otherwise considered worthy of preservation.

The system in Scotland is similar, though uses the designations A, B and C respectively. The respective proportions of the total are Grade I – less than 2 per cent; Grade II* – approximately 4 per cent; and Grade II approximately 94 per cent (for obvious reasons the exact figures are subject to frequent change.)

Entries are recorded in the 'Green Books' which normally give a location/schedule reference, the dates of construction and of listing, the name of the architect, and summary details of the building. These records may be inspected at the local planning authority or at the offices of English Heritage. It may be noted however that the only legal requirement of an entry is the building address, thereby allowing amendment and amplification of the descriptive material without statutory ratification. Similarly, although a developer is required to deposit drawings of a listed building before demolition or material alteration, there is no general statutory obligation upon an owner to provide an official record of his listed property. This is a situation which some conservationists regard as a weakness in current legislation in need of correction, though the problem of enforcement – if such a duty were imposed upon owners – would seem to be no less intractable than that of resources – if it were imposed upon conservation authorities.

In addition to the statutory list it is common for local councils to maintain 'Local Lists' of buildings in their area considered worthy of preservation, many of which will have been formerly classified Grade III under previous legislation – a grade that is now defunct. Such lists are not subject to the statutory controls covering nationally listed buildings.

The designation of Conservation Area status (now also covered by the 1990 Planning Act) imposes limitations on the freedom of property owners to alter or demolish buildings within the area. Any demolition requires consent, while enforcement and preservation notices may likewise be served upon errant owners for unauthorised alteration or persistent neglect. The number of Conservation Areas in Britain is now approaching 8 500.

11.5.4 The protection offered by listing consists in the 'stay of execution' that is provided by the requirement to obtain Listed Building Consent in the event of proposals to alter or demolish a listed building. Initiating such works without consent can render the owner (and even his agents and contractors) liable to criminal prosecution. Alternatively a local authority may issue an Enforcement Notice requiring restoration of the damaged fabric. Obtaining permission to alter a listed building normally involves submitting an application for planning consent to the relevant local authority, who, in Grade I or II* cases, will in turn remit the matter to English Heritage for expert opinion on whether the special architectural quality that caused the building to be listed is likely to be adversely affected. The ultimate decision on whether to permit the proposals to proceed rests with the planning authority. They are, however, unlikely to defy English Heritage's judgement, and applicants are well advised to precede a formal application with careful pre-consultation with both bodies.

11.5.5 Not surprisingly in view of its implications for potential development of property and land, the listing process can become 'politically' charged in certain circumstances. An intending developer may be able to obtain a Certificate of Immunity from Listing to prevent considerations of conservation obstructing his or her plans. Conversely, the Department has powers to 'spot list' an outstanding building if it considers it vulnerable to imminent disfigurement or demolition, provided it is at least 10 years old. Examples of these manoeuvres are, in the former case, Erno Goldfinger's office complex Alexander Fleming House in London, for which an immunity certificate was obtained by the building owners in 1988 very shortly before it would have become eligible for listing, and in the latter, Sir Norman Foster's Willis Faber Building in Ipswich, which was recently spot listed Grade I when unacceptable alterations were proposed. More recently a Certificate of Immunity was obtained to enable development within the Brunswick Square complex in Bloomsbury; while conversely Alexander Road Housing was spot listed Grade II* to arrest unsympathetic estate 'improvements'.

11.5.6 The architectural evaluation that informs the listing process is normally undertaken by English Heritage, although cases for additions to the list can also be presented by conservation and amenity bodies, pressure groups and individuals. English Heritage is an independent public body created by Parliament in 1984 to promote the cause of heritage conservation, advise the Department of National Heritage, and oversee properties and sites for which it has responsibility. It administers an annual budget of some £90 million, manages about 400 properties and enjoys a public subscription membership of 270 000 (1991 figures). The equivalent functions in Scotland and Wales are undertaken by Historic Scotland and Cadw.

11.5.7 English Heritage's role in public education and the control and care of historic buildings is

Figure 11.5 Britain's youngest protected building: Alexandra Road, London Borough of Camden, completed 1976 and spot-listed only seven years after becoming eligible under the ten-year rule. (Photograph: Martin Charles)

complemented by the work of the Royal Commission on the Historical Monuments of England, whose task is to survey and record them. Although most of this activity has, of course, been directed towards historic and archaeological buildings and sites, more attention is now being focused on modern architecture dating from the 1930s and into the post-war period.

11.5.8 Official recognition of Britain's modern architectural heritage began in 1970 with the statutory listing of 50 selected 'pioneer' buildings and some others of the period 1914–1939 following recommendations by the distinguished historian Sir Nikolaus Pevsner. The majority were accorded a Grade II or II★, although a select few achieved the Grade I rating – among them Lubetkin's Penguin Pool at London Zoo and Mendlesohn and Chermayeff's De La Warr Pavilion at Bexhill-on-sea. Since then a further

150 pre-war works have been added and several already on the list have been upgraded.

11.5.9 The so-called '30-year rule' adopted in 1987 enabled the government to list buildings started at least 30 years previously, and thus allows the vast output of the post-war decades to become eligible for consideration. Not surprisingly, it is this period which has generated the most controversy. The now-notorious episode in 1987 when a list of 70 buildings advanced for statutory protection by English Heritage was pruned down to 18 by the Secretary of State has underlined the difficulty of achieving consensus on the architectural values of a period still so relatively close and socially sensitive as the 1950s and 1960s. Since this particular experience English Heritage – rather than simply submitting assorted new candidates for listing on an annual 'roll-forward' basis – have been developing more 'objective' criteria for listing, approaching the period thematically by concentrating on specific building types (e.g. educational buildings, industrial buildings) in the endeavour to establish a more systematic basis for their proposals that will be less vulnerable to the vagaries of political favour or ministerial taste.

11.5.10 The attempt to extend the existing listing system to cover modern buildings has also highlighted another difficulty perhaps peculiar to the post-war period. This concerns the prevalence of large developments embracing multiple groups of buildings and extensive sites. (The recent batch of educational building entries approved for listing in March 1993 relates to 47 sites but actually comprises 93 structures.) To list individually all the buildings and ancillary structures making up a typical post-war housing estate, for example, could involve the inclusion of many items of marginal significance in themselves. The protection afforded by Conservation Area legislation, which has been suggested as a more appropriate alternative for such cases, may, on the other hand, be inadequate to deal with the multiplicity of individual alterations that cumulatively could significantly affect the overall ensemble – especially since the introduction of 'Right-to-Buy' housing policies and the ensuing 'personalization' of entrances, windows and parts of facades.

11.5.11 A similar problem for the listing process is presented by the modern phenomenon of 'serial building' in such fields as education and, to a lesser extent, housing. Where a prefabricated system of construction has been developed and progressively refined over a long-term programme it is as much in the series as a whole as in any of its individual exemplars that the architectural interest resides. The very process of improvement may have left the prototype looking primitive and less capable or worthy of preservation. Here, as elsewhere, difficult choices will have to be made.

11.5.12 It will be readily seen from the above that the acquisition of listed status is not always welcomed by property owners, as it imposes constraints upon their freedom of action in altering or disposing of a building. Cases of 'weekend demolition' have occurred when it was suspected that imminent listing was in the offing – the most notorious example being the disappearance of the famous Firestone Factory in west London over an August Bank Holiday. Unauthorized alterations can be stopped by a local planning authority, who in extreme cases of neglect can likewise serve a Repairs Notice obliging an owner to attend to a listed building,

or, failing this, compulsorily purchase a threatened property and present the owner with the bill for undertaking the work itself.

11.5.13 At the time of writing, with fewer than a hundred post-war buildings listed, it is clear that the application of statutory protection and special area status as conservation measures for modern architecture of this period is still in its infancy. (The national total number of listed buildings is a little above 400 000.) But it is also already evident that unless these initiatives can carry with them a significant degree of popular and political acceptance they will be unlikely to prove very successful on their own. Since most of the buildings under scrutiny remain in active use, only intelligent selectivity on the part of the listing authorities and the inculcation of voluntary conservation on the part of their users is likely to provide the necessary consensus. Policy can ultimately only flourish in practice if it finds sufficient public support, which in this case seems likely to depend upon a profound change in attitudes. This once again underlines the need for modern conservationists to seek a wider audience for their cause, to promote a greater understanding and appreciation of modern architecture, while not forgetting that any successful educational endeavour is usually a two-way process.

11.6 Grant aid

11.6.1 An important element in securing public support in the implementation of conservation policy is the provision of official financial aid in undertaking restoration and repairs. Building owners constrained to treat their properties according to certain architectural guidelines are more likely to be cooperative if they can see that the conservation authority is prepared to 'put its money where its mouth is'.

11.6.2 The system of grading referred to above is closely connected to the eligibility of buildings for grant aid. The principal agency for such aid is English Heritage, who administer several types of grant, including those for works to scheduled ancient monuments, Historic Building Grants, Conservation Area Grants, London Grants and Buildings At Risk Grants. Although circumstances vary, the most common type of grant (or loan) in the case of modern buildings is likely to be made available under Section 3A of the Historic Buildings and Ancient Monuments Act 1953, as inserted by the National Heritage Act 1983.

11.6.3 Cases normally depend upon the property being demonstrably 'outstanding' and the applicant's inability to meet the whole cost of repairs from his or her own resources. The first criterion is usually satisfied if the building is listed Grade II* or higher although the strict requirement is that the property is 'outstanding'; the latter involves a confidential means test. Grant aid is normally only applied towards the costs of structural repairs or restoration works and the reinstatement of special architectural features, and is not given for alterations or improvements.

11.6.4 The standard rate of grant is 40 per cent of the costs of eligible works although this can vary in particular cases. Small-scale works costing less than £10 000 do not usually qualify for grants. In projects involving a variety of types of work, valuations must identify those elements for which assistance is applicable. Other typical conditions involve obtaining all approvals before commencement of work, retrospective payment after satisfactory completion, and partial or full repayment if the restored property is sold within a certain period.

11.6.5 Having said this, it must also be pointed out that the availability of grant aid is ultimately dependent on the financial resources of English Heritage, and applicants whose buildings are eligible in every respect may still be turned away empty-handed. In such circumstances it is well worth approaching other sources of funding, such as the local council and/or county authority. The Architectural Heritage Fund receives DOE support to provide resources for feasibility studies and low-interest loans, and, at the time of writing, moves are afoot to establish a Modern Architecture Preservation Trust under its auspices to address cases of modern buildings in need of aid. English Heritage itself publishes a directory of public sources of grants for the repair or conversion of historic buildings.

11.6.6 The current scarcity of resources for grant aid makes it especially difficult to attract assistance for restoration work to Grade II buildings – the vast proportion of those listed. Although it is possible to base an application on the premise that the building is 'outstanding', it may be easier to take the indirect route of first applying to have the building upgraded to Grade II* or better so as to facilitate a subsequent grant application.

11.6.7 One further point to note in connection with the financial arrangements affecting conservation work concerns the application of VAT. As the regulations currently stand, only approved alteration work that has been the subject of a Listed Building Consent may be zero-rated. This state of affairs has been widely criticized for

acting as a disincentive to the conscientious owner who might otherwise be encouraged to maintain a listed property in good repair. It does indeed seem illogical to impose a statutory protection on a building, then reward an owner for breaking it, while penalizing him or her for seeking to uphold the intention of that protection. Perhaps a rational Chancellor of the Exchequer would reverse this anomaly.

11.7 Technical background

11.7.1 In much the same way that modern architecture presents particular problems in the application of statutory protection, so the technical background of these pioneering buildings now poses special challenges for the conservationist.

11.7.2 A key theme in the origins and development of modernism was the determination to address contemporary social needs by exploiting new materials and constructional techniques. Indeed, the new aesthetic of modern architecture, derived partly from the artistic discoveries of Cubism and Purism, partly from the imagery of ocean liners, cars and aircraft, was greatly influenced by an idealized role for technology that, at least in the early period, was often well in advance of the realities of building construction.

11.7.3 The new freedoms sought in planning, daylighting and architectural expression relied on new frame techniques, larger window openings and sheer planar effects. The evolution of two-way spanning panel and slab structure liberated architects from the inhibiting constraints of traditional load-bearing masonry construction. In place of stone, brick and timber came the materials of the modern age – steel, glass and, above all, reinforced concrete. Combining the tensile strength of steel with the compressive strength of concrete, this formless composite cheap and universal material promised almost limitless versatility.

11.7.4 The application of this technology to achieve the desired imagery of modernism produced, especially in the period of the International Style, one of the most consistent building 'vocabularies' in architectural history. Much of its visual impact depended upon the impression of lightness, thinness, whiteness and geometric purity attainable in fresh concrete. The rejection of traditional embellishment in the drive for formal clarity tended to lead to the omission of conventional details such as copings, sills, drips and overhangs, weathering falls and surface relief generally. Steel windows, another major mass

(a)

(b)

Figure 11.6 The early modern style: sheer planes, flat roofs, geometric purity and minimal trim. An architecture heavily dependent for its intended aesthetic impact upon conscientious upkeep and perfect weather – both rare in Britain. (a) Bexhill Pavilion 1936. (Photograph: Architectural Review) (b) Highpoint One. (Photograph: John Allan)

product development of the inter-war period and initially ungalvanized, were introduced instead of timber and frequently positioned at the outside edge of their building openings to emphasize the effect of weightlessness, thereby also being exposed to maximum risk of corrosion. By the mid-1930s the earliest attempts at curtain walling were being introduced. Large areas of plain concrete and flat terraces, the latter often without adequate falls, insulation or provision for thermal movement, replaced modelled facades and pitched roofs.

11.7.5 Although reinforced concrete was relatively familiar in the field of civil engineering, its usage in architectural design and by domestic building contractors was almost unknown in the early 1930s. While *avant-garde* architects like Lubetkin, Wells Coates and Connell, Ward & Lucas and innovative engineers such as Ove Arup and Felix Samuely could demonstrate the architectural and

structural potentialities of reinforced concrete it is clear that its long-term behaviour was not fully understood.

11.7.6 To this inexperience on the part of designers was added an equal inexperience on the part of contractors in the preparation, placing and curing of reinforced concrete, which was at that time invariably mixed on-site rather than delivered pre-batched with all the quality control procedures available today.

11.7.7 It is in this combination of a minimalist aesthetic with a young technology, not to mention a degree of professional naivety, that lies the origin of many of early modern architecture's technical shortcomings. The extensive innovation in building design and construction, deployed across such a wide front, makes it impossible to embark on conservation studies of all the new techniques within the confines of a single chapter. However,

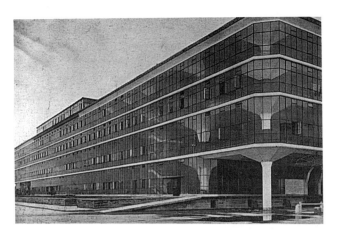

Figure 11.7 Dramatic early use of steel glazed curtain walling at the Boots Factory, near Nottingham, by Sir Own Williams, 1931 Restoration in progress 1993. (Photograph: Architectural Review)

(a)

(b)

Figure 11.8 (a) and (b) Typical window 'details' of early modernism, incorporating ungalvanised frames, almost flush mounting and no cill projection. (Photographs: John Allan)

the abiding and predominant association of modern architecture with reinforced concrete makes it appropriate to include a special section devoted to its problems and repair.

11.8 Modern restoration – principles, methodology and technique

11.8.1 Towards a modern restoration methodology

The preceding sections have outlined the historical, theoretical and technical background to the subject of modern conservation. The following discussion looks more closely at the practical procedures involved and concludes with a section on the particular problems and techniques of concrete repair. Since however the practical challenge of dealing with ageing modern architecture cannot be divorced from the historical tradition of conservation, of which some might say it is but the latest manifestation, it seems appropriate to preface the section with a summary of the generally accepted principles of conservation now established by many years of traditional application.

11.8.2 Principles of conservation

The following abbreviated summary is not offered as a definitive statement of principles, though most are represented or implicit in the manifestos or literature of established conservationist groups – often in a manner that illuminates the ethical dimension of serious conservation. Moreover, while some may be considered uncontentious, others are susceptible to differing interpretation and remain the subject of debate among afficionados. This serves as another reminder that 'universals' can become problematic when applied to the specific case, and that even seemingly self-evident propositions may well need to be re-argued in the new technical context of modernist conservation. It will also be evident that the predisposing decision as to a building's intrinsic architectural importance will determine whether some of these principles will apply at all. The reader is referred to the decision 'flowchart' of H-J Henket illustrated in Figure 11.9.

1. Pro-active maintenance. The principle of protecting and preserving a building's historic fabric from deterioration so as to avoid or minimize the need for repair or replacement.
2. Minimum intervention. The principle of restricting conservation operations to the minimum necessary for preservation of the fabric. The first premise of responsible

conservation is sensitivity and respect towards the original building or artefact.

3. Conservative repair. The principle of determining the quality of repair to accord with the quality of the original fabric. The avoidance of distortion of the original evidence is emphasized and the use of 'like for like' material is usually recommended in replacement.

4. Explicitness of alteration or addition. The principle of clear differentiation between the genuine fabric and any necessary modifications. This is sometimes expressed as an injunction to avoid pastiche.

5. Reversibility of alterations and extensions. The principle of conceiving any necessary modification in such a way as to enable the original design to be reinstated at a future date. Implicit in this principle is the need to record the details of any genuine fabric before it is covered or removed.

6. Compatibility of use. The principle of maintaining or introducing a use for the building that involves no, or least, change to the culturally significant fabric. In extremis this may be construed as an injunction to build anew rather than refashion the original to accommodate an incompatible function.

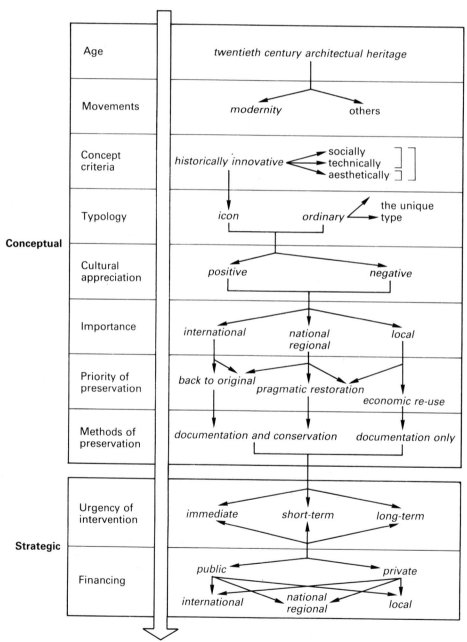

Figure 11.9 Decision flowchart devised by Hubert-Jan Henket to systematize selection of Modern Movement buildings and neighbourhoods for registration, documentation and restoration.

11.8.3 Conservation methodology

The discussion of methodology preceding Figure 11.9 is in the form of a checklist offering a six-point approach to the conservation of modern buildings from the viewpoint of the practising architect. This is not intended as definitive. It is more a distillation of themes which, in the author's experience, seem to recur in most projects albeit with differing emphasis depending upon the circumstances of the case, but which the practitioner can expect to have to address in achieving a successful result.

11.8.4 Obviously, it is based upon the premise that the building in question is retained, even if it is modified, for otherwise there would be no conservation project. As already indicated, this assumption may be the most difficult and controversial factor to establish. Indeed, even modern conservationists need not always advocate the preservation of every building in distress. In many cases documentation may be the appropriate response. A useful flow chart of decisions and choices to be made in reaching these initial conclusions has been prepared by DoCoMoMo and is illustrated in Figure 11.9.

However, as this book is on *preservation* and *maintenance* it is necessarily upon the problems and tasks encountered in *retaining* modern buildings that this study is focused.

11.8.5 Research into the building's original design, construction, materials and components, appearance and setting

Before embarking on *any* physical intervention in, or design strategy for, a potential conservation project it is essential to assemble all the available evidence on the building and/or site that is to be conserved. Library and archival sources should be consulted for drawings, documentary records, contemporary press coverage and photographic evidence. Contemporary Ordnance Survey maps may be valuable in providing information about the original setting. Trade journals of the period may also contain valuable information on particular products or components used in the building, and can often lead to 'rediscovery' of the original manufacturers or their successors. This may be vital in obtaining replacement items. Further technical record information requirements prior to physical survey work are listed in paragraph 11.11.5.

One element of design information that may not be available from photographic records (which, depending on their date, may only be in black and white) concerns the original colour scheme. Here it may be necessary to take surface scrapes at relevant points on the building to establish authentic data.

11.8.6 Understanding the intentions of the original designer/s, the circumstances of the building commission and of its original cultural context and significance

In cases where the original designers, or design practices, still survive, the task of establishing their operational and architectural purposes is made somewhat easier. However, access to 'original' sources is more often not available, and the conservator is accordingly committed to indirect research to gain insights into the circumstances surrounding the commission and the intentions of those involved. Obviously, any substantial biographical studies should be consulted.

An investigation of other works by the same author/s may also provide valuable information on the aesthetic, technical and social themes with which they were preoccupied. Comparative study of their contemporaries may elicit consistencies or differences of intention. It is necessary to be able to understand clearly, in the absence of other physical evidence, what the original designer was attempting to achieve, even if the result was otherwise, for this insight can serve as guidance in approaching the restoration or adaptation of certain details.

Having said this, it must be added that it is the work *as achieved* to which the conservator owes the primary loyalty, not what the designer may have wished to achieve. Equally the conservator must always be clear about which elements of his proposals constitute restoration on the basis of knowledge, and which constitute 'best estimates' on the basis of conjecture. The latter should only be adopted after objective evaluation of their necessity, and confirmation that they will not detract from the special interest and cultural significance of the work.

11.8.7 Discovering the history of the building's subsequent use, maintenance and alteration

It is essential in formulating an appropriate and informed conservation strategy to establish clearly which elements of the surviving building are original, and which have resulted from subsequent modification. The condition and therefore perceived value of the building 'as found' may be strongly influenced by later alterations to its fabric (whether by addition, removal or substitution), by the effects of an unsympathetic pattern of use, or by persistent neglect.

It cannot be assumed that all subsequent

(a)

(b)

Figure 11.10 Transformation of an original design by illiterate and misconceived alteration. (a) Tecton's restaurant at Dudley Zoo in 1937. (Photograph: H. Felton) (b) 1989. (Photograph: John Allan)

changes are 'foreign' to the original concept; in some cases they may add to the architectural or cultural interest. Buildings even as young as 50 years must be interpreted as palimpsests. An interesting example is the house in Willow Road, Hampstead, design by Erno Goldfinger for his own use and inhabited by the architect from the late 1930s until his death nearly 50 years later. The cumulative alterations to the house, all of which were also designed by the architect in response to changing family circumstances, provide an additional source of interest that greatly enhances the significance of the building.

11.8.8 Evaluating the building's architectural significance, identifying its capability for adaptation, and determining the interface between authentic restoration and legitimate intervention

The significance of a building from a conservation standpoint may reside in any or many aspects of

its design. Buildings have been preserved simply because they exhibit the earliest or only surviving example of a specific constructional technique. In others the value may reside in no single feature but in the completeness of all the original surviving fabric in its totality. The 1930s suburban house recently acquired by the National Trust with all its contents intact is an example of the latter.

The Burra Charter defines cultural significance as the 'aesthetic, historic, scientific or social value for past, present or future generations'. To some extent, if it has been listed, these features will have been identified by the criteria applied to determine its 'special architectural interest'. However, to the architect faced with the practical task of conservation the information contained in a listing entry will seldom provide sufficient detail to form the basis of an architectural strategy.

Such a strategy, especially if changes in use are contemplated, must be based on an *architectural analysis* of the building that identifies the essence

of its design. This will include making an
'anatomical' study of the types and disposition of
spaces it contains, its mode of access, its structural
logic, lighting, the relationship to its immediate
setting, and so on. The objective is to establish the
criteria and constraints to be adopted for *any*
restoration or conversion, so that alternative
options may be evaluated within an objective
conservation framework.

A similar exercise may be undertaken as a
means of managing an ongoing process of change
within a listed building to ensure that conservation
principles are not neglected or breached. Such a
dialogue has been established between the
occupants of the Grade I Willis Faber building in
Ipswich and the local conservation authorities to
regulate the types of alteration for which sanction
is needed and differentiate these from routine
modifications deemed to adhere to the original
design. Another example concerns Lubetkin and
Tecton's Highpoint flats, where the desire of
individual householders to replace their windows
could result in progressive erosion of architectural
integrity. Here the involvement of English
Heritage was necessary to establish a standard that
will ensure that any new windows respect the
original uniformity of overall fenestration.

11.8.9 Formulating a viable programme of future use

In many cases modern buildings may have
reached a state of distress not only through
material decay but also because the original use
has discontinued or is no longer required. Indeed,
it is often the loss of use that leads to decay rather
than the other way around, although the two
factors are usually compounded. In such cases the
conservator must be prepared to seek alternative
viable uses for the building as a means of
extending its life. It is a truism that the best way
of conserving a building is to use it effectively.

New uses for old buildings is now a
well-developed discourse within the design
professions. Many excellent examples exist of
conversions of Georgian and Victorian structures
to serve different needs, whether it be
transforming Billingsgate Fish Market into a
financial dealing office or Gare d'Orsay into an art
gallery. But finding new uses for 'new' buildings is
not yet so well established. However, there are
likely to be many cases where the salvation of a
modern building may only be achieved if it can be
made to serve a new and necessary purpose that
attracts the support and commitment of a new
client. The Rubber Factory at Brynmwr and the
huge Power Station at Battersea are two cases in
point.

The Burra Charter defines 'compatible use' as

(a)

(b)

Figure 11.11 The integrity of an original facade may reside
in the uniformity of all its windows, for example (a)
Highpoint, north London by Lubetkin, 1935. (b) At his
earlier apartment block 25 Avenue de Versailles in Paris note
how the otherwise immaculately maintained facade is slightly
spoilt by individual window modifications. (Photographs:
John Allan)

Figure 11.12 The Rubber Factory, Brynmwr, South Wales by The Architects Co-Partnership (1951) now derelict and dependent for its survival on a new but compatible use. (Photograph: Alan Powers)

that which 'involves no change to the culturally significant fabric, changes which are substantially reversible, or changes which require a minimum impact'. The identification of an alternative use consistent with the 'essence' of a building is, or should be, a skill at which architects excel, and the representation of such possibilities may be seen as an integral part of the practice of conservation. Agreement as to what constitutes a 'compatible' use may be easier to achieve in abstract terms than in relation to the particular circumstances of an actual building. This is why the prior analysis of the building's fundamental essence, as indicated above, is of such importance.

11.8.10 Establishing the conservation strategy and attracting support

The results of all the above investigations should ideally be brought together as a single report, supported by all the relevant illustrative material, to provide a coherent conservation strategy. This, and/or an abbreviated synopsis, may then be used both as a fund-raising tool, and as a reference sourcebook throughout the project. An example from the author's experience is illustrated below.

A recent attempt to produce such a document was the Feasibility Study for restoration and re-use of the buildings designed by Lubetkin and Tecton at Dudley Zoo (Avanti Architects and others, 1990). This study sought to provide a way forward *both* for the zoo operators, constrained by the problems of working with buildings which in many cases had been superseded by zoological practice or legislation, *and* for English Heritage and others concerned with the conservation of this unique ensemble of twelve listed structures by one of Britain's leading modern architects.

The thrust of this report was that only by synthesizing the legitimate claims of restoration *and* development could a viable future for these buildings be secured. The proposal that this must entail simultaneously repairing and judiciously adapting sought to grasp the familiar nettle of conservation – the conundrum of changing something in order to save it.

It may be useful for the further development of such studies to summarize the contents of The Dudley Report. Introductory material included sections on the historical background and social significance of the Zoo; architectural intentions in Lubetkin and Tecton's work; and issues raised by the restoration of 1930s buildings. The main study presented a building-by-building survey covering the following topics:

- *Building description and architectural assessment* A factual account of the original structure, illustrated with contemporary photographs, coupled with a qualitative assessment of the design and its relative position within the architect's canon.

- *Synopsis of existing condition and alterations from original* A synopsis of the condition survey notes compiled during site inspection. The 'clipboard' notes themselves are usually too voluminous to include in a strategy document, but must be kept as the detailed record for further reference when specifying works. Similarly, the detailed photographic survey may need to be edited down to give a representative picture of the building's condition.

- *Building capability and adaptation criteria* An evaluation of the building's potential from the point of view of possible re-use for other functions, together with criteria derived from the architectural assessment above to be adopted in weighing the merits of any specific proposal.

- *A summary of restoration criteria* A series of normative *do's* and *don't's* defining essential restoration that should be undertaken and incompatible alterations to be rectified, and for subsequent use as an estate maintenance reference manual. The section included information on the original colour scheme, expressed in terms of either British Standard references or the Dulux Colour Dimension system. It is worth noting in this context that the conservator cannot assume that he or she will always be either invited or available to offer guidance in the future after the project is completed. It is highly desirable that the recommendations of any systematic conservation study are, so to speak, 'attached' to the building, like the instruction manual in a car's glove compartment, for the benefit of new owners or changed circumstances, and are not solely reliant on the surviving involvement of the original personnel.

The next section of the Dudley Report presented specific proposals for architectural, structural, services and, where relevant, zoological works to be undertaken, including compatible alterations and options. All the options were also costed, with the findings coded to indicate the status of each work item according to whether it consisted of Restoration and Repair (R) or Conversion for new/improved use (C), and whether the work itself was deemed Essential (E) or merely Desirable (D). By this means it was possible to establish the broad-brush costs that could certainly be expected to attract grant assistance (i.e. for essential restoration works), those that might attract aid (desirable restoration), and then those for which the building owner would probably have to raise the necessary funding.

Following individual studies of each of the Tecton buildings, the Report proceeded to examine their settings, which were integral to the original design conception and of considerable interest and importance in relating the ensemble to pre-existing features of the site, including an outstanding medieval castle crowning the central hill. This study revealed the considerable changes that had taken place over the 50 years since the zoo opened, the combined effects of which had radically altered the context in which the buildings were perceived. On the one hand, was the addition of numerous additional structures in a heterogeneous variety of styles, and, on the other, a savage reduction of the original oversite coverage of mature forest trees.

A series of general environmental policies were

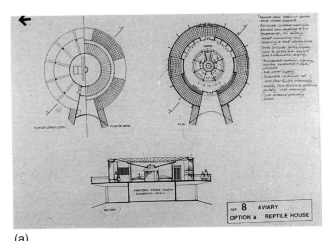

(a)

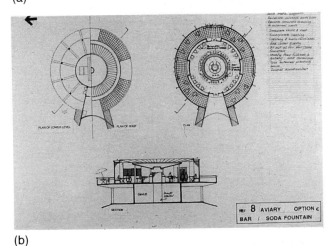

(b)

Figure 11.13 Alternative proposals for re-use of the Tecton Aviary at Dudley Zoo, by Avanti Architects, 1990. (a) Introducing a new zoological exhibit. (b) Converting the building to social use, but both schemes maintaining the architectural logic of the original design.

formulated to restore visual order through such details as signage and outdoor furniture, and retrieve the drama of the original contrast of abstract modern buildings in a natural woodland setting. This was then developed into a set of detailed studies of sub-zones of the site, identified by virtue of their environmental coherence.

A further section of the report addressed the subject of maintenance, both as a holding measure before restoration works were initiated and following works, to protect the value of the primary investment. Recommendations were included for the instigation of 'Log Books' for each building and for the site as a whole, and for the maintenance of a 'good housekeeping roster', noting such points as concealed gutters requiring routine pre-emptive cleaning, etc.

The final section sought to derive a programming logic from the prioritization of different restoration works and put forward a possible implementation strategy for the project as a whole.

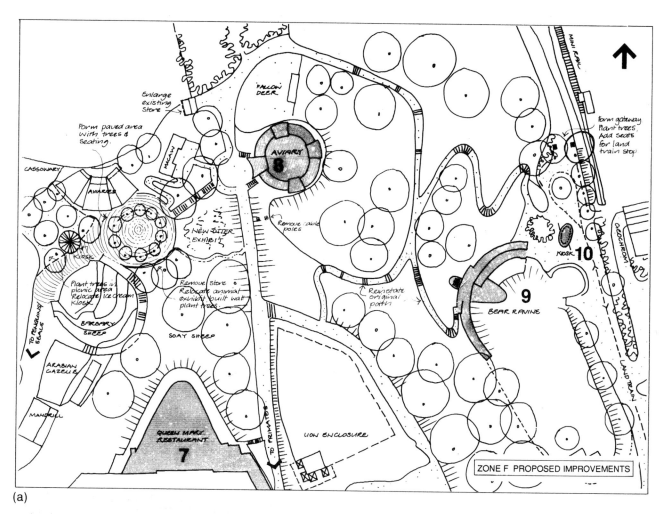

(a)

(b)

Figure 11.14 Zone study of part of the Dudley Zoo site. (a) Showing policies for improvement. (b) A view of the above area suggesting the importance of landscape setting to the architectural design. (Photograph: John Allan)

11.9 The technology of concrete failure and repair

11.9.1 Concrete failure

To understand the causes of concrete failure, it may be useful first to rehearse the reasons for its

appeal as a modern building material. As already noted, the unique potential of reinforced concrete derives from the complementary virtues of its two constituents – steel in tension and concrete in compression. To perform its task effectively, the steel as a ferrous metal must be adequately protected from corrosion, while to perform its task correctly, the concrete must not be subjected to undue tensile pressures. So long as these materials are 'playing to their strengths', so good. However, if and when their roles are altered or reversed, problems quickly ensue.

11.9.2 One of the important characteristics of fresh concrete of which the pioneers appear to have been unaware is its alkalinity. The pH value (i.e. the acid/alkali measure) of newly cast concrete is normally in the region of 12 to 14, that is, highly alkaline. This produces an extremely favourable environment for the reinforcement steel, which is passivated by the formation of a protective surface oxide. Over time, however, this environment is altered by the invasion of atmospheric carbon dioxide and water through the pore structure of the material, forming carbonic acid which, reacting with the calcium hydroxide in the mix, produces calcium carbonate.

Figure 11.15 Aviary, Dudley Zoo. Pre-emptive cleaning of concealed roof areas may be a simple and economic means of avoiding insidious damage, and should form part of a conscientious management and maintenance regime. (Photograph: John Allan)

(a)

(b)

(c)

Figure 11.16 Classic examples of corrosion and spalling. (a) Showing fracture of a thin concrete member through pressure of corrosion. (b) Eventual loss of cover and exposure of rusted reinforcement. (c) The damage appears similar, but is due to frost attack forcing delamination of an inadequately constituted parapet top stratum. (Photographs: John Allan)

11.9.3 The result of this process – known as carbonation – is a reduction in the pH value, which leads to breakdown of the passive oxide protective film on the reinforcement steel which in turn renders it vulnerable to corrosion from exposure to oxygen and water. The product of corrosion – ferric oxide, or rust – greatly increases (by factors of 2 to 6) the volume of the reinforcement, setting up tensile stresses in the adjacent concrete which will usually lead to spalling. As this exposes more of the interior fabric it is easy to see how deterioration can become progressive. In effect, the benefits of the two materials have been reversed – as more unprotected steel suffers corrosion and more concrete fractures under tension.

11.9.4 The rate of carbonation and therefore its adverse consequences can vary considerably depending on the quality and density of the original concrete, particularly its water/cement ratio and the relative humidity of the pore air. In general, the lower the water/cement ratio, the slower the rate of carbonation. As the reaction begins at the surface of the concrete and proceeds inwards, it is easy to see how the depth of cover given to the reinforcement steel is also a key factor in determining the point at which carbonation can become destructive.

11.9.5 Equally, as the rate of carbonation, being a square root function, also decreases over time, in well-designed and correctly constructed concrete the carbonation front may never reach the reinforcement steel, which therefore remains in an alkali-rich environment. This emphasizes the fact that it is not carbonation of itself that weakens concrete but the consequential corrosive effects on reinforcement when air and water reach areas of alkali depletion.

11.9.6 While the adverse effects of carbonation constitute the most common problem in ageing exposed reinforced concrete buildings, the presence of chlorides can set up reactions within the material which also give rise to corrosion of the reinforcing steel in advance of the arrival of the carbonation front. The detrimental effect of unstable chloride compounds varies, depending on the pH value of their environment. The presence of chlorides in concrete may be due to a number of factors, including the use of unwashed sea-dredged aggregates or contaminated mixing water or even their deliberate inclusion in the form of rapid-setting agents.

11.9.7 Chloride attack is more usually a problem in civil engineering construction, where more rigorous concrete specification and better compaction reduces the likelihood of carbonation becoming the principal danger. Typical examples include motorway bridges, car parks, viaducts and other structures in aggressive or marine environments, where saline spray or the persistent use of de-icing salts can cause serious damage to the concrete fabric. Chloride attack can greatly impair the integrity of reinforcement without the rust expansion factor normally associated with spalling, and pre-testing of any concrete structure, even a domestic building, should always include an analysis of chloride content (see also Chapter 5, section 5.7).

11.10 The development of concrete repair

11.10.1 As the understanding of concrete deterioration has developed, so has the technology of concrete repair. Even by the early post-war years some of the International Style icons of the 1930s were beginning to lose their pristine appearance and require remedial attention. But until the 1960s the approach to repairing spalls was seldom more sophisticated than simply removing the loose or flaking material in areas of visible damage and making good with cement mortar. From the 1960s to the mid-1970s techniques progressed slowly with the incorporation of bonding agents and better coatings. However, full appreciation of the significance of carbonation and the development of scientific repair products and systems does not appear to have really become established until the mid to late 1970s, to the extent that now the field has effectively become the province of a specialist industry. Although, at the time of writing, there is still no British Standard on concrete repair there is now an established cadre of firms, a steady stream of exhibitions, conferences and journals, and a veritable flood of trade literature devoted more or less exclusively to the subject of concrete repair.

11.10.2 Within or in close association with the materials and contracting industry there are a number of agencies promoting good practice in the use and repair of concrete, offering advisory services, or producing technical publications and newsletters of which architectural practitioners and modern conservationists should be aware. These include the Concrete Society, founded in 1966, which since 1987 has established a regional Concrete Advisory Service staffed by Chartered Engineers for the benefit of subscribing members (currently numbering some 7500), superseding the technical advisory service previously provided by the Cement and Concrete Association. The British Cement Association, the latter's successor, maintains a vigorous programme of conferences, lectures, technical seminars and other events, and until recently has published the well-known periodical *Concrete Quarterly*, which had appeared continuously since 1947.

11.10.3 Another more recently founded organization of specific relevance to the emergent field of modern conservation is the Concrete Repair Association, which brings together a number of leading companies to promote a series of common aims, including progress in the practice of concrete repair, advancement of education and technical training within the industry, and improved liaison with professional bodies, local authorities and specifiers. Members of the Association are enjoined to fulfil a number of standards in such matters as technical expertise, employment conditions, performance records and Quality Assurance. The Association's promotion programme includes exhibitions, seminars and publications. In 1990 it promulgated its Method of Measurement for Concrete Repair in an endeavour to introduce consistency into the itemization and pricing of this difficult area of tender documentation. Details of the above and other organizations are in Chapter 19.

11.10.4 Before engaging in the detailed techniques of concrete repair it may be useful to note how much of the initiative behind its development has come from contractors and the specialist product-manufacturing industry. This is perhaps due to the fact that defects in concrete buildings and structures became a matter of concern to their owners – often local authorities grappling with daunting problems of backlog maintenance – long before architects and conservationists became engaged with the 'heritage' aspects of the subject.

11.10.5 As a consequence, there is within the industry what might be described as a presumption in favour of achieving the most effective technical result in repair work, which

Figure 11.17 Introduction of non-insulating replacement wall panels at Finsbury Health Centre contradicts the servicing strategy of the external ducts. Repair interventions should be based on an informed understanding of the original design. (Note also progressive tile loss). (Photograph: John Allan)

(a)

(b)

may not necessarily represent what, from a conservationist stance, would be desirable as the most historically authentic result. A contractor, understandably concerned to minimize his or her liability for latent defects, will – in the absence of conservationist restraint – be more inclined to obliterate an original feature or finish if this appears to facilitate the achievement of a more durable result.

11.10.6 Meanwhile, from the viewpoint of the client, whether in the public or private sector, the responsibility of maintaining 'modern' property in a viable state of repair or even altering it to accommodate new requirements may generally not be regarded as one with *architectural* considerations or consequences. It is more usually seen as a pragmatic affair in which if troublesome features of the original design need attention they are better eliminated with altogether – if possible.

11.10.7 The removal of the cornice rails from Erno Goldfinger's Trellick Tower in Kensington, west London, regardless of their architectural significance (and weathering benefits) is a typical case. The replacement of the original Thermolux curtain wall panels in Tecton's Finsbury Health Centre with non-insulating substitutes effectively contradicts the whole logic of that building's externally accessible service strategy. Corroding painted original steel windows will often be replaced by aluminium or PVC substitutes on the assumption that this is the only way to reduce future maintenance, without considering the more

(c)

Figure 11.18 a and b The correct conservationist response to irreparable corrosion of original ungalvanized steel windows (a. shown causing glass fracture; b. exposed on hidden surfaces after removal) is to replace with like steel windows made to current standards, not with uPVC or aluminium. c. Example of replica replacement at Highpoint. (Photographs: John Allan)

Figure 11.19 Overcladding – a popular non-conservationist approach towards upgrading troublesome post-war blocks of disputed architectural appeal. (Photograph: Makers Ltd)

authentic result that could be achieved by using equivalent steel windows made to current manufacturing standards.

11.10.8 More generally, one has only to look at the way in which modern hospital, educational and industrial complexes have tended to expand by incremental accretion rather than through any extension of the original design concept to see how little considerations of modern conservation have figured in the thinking of many commissioning authorities.

11.10.9 Alternatively, the 'change of image' entailed in a particular mode of repair or upgrading may be a deliberate objective, and an additional feature of its 'customer appeal'. The widespread adoption of overcladding by housing authorities from the 1980s must be attributable at least as much to its camouflage of an unpopular aesthetic as to its insulation and weathering benefits and the convenience of being able to

undertake estate improvement with tenants in occupation.

11.10.10 Such examples are a strong indication of the extent to which the repair of modern buildings is still a contractor- or surveyor-led field. Unlike Georgian, Victorian or even Edwardian buildings, modern architecture is not yet generally perceived as a subject in which conservation criteria arise when maintenance or refurbishment is undertaken. Over a hundred years since the SPAB launched its crusade for the 'gentle, careful and conservative' repair of ancient buildings such principles have yet fully to permeate the debate on modernism.

11.10.11 Of course, in many cases where the properties are of no particular architectural interest this state of affairs may be of little or no consequence, but architects and others concerned with preserving the authenticity of *significant* modern buildings must quickly catch up with the technical developments within the industry while at the same time seeking to stimulate a wider awareness of the aesthetic and cultural implications of modern conservation. This point connects directly with the issues of education, statutory protection and popular consensus raised earlier in this chapter.

11.11 Concrete repair procedures – testing and diagnosis

11.11.1 Concrete repair of any kind starts with testing and diagnosis. Unless a clear picture is gained of the nature, causes and scope of deterioration, it will be impossible to be certain of the appropriate remedies to be applied and equally difficult to predict costs. The aim must be to establish where and to what extent carbonation has taken place, what cover the reinforcement has, and where bars lie in or near carbonated zones. It is also important to establish the concrete strength and mix quality, and discover any deleterious substances such as chlorides which may be causing internal debonding or delamination not detectable from surface inspection.

11.11.2 Testing must usually be undertaken in two stages. The first (pre-contract stage) is necessary to gain sufficient knowledge of the problem to enable documentation to be prepared for the invitation of competitive tenders for carrying out the repair work proper. This preliminary testing may be undertaken by a specialist contractor who is subsequently included in the list of tenderers. The second stage of testing ensues after a contractor has been appointed and has established full control of and comprehensive access to the works, and forms part of the detailed project record on-site.

11.11.3 The particular importance of prior survey work in the context of reinforced concrete repair must be carefully explained to clients, especially private domestic owners used to offers of 'Free Estimates Without Obligation' from traditional builders. As the testing procedures (explained below) usually involve assisted access to and invasive examination of the structure by a specialist contractor, followed by laboratory analysis of specimen samples, it is common practice for a charge to be made for this service. Clients must be persuaded that it is in their interest to fund this preliminary study as a means towards obtaining more accurate knowledge of, and therefore costings for, the eventual work. It is perhaps analogous to commissioning an exploratory soil survey in order to inform the design of an economic substructure. In both cases, unless a fixed price offer is given, it is normally necessary to re-measure the job on completion, but without the preliminary survey there could be no systematic basis for designing and costing the work at all.

11.11.4 Depending on the type of structure under consideration, the mode of access for testing (and subsequent repair) can become a significant factor in the cost equation. Scaffolding a tall building may be slow and expensive, while cradles may not enable access to be gained to all parts of the structure. Prefabricated towers may usually be taken only to prescribed heights, while hydraulic platforms, although flexible, have limited reach. Abseiling has become an accepted method of providing access for inspection in difficult high-rise cases. At the other extreme, the use of binoculars, a telephoto camera and an extension ladder may be sufficient to carry out the preliminary survey of a typical 1930s domestic reinforced concrete house. The professional adviser must consider the nature of the specific task to be undertaken and determine the most cost-effective mode of access.

11.11.5 The survey, inspection and testing programme should also be designed to suit the structure being considered, and undertaken systematically to provide an 'auditable trail' for future reference. Before embarking on sitework it is desirable to assemble as complete a database as possible as a desktop study. Useful information can include:

- Dates and details of the original design commission
- As-built drawings and specification
- Photographs of the building in construction, at completion, and thereafter
- Details of subsequent alterations and extension
- Details of major or persistent defects
- A record of previous repairs and maintenance
- Information on the pattern of ownership and use over the building's life cycle

It is easy to suggest that all these data should be compiled and readily available for every modern building as a matter of routine or good stewardship, rather as a logbook covers a car over its lifespan, or doctors' medical records accompany a patient. But it is surprisingly difficult in practice to assemble such information on even a recent building on a fully comprehensive basis. Much original data on pre-war modern buildings have been lost or dispersed, although important deposits of material are held at such centres as the British Architectural Library Drawings Collection and Photographic Archive. Information on post-war projects is more likely to be retained by the original commissioning authorities, though the demise of many Architects' and Works' Departments in the public sector over recent years makes this uncertain.

11.11.6 The range and extent of survey/testing will depend upon the nature of the building and size of project, but typically will include the following:

- *Pertinent visual observations* Of course, visual observations on their own are insufficient – for the very reason that concrete deterioration, as already indicated, is a phenomenon in which appearances may be deceptive – but they can be of great benefit in relating damage and decay to the contextual and architectural characteristics of the subject. Such factors as the location of the building, prevailing wind direction, proximity of other structures that might affect weathering, in addition to the architectural design and constructional features of the building itself, should all be considered. All patent visual damage, including obvious spalls, cracks, evidence of frost attack, loss of coatings, etc., should be recorded. It is desirable to annotate a set of as-existing drawings with the findings of a visual survey, and coordinate this with as complete as possible a photographic record of the building 'as is', before the commencement of any works. A series of relevant fixed viewpoints should be established from which to take progress photographs at regular intervals throughout the contract. This can be written into the Preliminaries as a contractual requirement.

(a)

(b)

Figure 11.20 Original surface texture should be noted as it provides vital evidence of the method of construction to be retained in authentic restoration. (a) The Gorilla House 1933. (b) Chapelle de Notre Dame du Haut, Ronchamp 1955, (both photographed in close-up.) (Photographs: John Allan)

• *Hammer survey* A simple test (for a skilled operative) to establish the extent of delamination and hollow areas within the concrete fabric may be undertaken by tapping or 'rubbing' the surface with an ordinary club hammer. Care must be taken not to damage sound concrete unnecessarily. Carried out systematically over the survey area, such a test will indicate a series of live and suspect zones which should be mapped on the record drawing for further attention.

• *Crack survey* Careful inspection of the structure for cracks can yield valuable information about the condition of reinforcement, quality of the concrete, and possible settlement or differential movement of parts of the structure. The correct interpretation of cracks is of great importance, as their causes and significance can differ considerably. Pullar-Strecker (see Bibliography) has identified seven types of crack, a 'generic' understanding of which must be linked to specific knowledge of the structure under investigation and an appreciation of its probable behaviour. Locations and dimensions of all visible cracks should be logged on the survey record drawing.

• *Covermeter survey* Various types of electromagnetic covermeters have been developed to enable the location depth, size and direction of reinforcing bars to be determined from working the sensor across the surface of the concrete. This is of great importance in establishing a picture of what was actually built (especially in the absence of as-built drawings and, on occasion, even with them) and also in conjunction with the carbonation test (see below) in assessing the extent of reinforcement at risk. Readings are normally taken on a representative grid basis, and should be recorded on the survey drawings.

• *Carbonation test* The purpose of this test is to establish the extent and severity of carbonation relative to the reinforcement locations as determined above. The normal procedure consists of applying the indicator solution phenolphthalein to an area of freshly broken concrete and observing the colour as it discloses the degree of alkalinity. This will range from neutral to bright pink as the pH exceeds values of 10. The quality of concrete compaction will also become apparent, as greater porosity presents less resistance to the advancing carbonation front. Again, representative positions should be tested and the average carbonation depths marked on the survey drawing. These should then be correlated with the covermeter readings to establish areas where the reinforcement may be considered to be at risk. For repair purposes, steel within 5 mm of the carbonation front may be so classified, and the number of locations of potential latent damage calculated accordingly.

• *Strength testing* Depending on the size of project and type of damage it may be appropriate to take core samples of the concrete from key positions. These can then be subjected to a variety of tests to produce valuable data on the strength, compaction quality and mix

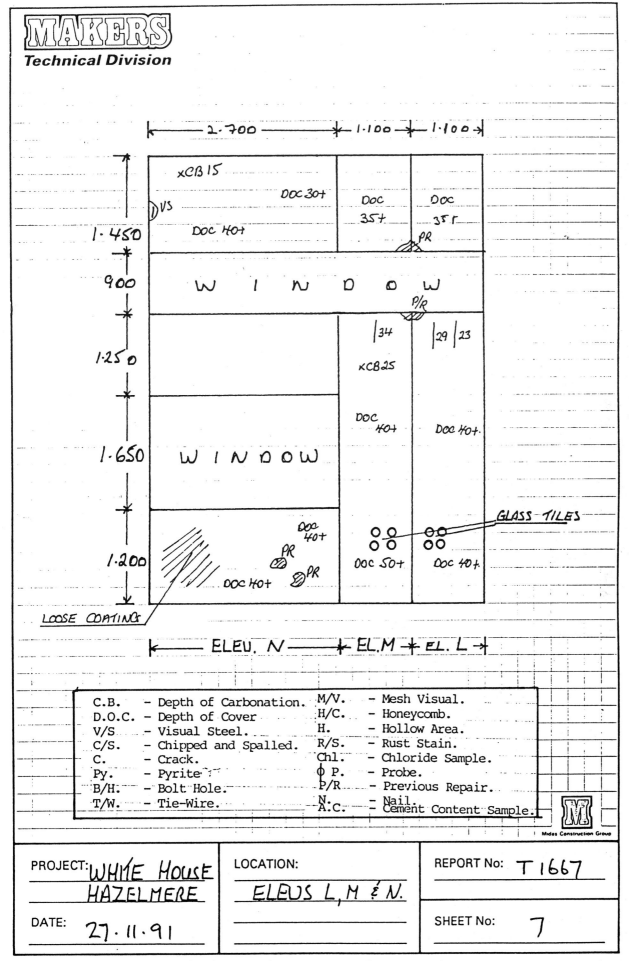

Figure 11.21 Typical pre-contract concrete condition survey site worksheet showing wall elevation marked up with coded diagnostic notes. (courtesy of Makers Ltd)

constituents of the concrete as well as information on the extent of carbonation and corrosion, if reinforcement is included in the sample. The location of core samples and possible effects on the stability of the adjacent structure must be very carefully considered. Alternatively, a non-invasive and cheaper method of obtaining indications of concrete strength is to use a Schmidt rebound hammer which will measure the resistance of the material to a known impact.

● *Concrete dust sampling* The cheaper and more usual way to gain information on the mix quality of the concrete, and the presence of chlorides, is to collect a sample of dust from drilling. This is then sent to a laboratory for investigation. The dried, crushed and homogenized material may be analysed for insoluble residue, soluble silica and calcium oxide, from which the probable mix proportions, including cement content, may be calculated. Results must be averaged from several samples. The chloride content of the sample is determined by a potentiometric titration procedure specified by BS 1881: Part 124.

11.11.7 The collected information from all pre-contract testing undertaken will provide the basis for a diagnostic report and specification of remedial works. If Bills of Quantities are prepared for tendering the works the desired survey methods and documentation requirements should be fully specified. The model document Method of Measurement for Concrete Repair published by the Concrete Repair Association, referred to earlier, contains a useful schedule of survey methods and recording standards for this purpose.

11.12 Concrete repair techniques – introduction

11.12.1 On the assumption that the survey and testing procedures described above will have produced a remedial works strategy, this chapter now looks at typical repair methods. Although, as mentioned above, there is now a wide variety of proprietary products and techniques available for achieving high-quality repair of structural concrete in buildings, we will concentrate on the 'generic' approaches most commonly adopted in current practice. All have as their end objective the restoration of an alkaline-rich environment within the concrete fabric or the re-passivation of the reinforcement.

11.12.2 The first method is usually referred to as the 'traditional repair technique', although the materials and procedures now involve 'state of the art' practice and products. The second is cathodic protection and exploits the electrical conductivity of reinforcement to reverse its anodic behaviour when corroding. The third and more recent technique is commonly termed 're-alkalization', and uses an electrochemical process to achieve the desired aim. The fourth, known as 'desalination', is, in effect, a variation of the third in using similar means to extract chlorides. These techniques are not to be understood as mutually exclusive. Their application will depend upon the nature of the diagnosis and the balance of advantage relative to cost, and it is frequently necessary to combine traditional repairs in selective locations with another of the techniques in a single project. A brief note on the treatment of cracks is added in conclusion.

 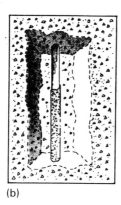 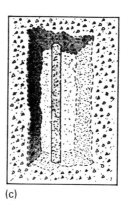 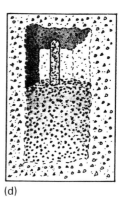 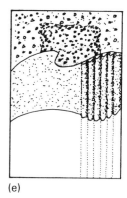

(a)   (b)   (c)   (d)   (e)

Figure 11.22 Sequential diagram of the traditional concrete repair process. (a) Test, expose and remove defective concrete. Expose reinforcement beyond corroded length. Replace rods if necessary, splicing to provide continuity. (b) Blast clean steel and protect with polymer modified cementitious slurry, two coats, second coat blinded with quartz sand. (c) Apply bonding bridge slurry, (sand/cement/polymer) as primer for whole area to be repaired, while still wet. (d) Place and compact repair mortar in layers not exceeding 25 mm. Apply further bonding bridge between layers if greater depth required. (e) Apply thin film (alkaline rich) levelling render (2 mm) to fill blow holes and provide substrate for protective coatings. Profile as required to match original. Apply coatings with a minimum three applications, thinning first coat.

11.12.3 It should already be apparent from the description of the way concrete deteriorates that a thorough scientific understanding of the problem is required if remedial works are to be more than merely cosmetic. The repair techniques described here must therefore be seen as elements within a comprehensive strategy properly only carried out by fully qualified firms and personnel. Concrete repair is emphatically *not* a DIY process. Several proprietary systems of concrete repair carry an Agrément Certificate and certain manufacturers of such systems will supply their products only to specialist contractors with BBA recognition. Clients, architects and specifiers should carefully check the credentials and track record of potential contractors when compiling tender lists.

11.12.4 Traditional concrete repair

This approach to repairing concrete proceeds from the premise that all areas of actual *and* latent or potential damage (i.e. where the carbonation front is unacceptably close to reinforcement even if corrosion has not yet occurred) should be replaced with new high-quality mortar and, if necessary, new reinforcement.

The areas for treatment identified from the survey are marked out on the structure. An assessment of the need for temporary support must be made where defective concrete is to be removed from load-bearing elements. The defective concrete is then removed and the repair area cleaned. Deeper damage may be drilled out or cut out by pneumatic hammer, while surface cleaning, including the removal of previous coatings, is normally dealt with by grit blasting, though high-pressure water or suction blasting may also be used. An indicator solution test should be applied to check that one has got back to sound concrete.

In removing damaged material care must be taken not to cause unnecessary damage to the adjacent concrete. Drilling out large sections will transmit vibration through the structure. In the restoration of Lubetkin's Penguin Pool at London Zoo (1987) a high-pressure water lance was used to cut out the old diving tank to avoid disturbing the delicate adjacent canopies. Costs of alternative preparation techniques can vary significantly, but the decision as to which method to adopt must also take into account the possibility of consequential damage as well as the nuisance factor of noise and debris to occupants, neighbours and the public.

The thorough removal of the products of corrosion from steel reinforcement is most important in achieving a sound result. Damaged bars should be exposed beyond their corroded

Figure 11.23 Drilling out defective concrete with a pneumatic hammer. (Photograph: John Allan)

length and at the rear, with portions being replaced where weakened beyond repair. Structural advice should be sought in cases of replacement. Rusted reinforcement is best cleaned by grit blasting. The Swedish Standard SA $2\frac{1}{2}$ for painting preparation is commonly specified for this process.

As soon as possible after cleaning, the prepared steel is primed in two coats using an epoxy coating or polymer-modified cementitious slurry – the second coat being applied while the first is still tacky, and including a quartz sand blinding to optimize keying of the ensuing repair. Site procedures will differ depending on the product system specified.

The constituents of the mortar will depend on a number of circumstances including compatibility with the parent concrete, whether the repair is structural (e.g. strength over $30 \, \text{N/mm}^2$), whether the material is pre-or site-batched and whether the repair is thin or thick section. A broad distinction in mortar types is that between epoxy systems showing rapid initial strength and low shrinkage characteristics and cementitious mortars using low water/cement ratios and resembling more closely the properties of the parent concrete. The

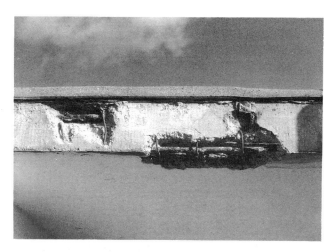

Figure 11.24 Repair zone exposed prior to cleaning of reinforcement. Note concrete behind bars has also been removed to enable repair mortar to completely envelope and protect the steel. (Photograph: John Allan)

different and contending advantages of each generic type are vigorously promoted by their respective manufacturers.

The whole repair area should be dampened to prevent premature setting, and a bonding bridge slurry applied to enhance adhesion of the repair mortar, which should be placed in position while the bonding is still wet. Thorough compaction is essential, and to achieve this the repair mortar is best 'punched' home by hand rather than being rendered over the area with a plasterer's float. Further bonding bridges should be applied between layers if a number of applications is needed to complete the repair. Whichever proprietary materials are used, it is inadvisable to mix products from the repair systems of different manufacturers. Close attention should also be paid to the ambient air and substrate temperatures (which should normally be at least 5°C and rising) before starting repair work.

When all the areas designated for concrete replacement have been thus completed and cured, and the remaining surfaces have been cleaned of previous coatings, an important architectural/technical decision must be made. This concerns the quality of finish desired for the eventual result. This in turn will depend upon the regularity or otherwise to the substrate to which protective coatings are applied. As the finished repair work is likely to be slightly uneven and also exhibit different surface characteristics from the adjacent original concrete, it is common for a thin-film levelling render or 'fairing coat' to be applied to achieve a consistent substrate. This also has the important technical advantages of filling any surface pores and blowholes that coatings could not bridge on their own, and providing a dense alkaline-rich layer of resistance to further

carbonation. Levelling renders should, of course, be properly cured and are ideally finished with a wood float or rough sponge to provide a good key for final coatings.

However, it will be evident that this overall levelling process produces superficially a wholly new artefact by obliterating any irregularity in the original surface. This may not be inconsistent where the building was brought to a high state of finish originally – for example, Lubetkin's Penguin Pool (where levelling coats were accordingly used in restoration). But in other cases it may be thought to detract from the authentic 'roughness' of much early Modern concrete construction, the retention of which – as an original feature – would be a prime objective according to traditional conservation principles.

The same architect's celebrated ensemble of pavilions at Dudley Zoo, for example, very clearly displays the fact that the construction process was quick, cheap and unsophisticated. The resulting

(a)

(b)

Figure 11.25 Making the repair following priming of reinforcement. (a) The mortar is 'punched' home to ensure compaction. (b) Trowelling off should only be carried out after the repair zone has been filled and compacted by hand. (Photographs: John Allan)

(a)

(b)

(c)

Figure 11.26 The Penguin Pool Restoration 1987. (a) General view of works including reformation of ramp edges, application of levelling mortar, and curing of finished repair before coatings (wrapped beam). (b) and (c) Studies of repaired wall before and after fairing coat (Photographs: Sir Robert McAlpine & Sons Ltd)

irregularities could be regarded as original, indeed historic, architectural features worthy of preservation. Conversely, and perhaps ironically, the recreation of some of the original surface effects, e.g. the highly characteristic corrugated formwork pattern (employed as a cheap device to control weathering), can now only be achieved by an extremely labour-intensive manual 'combing' operation.

This is an issue likely to divide opinions as between the conservationist and the contractor. The former will be inclined to seek authenticity of result where the latter will tend to place a higher priority on comprehensiveness of repair. The client, unless he or she is unusually interested in the constructional foibles of early modernism, is also likely to prefer the security of a more permanent remedy, provided the additional costs of levelling coats can be afforded. It is, however, an interesting area where the different repair technology of modern construction may impinge upon a traditional conservation principle. The conundrum arises from the fact that, in preventing

Figure 11.27 The Polar Bear Pit, Dudley Zoo restored 1988–9 showing replication of original surface pattern achieved by manual combing. (Photograph: Makers Ltd)

(a)

(b)

Figure 11.28 (a) The White House restoration 1993 showing brush application of pore stopping mortar as a substrate preparation for protective coatings. (b) View of finished wall after coatings showing retention of surface roughness. (Photographs: John Allan)

carbonation or chloride attack in reinforced concrete, the surface coatings are effectively playing a 'structural' role.

An acceptable compromise can sometimes be achieved by using only a brush-applied pore stopper as the substrate preparation prior to the application of surface coatings, thereby keeping the structure dry while not unduly modifying its surface texture. At The White House (formerly New Farm) in Surrey designed by the pioneer modernist Amyas Connell this approach was used both to control costs and, architecturally, to retain the characteristic roughness of the original surface. The objective, strongly supported by the conservation authorities, was not to produce the appearance of a modern style house *built* in the 1990s, but of a 60-year-old building *restored* in the 1990s. The traditional conservation principle of explicit repair was accordingly upheld.

Having obtained an appropriate substrate, the final stage of the concrete repair process is to apply suitable anti-carbonation coatings, (normally in not less than three coats, but, in any case, to a prescribed thickness in micrometres). In specifying these, consideration should be given to whether the area to be covered is likely to be subject to cracking or not. In the former case a coating system with elastomeric properties should be used. In all cases it is important to use coatings with sufficient resistance to freeze–thaw cycles and vapour-permeable properties that allow any moisture within the fabric to migrate to the outside.

The unscientific use of non-permeable coatings in the past has frequently led to moisture becoming trapped behind the surface and continuing to cause damage. It should be understood that vapour permeability is a function of the molecular structure of the coating, and in properly formulated systems does not compromise the water-resistant capability of the material.

The other purpose of coatings is, of course, to provide a decorative finish, where again the question of authenticity arises. Clear coatings are available for cases where the original concrete was left undecorated. But in most instances the architect must seek to retrieve the original colour scheme. This may not be as simple as the black and white photography of the 1930s makes it appear. The fact is that external colour was widely and adventurously used by many modernists, and in the absence of survivors' memories or any documentary information it is important to take the opportunity during the initial building survey of carefully removing coatings in selected areas to establish the original colours by 'archaeological' means. The dramatic blue undersides of the Penguin Pool ramps at London Zoo had escaped Lubetkin's recollection but were rediscovered during a site inspection.

11.12.5 Cathodic protection

This well-established technique could perhaps be regarded as the 'traditional' response to the problem of chloride attack and makes use of the conductivity of reinforcement to continuously

Figure 11.29 The restored blue ramp undersides at the Penguin Pool are a reminder of the vital role of colour in early modern buildings concealed in the black and white photography of the period. (Photograph: John Allan)

transmit a small d.c. current through the structure to prevent the steel reaching the electric potential at which corrosion occurs. A positive anode or anodic covering is installed in or attached to the structure and connections made to the reinforcement, which functions as a cathode when the system is activated, attracting positively charged ions.

This technique cannot repair areas where chloride attack has already corroded reinforcement – these must still be remedied by replacement as described above. However, cathodic protection is widely used to arrest potential damage in other parts of the structure where chloride contamination has not yet begun to corrode the steel. Typical applications include motorway bridges and civil engineering structures where latent chloride attack rather than carbonation presents the main threat to the integrity of the structure.

Although cathodic protection can offer a cheaper initial outlay compared with full replacement of contaminated concrete, this must be balanced against the ongoing costs of monitoring and regulating the installation to ensure that the desired result is being maintained.

Cathodic protection is currently adopted in Britain by the Department of Transport on various motorway bridges and has been used in the USA for many years.

11.12.6 Re-alkalization

Although the traditional method of concrete repair described above, properly applied, can be relied upon to restore a structure to a 'better than new' condition, it will be evident that the need to remove and replace not only the concrete already

damaged by spalling but also all carbonated concrete that *could* threaten adjacent reinforcement may – depending on the type of project – involve considerable invasive remedial work, with its attendant disturbance and unpredictable cost (as more and more latent damage is discovered when a contractor is on-site). The relatively new technique of re-alkalization proceeds towards the same end objective as traditional repair – the restoration of an alkaline fabric – but employs a non-destructive method of achieving this for the areas of latent damage.

Re-alkalization cannot be used to repair areas of patent damage where reinforcement corrosion and spalling have already taken place. These must be identified and dealt with in the traditional way. The decision to adopt re-alkalization as a more economic restoration strategy will therefore depend upon the relative extent of patent damage as a proportion of the total area of concrete structure, where invasive repair would otherwise not be necessary. Where the preponderance of work involves treating carbonated but otherwise sound concrete then re-alkalization becomes an extremely attractive restoration technique, from both the financial and the conservationist viewpoints.

The principle behind re-alkalization, which was pioneered by Norwegian Concrete Technologies in the early 1980s and introduced to Britain by the firm Makers Industrial Ltd, is to induce into the concrete fabric by means of an electrochemical process an alkaline solution that will permanently raise the pH value above 10, passivating the reinforcement steel and resisting future carbonation.

The initial operations prior to re-alkalization, including surveying, testing, diagnosis, etc. and the general removal of surface coatings, must all be undertaken as in traditional repair. A series of $25\,mm \times 25\,mm$ timber battens (spacers) supporting a $50\,mm$ steel mesh grid is then fixed (temporarily) to the concrete with plastic plugs. This provides the matrix for an alkaline poultice, which is created by spraying a mixture of fibrous material (usually shredded newsprint) and sodium carbonate through twinned hoses.

Connections are then made to suitable points on the reinforcement (for example, at a spall position where the concrete is already fractured) and to the external 'loom' and a low-voltage d.c. electric current applied. With the reinforcement acting as a cathode and the external net as an anode the sodium carbonate solution is drawn into the pore structure of the concrete by a process of electro-osmosis. When it reaches the reinforcement electrolitic action produces hydroxyl ions, which greatly enhance the alkalinity of the reinforcement environment.

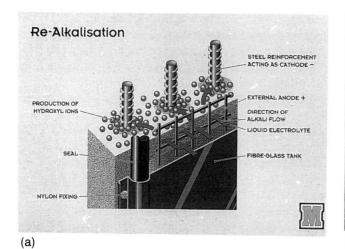

(a)

(a)

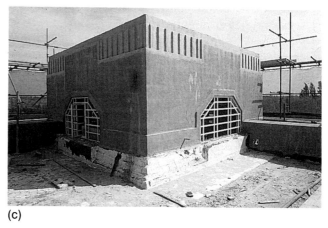

(b)

(b)

Figure 11.30 (a) The principle of re-alkalization with applied anodic poultice supplying alkaline solution into the structure towards the cathodic reinforcement. (b) Diagram of the typical management and control system for re-alkalization. (Illustrations: Makers Ltd)

At suitable intervals during the treatment checks are made to confirm the increased pH. One method is to monitor the resistance, which is high while the fluid is still travelling within the pores but drops when it reaches the steel and finally stabilizes at a low level when maximum re-alkalization is achieved. Double checking is carried out by taking and fracturing core samples at completion and applying indicator solution to the exposed surfaces.

The re-alkalization technique is concluded by removal of the poultice, cleaning down, and applying fairing and finishing coatings – as before described. Advantages claimed for re-alkalization include its non-destructive and environmentally safe processes, its comparatively non-disruptive nature in reducing noise and dust, thereby allowing uninterrupted use of a building during the work, and its long-term durability.

Good case studies now exist for this technique from completed projects in Norway, Belgium and also in the UK – most notably the recently

(c)

Figure 11.31 Site operations. (a) Spraying the alkali electrolyte. (b) the completed loom ready for connection, and (c) the treated concrete following removal of the loom prior to application of coatings. (Photographs: Makers Ltd.)

refurbished Hoover Building, a 1930s landmark in west London.

11.12.7 Desalination

This technique makes use of procedures similar to those of re-alkalization except that it is applied in cases where the concrete damage has occurred

Figure 11.32 The Hoover Building, West London, a major restoration project involving the re-alkalization process. (Photograph: Makers Ltd)

through chloride attack and the ingress of harmful salts. Here the negatively charged chloride ions in the cement matrix of the reinforcement are attracted to the positive external anode formed by the temporary wire poultice/loom.

This technique, also promoted in Britain by the firm Makers Industrial Ltd, is likely to be more appropriate for use in marine locations or civil engineering structures affected by de-icing salts. Depending on the severity of the attack, the process may take several weeks to achieve complete desalination, with intermediate stop–starting to restimulate chloride ion retraction. Areas of patent damage will still require traditional repair, and finishing procedures will similarly be as already described. Unlike cathodic protection, which relies on the continuous application of the electric charge, desalination promises a once-and-for-all remedy, provided finishes and coatings are used to prevent the ingress of further chlorides.

11.12.8 Crack repairs

The variety of possible causes of concrete cracks and the differing circumstances of individual cases make it impossible to provide here any complete account of the available remedies. However, certain general principles may be rehearsed as follows.

It is obviously of paramount importance to diagnose cracking in concrete correctly if the appropriate remedy is to be specified. Young shrinkage cracks, provided they are not related to structural movement still actively taking place, may be cleaned of contaminants and grouted up for redecoration. If, however, the crack has initiated reinforcement damage and presents an incipient spall a full traditional concrete repair is indicated.

Resin injection techniques may be suitable to bond and seal larger cracks provided they are not subject to further movement. If the crack still appears to be active the attempt to prevent movement by resin bonding will probably prove fruitless and it may be more realistic to cut out and re-form the crack as a deliberate movement joint of sufficient width to provide adequate elastic capacity in the flexible sealant. Naturally, any deep damage resulting from the ingress of water or salts or the advance of carbonation via the crack must be repaired first.

If an active crack is symptomatic of more serious structural problems even conscientious remedial work at the crack site will only be successful if undertaken in conjunction with appropriate structural engineering intervention. The rectification of cracks should, in any case, form part of the overall restoration strategy, and will normally accompany one or more of the repair procedures already described.

11.12.9 Repair records

Having emphasized the value of researching a building's original construction and history in use as a prior database to inform a restoration strategy,

(a)

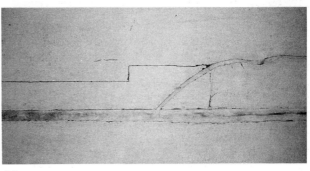

(b)

Figure 11.33 (a) Cracking and delamination of granolithic paving re-bonded and sealed by resin injection. (b) A live crack re-formed as a permanent movement joint.

it should go without saying that a record of all repair works carried out and materials used should be compiled during and/or at completion of the project. Even comprehensive restoration schemes cannot be regarded as 'the last word' on a building, especially a purist modern one. An ongoing regime of monitoring and maintenance is essential for a client to protect the value of the principal investment and to provide the industry at large with essential feedback on the many new techniques and products appearing in the field.

Ideally, a maintenance manual should be prepared to give clear guidance to future personnel who may not have been involved in the restoration project. The document should include details of product life-expectancy and renewal requirements, specifically in the case of protective finishes which may require overcoating after periods of 8 to 10 years, depending on the conditions to which they are subjected.

Finally, the originals of any warranties and product guarantees should be lodged safely by the building owner, preferably with copies included in the maintenance manual where their expiry dates should be related to the recommended inspection cycle.

11.13 Challenge

One of the most compelling aspects of the Modern Movement in architecture was its attempt to integrate its constituent themes – social betterment, technical progress and aesthetic innovation – into a single vision. It may be easy, looking back after more than half a century, to deride the whole endeavour on account of its later failures. Yet it is also difficult not to admire the

optimism behind this vision, the sense of hope and responsibility of a generation emerging from the slaughter of two world wars determined to fulfil its obligation to the twentieth century. As we move towards the next century it will become the task of a new generation to determine an appropriate response to the now-historic legacy of modernism while simultaneously facing unknown but probably even greater challenges with comparable hope and courage.

Selected bibliography

Introduction

The following references give a selective guide to further reading in the main areas covered by the text. Very few books yet exist on the specific subject of modern conservation – an indication of how recently this field has emerged – and the general material available is more often in the form of articles in the professional or national press, exhibition catalogues, reports or collections of conference papers. The list opens with a short summary of the relevant charters and official publications. There are a number of valuable technical publications dealing with concrete repair on a purely scientific basis (i.e. without regard to conservationist considerations). Trade literature from particular manufacturers and specialist contractors is not included, though there is often useful technical information available in such publications and in the Agrement Certificates covering certain products and repair systems. Readers are referred to the trade associations of the relevant industries (see Directory of organizations) for details of firms who may be approached directly. Articles dealing with individual restoration projects are included within the technical section. There are also select lists of leading architects of the Modern Movement in Britain. The bibliography is accordingly arranged in four sections as follows:

1. Conservation charters and official publications.
2. General studies, exhibition catalogues and articles on modern conservation.
3. Technical publications, articles and project studies.
4. Monographs on leading modern architects in Britain.

For architectural surveys of the Modern Movement in Britain please refer to the bibliography at the end of this book. Readers are recommended to consult the references and bibliographies within these titles as a guide to further reading.

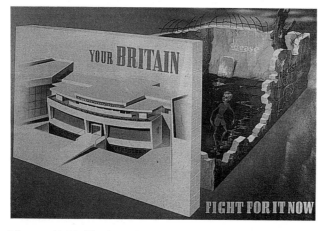

Figure 11.34 Finsbury Health Centre – the most complete pre-war realization of modernism's constituent themes, a symbol of Britain's progressive tradition worth fighting for. (Poster by Abram Games 1943. Copyright: The Trustees of the Imperial War Museum)

Conservation charters and official publications

English Heritage, *Buildings at Risk – A Sample Survey*, English Heritage (1992).

The Building Conservation Directory, Cathedral Communications Ltd (1st edition, 1993, 2nd Edition, 1994). This new directory concentrates on historic conservation but does contain some useful general information.

English Heritage, *Conservation Bulletin* (published three times a year) English Heritage. See especially Issue 17, June 1992, 'Modern architecture reappraised' by Andrew Saint and 'Modern architecture restored' by Elain Harwood; also Issue 20 July 1993, 'Post-war listing – An update' by Diane Kay.

Context, Quarterly Journal of the Association of Conservation Officers, Hall-McCartney Ltd, Letchworth, Herts.

English Heritage, *Directory of Public Sources of Grants for the Repair and Conversion of Historic Buildings*, English Heritage (1990).

DoCoMoMo, *The Eindhoven Statement*, Eindhoven University of Technology, The Netherlands (1990).

DOE, *PPG/15: Draft Planning Policy Guidance Note 15 – Historic Buildings and Conservative Areas*, Consultation Paper available from HMSO, including specific reference to post-war buildings, due to be adopted 1994.

English Heritage, *Framing Opinions*, A national campaign to promote the retention and replication of authentic windows and doors, English Heritage (1991). See also supplement to *Conservation Bulletin* **14**, June (1991).

HMSO, *Circular 8/87: Historic Buildings and Conservation Areas – Policy and Procedures*, HMSO (25.3.87). Official advice and criteria for listing. Due to be superseded in 1994 by PPG 15 (see above).

HMSO, *Planning (Listed Buildings and Conservation Areas) ACT 1990*, HMSO. The principal act concerning protected buildings and sites, consolidating previous legislation.

ICOMOS Australia, *Charter for the Conservation of Places of Cultural Significance* (The Burra Charter), Australia ICOMOS Inc. Sydney (1981).

ICOMOS, *International Charter for the Conservation and Restoration of Monuments and Sites* (The Venice Charter), Venice (1966).

English Heritage, *Repair Grants*, information leaflet, English Heritage, May (1993).

HM Customs and Excise, *Value Added Tax – Protected Buildings* (Listed Buildings and Scheduled Monuments) VAT Leaflet 708/1/90, HM Customs and Excise, 1 July (1990).

Modern conservation theory, studies, exhibitions and articles

'The Age of Optimism – Post-war architecture in England 1945–70', Exhibition by the Royal Commission on the Historical Monuments of England and English Heritage, 1994.

Allan, J., Instruments or Icons? *Architectural Review*, pp. 8–9, November (1990).

Allan, J., Modern Theory of Repair, *Architects Journal* – Renovation, 22 March 1989.

Barrie, G., 'Replicas' plan to save modern listed buildings, *Building Design*, **20** August (1993).

Benton, C., Endangered Species? *Building Design*, 28 September 1990.

Cadogan, G., Confusion clouds the listing regulations, *Financial Times Weekend*, 9/10 January 1993.

Chandler, I., *The Repair and Refurbishment of Modern Buildings*, Batsford (1991).

'A Change of Heart – English architecture since the war, a policy for protection', Exhibition by The Royal Commission on the Historical Monuments of England and English Heritage, 1992. See accompanying booklet of same title with essay by Andrew Saint.

Conservation and Repair of Twentieth Century Historic Buildings, Conference organized by The Institute of Advanced Architectural Studies, York, 4–6 May 1993. Conference Proceedings due to be published 1994.

Conservation Today, exhibition by the Royal Fine Art Commission, 1989. See guidebook, *Conservation Today*, David Pearce, Routledge (1989).

DoCoMoMo, *1st International Conference Proceedings, 1990*, Eindhoven, The Netherlands (1991). This collection of papers, and the equivalent compilation from the second international conference at Dessau 1992, taken together, are the best and most comprehensive survey of the modern conservation initiative from an international perspective. Highly recommended as foundation reading for practitioners, academics and students embarking on the subject.

DoCoMoMo, *2nd International Conference Proceedings, 1992*, Dessau Germany (Due 1994).

DoCoMoMo, *International Newsletters*. Issues 1–8 (published since the foundation of DoCoMoMo by its International Secretariat, Eindhoven). From July 1993 (Issue 9) re-titled the *DoCoMoMo Journal*. The best regular source of international news on modern conservation activities, with theoretical articles, technical studies and detailed project reviews.

DoCoMoMo-UK Newsletters. Twice yearly

updates on the activities of the British branch of the international organization. Available from The Building Centre, London.

Cruickshank, D., Some Concrete Proof of Changing Tastes, *The Independent*, 25 August 1993.

Dunnett, J., Through a glass darkly, *Building Design*, 1 October 1993. (Report on English Heritage Metal Windows Conference of 15 Sept 1993.)

Fawcett, J., *The Future of the Past – Attitudes to Conservation 1147–73*, Thames and Hudson (1976).

Fiorini, L., and Conti, A., *La conservazione del moderno: teoria e practica. Bibliografia di architetture e urbanistica*, Alinea Editrice, Florence (1993). A valuable bibliography of over 500 titles (including press clippings) relating to modern conservation. Text (and emphasis of selection) is Italian.

Harvey, J., The Origin of Listed Buildings, The Ancient Monuments Society, *Transactions*, Vol. 37 (1993).

Henket, Hubert-Jan, Documenting the Modern Movement, *World Architecture*, pp. 84–5, August (1990).

Hewison, R., *The heritage industry – Britain in a climate of decline*, Methuen (1987).

Jurow, A., The Immaculate Conception – ageing and the modernist building, Part I, *Archetype*, Vol. 2, No. 4, pp. 10–13 (1982). Part II, *Archetype*, Vol. 3, No. 2, pp. 13–15, Winter 1983. An extremely valuable essay on decay and modern architecture.

Listing the '60s, *Architects Journal*, p. 12, 21 November (1990).

Modern Architecture Restored, exhibition organized by DoCoMoMo-UK, 1992. See accompanying booklet of the same title.

Modern Moves by Conservationists, *Architects Journal*, p. 9, 26 September 1990.

Moore, R., The Modern World Grows Old, *Blueprint*, pp. 58–60, Nov (1990).

Morris, N., Catching up with the 20th Century, *Architects Journal*, 10 April 1991.

Mostafavi, M., and Leatherbarrow, D., *On Weathering: The Life of Buildings in Time*, MIT Press (1993).

Pawley, M., PS In the grip of museum culture', *RIBA Journal*, p. 125, October (1990).

Pearman, H., 'Re-doing it in Style', *Sunday Times*, 23 Feb 1992.

Restoring buildings of the Modern Movement, special issue of *Architects Journal*, 16 February 1994, featuring project studies of The White House and Bexhill Pavilion.

Roberts, J., Group may step up pressure for listing', *Estates Times*, 5 October 1990.

Robertson, M., Listed Buildings – The national re-survey of England, The Ancient Monuments Society, *Transactions*, Vol. 37 (1993).

Scotland – The brave new world: Scotland re-built 1945–70, exhibition organized by RIAS and DoCoMoMo Scotland, Edinburgh 1993. See accompanying guide book of same title.

Sharp, D., Preserving the Modern, *World Architecture*, No. 9, p. 84, 1990.

Spring, M., (ed.) Restoring Modern Buildings, refurbishment issue of *Building*, 11 December (1992).

Thorne, R., The right conservation policy for listed post-war buildings, *Architects Journal*, p. 21, 13 October 1993.

Technical publications and project reports

(individual sections are devoted to concrete and metal windows)

Concrete

Allen, R. T. L., Edwards, S. C. and Shaw, J. D. N. (ed.) *The Repair of Concrete Structures*, Blackie/Chapman and Hall (1993).

Building Research Establishment Digests, *Durability of steel in concrete*; No. 263 – Mechanism of protection and corrosion; No. 264 – Diagnosis and assessment of corrosion-cracked concrete; No. 265 – Repair of reinforced concrete, Garston (1982).

The Concrete Society, *Cathodic Protection of reinforced concrete*, Technical report No. 36, Concrete Society (1989).

Coping with Concrete, *Architect, Builder, Contractor and Developer*, pp. 63–66, October (1990).

International Journal of Construction, Maintenance and Repair, Vol. 5, No. 4 July/August (1991). Special issue on concrete repair.

McGuckin, S., Visual Concrete – Making Good, *Architects Journal*, pp. 38–41, 2 Dec 1992.

Method of Measurement for Concrete Repair, The Concrete Repair Association, (1990).

Perry, A., Concrete Repair – Guidance from a specialist contractor, *Construction Repair*, Vol. 2, No. 3, May/June (1988).

Plum, D. R., Epoxy Resins in the Laboratory and In-situ, *Proceedings of the International Conference on Structure Faults and Repair 87*, July (1987).

Plum, D. R., The Behaviour of Polymer Materials in Concrete Repair, *The Structural Engineer*, September (1990).

Pullar-Strecker, P., *Corrosion Damaged Concrete: Assessment and Repair*, CIRIA/Butterworth-Heinemann (1987).

Plaster and Concrete, renovation issue of *The Architects Journal*, March (1989).

Re-alkalising Concrete, articles by M. Darby; D. Gooda; G. Jones, *Concrete Quarterly*, Spring (1993).

The Concrete Society, *Repair of Concrete Damaged by Reinforcement Corrosion*, Technical Report No. 26, The Concrete Society.

Metal windows

Blake, D., *Window Vision – Crittall 1849–1989*, Crittall Windows (1989).

English Heritage, *Metal Windows*, Synopsis of the Metal Windows Conference, 15 Sept 1993, English Heritage (due 1994).

Steel Window Association Fact Sheets (1989–92). Useful advice in leaflet form on a range of issues pertaining to steel windows. The series includes sheets on Security, Maintenance, Fixing, Repair and Replacement of Fittings, Replacement windows in Housing, Fire, Your Questions Answered, and Glazing.

Steel Window Association, *The Specifier's Guide to Steel Windows*, Steel Window Association, January (1993).

Windows, *Architects Journal* Focus Issues, September 1992 and October 1993.

Project studies

(See also DoCoMoMo *International Newsletters* for regular coverage of British and European restoration case studies.)

Allan, J., Landmark of the Thirties Restored, *Concrete Quarterly*, **157**, pp. 2–5, April–June (1988).

Allan, J., Lubetkin Legacy Assured (Listing at Whipsnade), *Building Design*, *21*, Oct (1988).

Allan, J., Renovation – Tecton's concrete at Dudley Zoo, *Architecture Today*, **13**, p. 91, (1990).

Allan, J., The Restoration of Holly Frindle, *A3 Times*, No **12**, Vol 5, p. 28, (1989)

Allan, J., Tectonic Icon Restored, *RIBA Journal*, pp. 30–32, Feb (1988).

Avanti Architects Limited *The Tecton Buildings at Dudley Zoo – A Feasibility Study for Restoration and Re-Use*, June (1990). Unpublished (available from Avanti Architects.)

Baillieu, A., Lack of funds threatens repair of Lubetkin building (Finsbury Health Centre), *Building Design*, 15 January 1988.

Barrick, A., Final curtain for Goldfinger block, (Alexander Fleming House), *Building Design*, 12 April 1991.

Chablo, A., Unfashionable Listing (Willis Faber), *Architects Journal*, 15 May, p. 19, 1991.

'Cladding – Health Centre – Tecton', (Finsbury Health Centre), *The Architects Journal*, pp. 63–5, 8 November 1989.

Clayton, H., and Pearman, H., Nuclear plant may be listed, *Sunday Times*, 1 Nov 1987.

Davies, C., Prescription for a Health Centre, (restoration proposals for Finsbury Health Centre), *The Architects Journal*, Renovation Supplement, pp. 12–21, 22 March 1989.

Dunnett, J., A Gorbals Requiem (Sir Basil Spence's Hutchesontown Flats), *Building Design*, 6 August 1993.

Glancey, J., It's very clever, but its still a concrete block (Alexander Fleming House), *The Independent*, 15 May 1991.

Greenberg, S., The Economist Building – 1. Modernism in the making', *Architects Journal*, 21 November (1990). 2. Going Club Class, pp. 53–8, 28 November 1993.

Hetherington, P., High-rise gems face explosive downfall (Gorbals) *The Guardian*, 28 April 1993.

The Hoover Building Conversion, *British Architectural Profile*, pp. 24–25, May (1993).

Leroy, A., Lo Zoo di Cimento – The Concrete Zoo (Regents Park and Dudley Zoos), *Area*, pp. 4, 46–53, June (1991).

McGhie, C., Homes for a New Age (The White House, Amyas Connell), *The Independent on Sunday*, 4 October 1992.

Mead, A., Restoring an early modern English house to exhibit art (High Cross House, Dartington by W. Lescaze), *Architects Journal*, pp. 32–33, 22 Sept 1993.

Mead, A., Balancing conservation with improved performance (The White House by A. Connell), *Architects Journal*, pp. 18–20, 16 February 1994.

Moubray, Amicia de, Life in the Round (Chertsey House by Raymond McGrath), *Architects Journal*, pp. 28–31, 26 Sept 1984.

North, R., Modern medieval church stands the test of time (St. Paul's Church, Bow Common), *The Independent*, 30 March 1988.

Pearman, H., Estates of grace and favour, *Sunday Times*, The Culture section, pp. 6–7, 26 Sept 1993.

Pearman, H., Modern Classics, *Sunday Times*, Style and Travel section, p. 23, 26 Sept, 1993.

Powell, K., Why is this house so special? (Whipsnade bungalow by Lubetkin), *Daily Telegraph*, 13 March 1989.

Powell, K. and Schollar, T., Restoring a milestone of Modernism (De La Warr Pavilion by Mendelsohn and Chermayeff), *Architects Journal*, pp. 35–44, 16 February 1994.

Purchase, H., Listed status splits leaking 'Lego Lane' (Alexandra Road), *London Evening Standard*, 12 October 1993.

Rowland, T., As safe as houses in the zoo? (Lubetkin's Whipsnade bungalows), *Weekend Telegraph*, 29 June 1991.

Russell, B., Mending the Modern Movement (Saltings, Hayling Island by Connell Ward and Lucas), *Architects Journal*, pp. 582–3, 30 March 1977.

Stirton, P., Designer blocks undermined by innovation, fashion and bad management, *The Scotsman*, 20 April, 1993.

Stungo, N., Goldfinger becomes a Hero, *The Independent*, 18 Sept 1991

When Quality Counts (Willis Faber, Ipswich), *Architects Journal*, p. 5, (See also 'AJ action saves modern classic', p. 11.) 1 May 1991.

2 Willow Road – Success in Waiting, *National Trust (Thames & Chilterns) Newsletter*, Autumn (1993).

Monographs/studies on leading modern architects in Britain

Allan, J., *Berthold Lubetkin – Architecture and the tradition of progress*, RIBA Publications (1992).

Cantacuzino, S., *Wells Coates, A Monograph*, Gordon Fraser (1978).

Dunnett, J. and Stamp, G., *Erno Goldfinger, Works 1*, Architectural Association (1983).

Frampton, K. and Kolbowski, S., *William Lescaze*, No. 16 in series, Institute of Architectural and Urban Studies/Rizzoli (1992).

Erich Mendelsohn 1887–1953, book to accompany exhibition on Modern British Architecture (ed. Jeremy Brook, et al.) with *A3 Times* (1987).

Fry, M., *Autobiographical Sketches*, Elek Books (1975).

Powers, Alan, *In the Line of Development, FRS Yorke, E Rosenberg and CS Mardall to YRM, 1930–92*, RIBA Heinz Gallery (1992).

Stevens, T., Connell, Ward and Lucas 1927–1939, Special Issue, *Architectural Association Journal*, Nov (1956).

Ward, B., 'Houses of the Thirties', *Concrete Quarterly*, **85**, pp. 11–15, April/June (1980).

Ward, B., Things remembered – Heroic Relics, *Architecture North–West*, pp. 12–16, June/July (1968).

Owen Williams, Works 3 Catalogue, The Architectural Association (1986).

12
Safety and security in accessibility for maintenance
E. Geoffrey Lovejoy

12.1 Legal requirements in the United Kingdom

12.1.1 Accessibility for maintenance means that the place to be maintained is capable of being reached or entered for maintenance to be carried out, and it may well be thought that how this is achieved is the responsibility of the factory or office occupier. The title of this chapter could be amended to read, 'accessibility for *safe* maintenance'. The following paragraphs set out the various legal requirements in the United Kingdom.

12.1.2 There have been several Acts of Parliament covering work carried out in factories since 1833 following the Reform Act of 1832. The latest is the Factories Act 1961, which governs work not only in factories and docks but also in other processes such as building and engineering construction. The Factories Act covers general requirements, while individual processes are dealt with in greater detail by Regulations covering the particular trade or process. In the Factories Act 1961, Section 1 deals with the need for cleanliness, including daily cleaning of areas such as floors and benches of workrooms, and the longer-term maintenance of walls and ceilings, by means of washing or painting.

12.1.3 Section 28 of this Act states that

all floors, steps, stairs, passages and gangways shall be of sound construction and properly maintained and shall, so far as is reasonably practicable, be kept free from any obstruction and from any substance likely to cause persons to slip.

Other sub-clauses require that means of exit shall have handrails, and that openings shall be protected.

12.1.4 A great deal of argument has taken place in the courts with regard to the meaning of Section 28, and various judgments have been given with regard to what is a floor or step, and what is meant by 'sound construction'. In this context only two terms will be considered. The first of these 'properly maintained', refers to the structural condition of the floors, etc. The criterion is safety, which is a question of degree, dependent on the particular facts. The requirement is absolute that the floors and stairs are maintained in an efficient state and working order, and in good repair. It is a result to be achieved rather than a means of achieving it.

12.1.5 The second of these terms, 'Obstruction', means something which has no business to be on the floor or stair, and which should not reasonably be there. It therefore follows that there is no obstruction, within the meaning of Section 28 of the Factories Act 1961, by the proper storage of objects on the floor, or by the presence of a trolley in a gangway in the ordinary course of work, or by part of a machine fixed to the floor.

12.1.6 Having determined what is meant by a floor or stair, then Section 29 of the 1961 Act requires that there shall, so far as is reasonably practicable, be provided and maintained, safe means of access to every place at which any person has at any time to work, and every such place shall, so far as is reasonably practicable, be made and kept safe for any person working there.

12.1.7 It would be possible for the whole of this chapter to be devoted to this section alone, as it is one that has occasioned a great deal of argument with a large number of judgments relating to it. Without trying to summarize the large number of such judgments it must be pointed out that it is the duty of the occupier to provide and maintain a safe means of access for any person who has to work on the premises, including an independent contractor and his or her servants. This is contained in *Whitby* v. *Burt, Boulton and Hayward Ltd*, among others, and is, of course, reinforced by the provisions of the Health and Safety at Work etc. Act 1974.

12.1.8 It must be pointed out that particular regulations, such as the Abrasive Wheels Regulations 1970, also contain a specific regulation (Regulation 17) as to the condition of floors around an abrasive wheel which is similar to those mentioned in the Factories Act, but where there is the requirement for the floor around the fixed machine to be maintained in good and *even* condition the floor should be, as far as is practicable, kept clear of *loose* material and prevented from being slippery.

12.1.9 In Section 6 of the Factories Act 1961 there is also the provision that where water is capable of being removed by effective drainage it shall be provided and maintained for draining off the water where any process is carried on which renders the floor liable to be wet.

12.1.10 The introduction of the Offices, Shops and Railway Premises Act 1963 introduced similar requirements in Sections 4 and 16 regarding the need for cleanliness in such premises.

12.1.11 The passing of the Health and Safety at Work etc. Act 1974 now brings all work, with the exception of domestic servants in private households, under the need for a duty of care while at work and the preamble of the Act is worth quoting: 'It shall be the duty of every employer to ensure, as far as is reasonably practicable, the health, safety and welfare at work of all his employees.' This requirement is also contained in Section 2 of the Act, and other sections require that same duty of care to persons not employed (e.g. passers-by); employees to themselves and one another; and to manufacturers and designers with regard to their work.

12.1.12 The Health and Safety at Work etc. Act 1974 is an enabling Act and in detail the requirements are contained in the Factories Act 1961 and other Acts which are also backed by a number of Regulations and Approved Codes of Practice which relate more specifically to industries and processes. Where such Regulations exist, then the requirements contained therein take precedence over the general requirements of the Factories Act with regard to the particular process. All this legislation is detailed in Redgrave, Fife and Machin's *Health and Safety* (Butterworths)

which is the successor to Redgrave's *Factories Acts*.

12.1.13 The first sections of Part 1 of the Health and Safety at Work etc. Act 1974 are the important ones as far as people at work are concerned, and attention is drawn to the fact that Section 2 places a duty of care on an employer for his or her employees' health, safety and welfare while at work. Section 3 covers the duties of employers and self-employed to persons other than their employees, while Section 4 deals with the duties of persons concerned with premises to persons other than their employees.

12.1.14 Section 5 is concerned with harmful emissions into atmosphere, and Section 6 deals with the duties of manufacturers as regards articles and substances for use at work. The last section referred to is Section 7 and this concerns the general duties of employees at work.

12.2 Legal requirements in the European Community

12.2.1 While a great change has taken place since the publication of the first edition of this book by the move towards European Community legislation the duty of care required in the 1974 Act has not changed. Such duty of care by professional staff may be rather more quantified by the publication of the Construction (Management and Miscellaneous) Regulations.

12.2.2 A European Community Council Directive seeks to clarify the responsibilities for safety in construction by means of the Temporary and Mobile Works Sites Directive. A means of implementing this Directive is by the publication of the Construction (Management and Miscellaneous) Regulations and this should be implemented by the member states of the EC by the end of 1993.

12.2.3 The requirements may be summarized as follows:

1. Arrangements for site safety management.
2. Appointment of a health and safety adviser.
3. Client to provide contractors with information about hazards or risks associated with the works.
4. Client to appoint contractors with competence in safety matters.
5. It is the duty of designers (Engineers or Architects) to take account of health and safety matters during construction, maintenance (including repointing, redecorating and

cleaning), repair and demolition and to pass appropriate information to the client for future reference.

6. Revision of notice to work provisions.
7. Revocation of parts of existing Construction Regulations requiring safety supervision on site which are replaced by the new Regulations 3 and 4.

12.3 The need for access

12.3 Access varies from the everyday need for a person to walk from one end of the building to another to that for a trained and experienced person to ascend, for example, a radio tower 122 m in height to deal with a particular problem. Access will include maintenance of a machine which if left 'live' may be in a position to damage, to a greater or lesser extent, the person at work within or adjacent to the body of the machine. At its simplest, floors and stairs should be properly designed and constructed, properly maintained and suitably cleaned or drained. It would not be expected to have all areas of floors finished with, for example, granolithic flooring, but with today's

prime cost-cutting requirements and the use of carpet or carpet tiles on screed, there is a likelihood of the carpet flooring wearing through use, and the subsequent possibility of a tripping hazard while the worn area is left unrepaired or unrenewed.

12.3.2 Very often buildings are taken over from a developer without the occupier having any choice in surface finishes. This leaves the occupier in the position that, in order to conform to legal requirements, repair and maintenance must be kept under constant review and action taken immediately where wear is likely to cause a hazard. The correct use of an appropriate finish for the expected use may be nullified by a consequent change of use. With the present trend towards the use of labour-saving devices it may well be that in a building of four storeys no lift has been installed, and the subsequent use of a large floor-cleaning machine raises the need for the transport of it between floors. Such a requirement may be lost in the welter of other design problems. Nevertheless, on pure economic grounds, it may be cheaper to install a floor-cleaning machine on each floor to obviate the cost of a lift. This, of course, will present a problem of storage and maintenance of a different kind.

12.3.3 Within a factory or storage warehouse it should be defined as to where pedestrians may and should be allowed, and, if possible, to what areas fork lift trucks or other vehicles should be restricted. Where areas are large enough for these to be kept completely separate, they should remain so. Entrance doors should be protected by railings and barriers so that no-one inadvertently enters a vehicle access. Sight glasses in doors, especially rubber doors used in warehouse division walls, are essential, and the size and position must allow for all heights of view.

12.3.4 Although it would be preferable for internal rainwater downpipes to be kept clear of the internal access areas of the building, this cannot always be so, and although downpipes can be accommodated within a steel stanchion section, often columns in reinforced concrete or concrete encased steel are used, and these do not offer this protection. In this case and also where the downpipe is housed within the steel section, protection around the downpipe needs to be strong enough to resist impact from vehicles. Sufficient space needs to be left between the pipework and its housing for painting of the pipe to be properly carried out after installation.

Figure 12.1 Reinforced concrete floor slab showing signs of acid attack

Figure 12.2 External access of scaffolding

12.3.5 Reference had already been made to the various Regulations concerned with the need to keep areas around machines from becoming slippery or congested. Sufficient thought must therefore be given to permanent and temporary drainage not only of water but also of possible engine oil or other liquids likely to make the surrounding areas a hazard for workpeople. It will also be necessary to ensure that if a machine needs to be dismantled there is adequate space for the parts from it to be withdrawn safely and laid out without causing an obstruction to others.

12.3.6 It may be thought that the term 'machines' is only applicable to the factory, but with the greater use of electrical equipment in offices, stores and laboratories, the servicing of these electrically operated appliances means that an effective and easily accessible means of isolating each machine is vital. Operators must be warned against interference with their equipment while it is still live and they should be required to obtain trained help when necessary.

12.3.7 Transport requirements within buildings will vary from access for the humble tea-trolley to the latest mobile crane. It is obvious that designers need to be aware of what is likely to need conveying around a building, and to allow accordingly. However, there are many buildings where the introduction of a short flight of steps, or the building of a fire lobby with too small a space allowed between the two sets of doors, has precluded the easy access that must surely be a prime requirement by the users. There is often a conflict between the requirements of the user and the Fire Prevention Officer; the illegal propping open of fire doors is a common sight. Pressure-pad operated doors would obviate many of the problems but the expense must also be considered.

12.3.8 While the normal conveyance of goods and materials will usually be considered, it may often be forgotten that, for example, carrying a ladder in order to complete normal maintenance may involve not only difficulties when going through doors but, once in use, the ladder or steps may impede access by other personnel, or be within the swing of a door and hidden from view. Ample clearance should be allowed in planning, and sight glasses should be built into solid doors.

12.3.9 Vehicles such as fork lift trucks should not proceed with their masts in the raised position, but some operators continue to travel with them raised with subsequent damage to door heads and lintels. Vulnerable areas should be strengthened to cater for these effects. The height of service pipe runs and lighting fittings need to be above the minimum to obviate any foreseeable damage. Where access is required by a vehicle, there should be no placing of ladders or access towers in these areas of maintenance work unless it is unavoidable, and then warning notices should be displayed or the particular route barred to vehicles.

12.3.10 With services it should never be possible for other than trained operatives to gain access to vulnerable or dangerous areas. Doors to lift motor rooms, transformer rooms and switchgear should be locked while in use, and when the plant is under maintenance, inspection or repair, such rooms should not be accessible to the general public in the building. Areas where there is machinery or electric current should be well lit and have adequate space for maintenance to be safely carried out. Lifting eyes or beams should be installed to cater for the lifted loads of machine parts. Where machinery may need to operate while under maintenance, adequate safeguards and interlocks should be installed and used to prevent accidents. An accident such as that which killed a

Figure 12.3 Battery-operated mobile platform. (Courtesy John Rusling Ltd)

man working in the bottom of a lift shaft was caused when the cage, operated by another man, descended and crushed the man at the shaft bottom could have been avoided. Multi-keyed padlocks are available, so that a number of operatives can work in areas dissociated from one another, and no one person can release lethal machinery to the detriment of colleagues.

12.3.11 The layout of service pipe and cable runs should be planned so that joints in liquid-carrying pipe runs are not placed over electrical equipment. BS 83131: 1989 (Code of Practice for accommodation of building services in ducts) gives recommendations as to which liquids and gases should be kept clear of one another, and this Code must be followed.

12.3.12 Where electrically operated overhead travelling cranes are in use these should be isolated once access is required in the area of the crane rails, for whatever reason this access may be necessary. If permanent maintenance access is provided, then this access must be barred to casual users.

21.3.13 Manhole covers to inspection chambers of surface water soil drainage are not usually

lockable, but when such covers are lifted for access to the shaft they should be placed in a secure position and the open hole fenced by means of a portable guard. Where the cover to an access shaft can be locked in the shut position it should also be capable of being locked in the open position, the key being in the possession of the person working below ground. A portable guard should be used around the hole and cover. Step irons will normally be used for descent and ascent and these should be at not greater than 300 mm centres.

12.3.14 Underground tanks holding effluent occasionally need to be entered for maintenance purposes. Access is usually by permanent ladder which can be of steel construction or, if the contents of the tank are harmful to metal, glass-reinforced plastics ladders are available which will resist most acidic liquids. Access into the effluent tanks should never be made without testing the atmosphere of the tank to determine if there is any harmful gas present, or if there is any oxygen deficiency. Purging of harmful atmospheres should always be carried out, and the atmosphere retested before entry. Approved breathing apparatus, together with safety lines and harness, should be used, and no entry to these tanks should be made without a further person being present at the tank top.

12.3.15 It may not always be appreciated that even surface water drainage shafts can have similar hazardous atmospheres. Carbon dioxide gas may be present or there may be an oxygen deficiency when the shaft has been sealed for some time, where the shaft is in chalk subsoil, or when there has been an appreciable change in the barometric pressure. Similar procedures for entry should be adopted as that outlined for the effluent tank.

12.3.16 External access to buildings is usually for the cleaning of roof glazing and gutters, or the repair or replacement of roof sheeting. General access will be necessary more often for the purpose of window cleaning. Designers need to pay more attention to the planning stage to the access requirements for these maintenance jobs, and not assume that access is not their responsibility, or that the provision of permanent walkways is too expensive. The provision of properly designed and constructed access in order to clean patent glazing runs and gutters will mean that the hazardous occupation of walking along valley gutters between two slopes of fragile roof sheeting will then be unnecessary; injury and death resulting from this exercise will then be reduced or eliminated. The provision of warning notices drawing attention to the fragility of roofing

Figure 12.4 Demountable staging. (Courtesy John Rusling Ltd)

material is often forgotten, and by law these must be provided.

12.3.17. Access for cleaning will normally not give problems in relation to daily cleaning, but the requirements for access for external window cleaning and internal wall and ceiling maintenance may often include the provision of appliances such as ladders, access towers or suspended or slung scaffolds. BS 8213: Part 1: 1991: Part 1 (Code of Practice for Safety in use and during cleaning of windows and doors) refers to the cleaning of windows and rooflights and gives maximum dimensions of reach, and the various dangers inherent in horizontally pivoted windows and low sills. This Code of Practice also recommends a limit of height of a ladder to 9 m and this limit is endorsed by the Health and Safety Executive.

12.3.18 Recommended sizes of pane and thickness of glass is given in BS 6262: 1982 (Glazing for buildings). It is a necessity for glazed screens and glass doors to be permanently marked in order to avoid accidents due to people not realizing that glass is present. The use of polycarbonate film applied to the outside surface of the glass will help to reduce solar gain and resist vandalism. The manual published by the Glass and Glazing Federation gives recommendations regarding types of glass to be used and their fixing. Internal fixing of glass for multi-storey buildings is a necessity, otherwise expensive temporary suspended scaffolds will be required each time a pane needs replacement.

Figure 12.5 Example of improperly connected staging sections

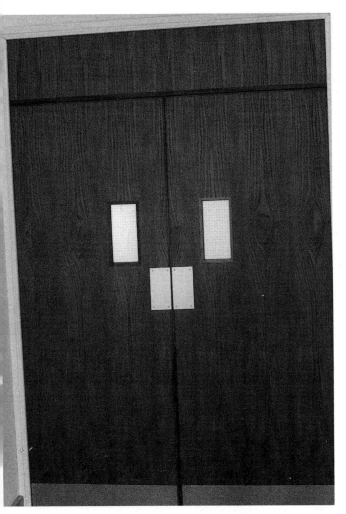

Figure 12.6 Doors with observation panels

12.3.19 If external painting is to be carried out from suspended scaffolds it is debatable whether a permanent system is the most economic method of supplying the access to the facades. Permanent installations, plus the cost of maintenance and repair of the equipment, can be more expensive than the relatively infrequent erection and dismantling of temporary suspended scaffolds. The provision of permanent fixing points in the structure for temporary installations also needs careful selection, detailing and maintenance, as often these fixings provide a hazard due to possible unseen corrosion. Great care should be exercised when determining whether or not to use them.

12.3.20 While most single-storey type buildings can be painted internally from slung scaffolds, mobile towers or mechanical equipment, there are a number of buildings of great height or complex section which may preclude the use of any or all these methods. Similarly, the layout or height of the machinery or plant contained within the building may make the task of painting or cleaning

a difficult one. Again, consideration should be given to the provision of permanent access, or at least to the provision of fixings to enable lightweight staging to be rapidly and safely installed and dismantled.

12.3.21 Maintenance of a building can be carried out either by personnel employed by the occupying firm or by an external contractor. While occupiers should have no doubt as to their legal responsibilities for the safety and protection of their own staff carrying out maintenance work, they may not realize that they also have duties under Section 4 of the Health and Safety at Work etc. Act 1974, with regard to persons on their premises who are not actually employed by them. It is the occupiers' duty to ensure that any plant and any means of access supplied by them is safe and without risks to health.

12.3.22 It may be difficult to determine a complete division of areas of work for a contractor working within a factory building. Where, for example, an extension of a factory can be fenced to prevent access by unauthorized personnel, the division of responsibility is clear between the factory occupier and the contractor. Generally, the contractor's work areas cannot be so easily kept separate from the occupier's areas of work. The need for prior discussion and agreement as to the provision and maintenance of safe systems of work on behalf of both parties cannot be overemphasized.

12.4 Methods of access

12.4.1 Permanent requirements

The need to conform to the provisions of the Building Regulations generally will normally provide safe access as far as stairs and landings are concerned. Reference has already been made to the need to maintain the various surfaces of stairs and floors in good order so that no tripping hazard exists. Roof access ladders are sometimes provided on a permanent basis and these should be in accordance with BS 4211: 1987.

12.4.2 Where fragile roofing materials are used such as fibre-cement or plastics sheeting then the provision and installation of 'Fragile Roof' notices is called for at the approach points to the roofs in accordance with Regulation 36(3) of the Construction (Working Places) Regulations 1966. Although glass is also a fragile material, where the roof is wholly of glass no legal requirement exists for such a notice, although this may be desirable.

12.4.3 Although maintenance work on and adjacent to such fragile materials can be carried out from temporarily supplied and erected ladders and walkways, greater care must be exercised by designers in the future with regard to the provision of permanent platforms and walkways so that regulation inspection and maintenance can be carried out from a properly installed and safe means of access. Maintenance will not then need to be attempted from inadequately provided, installed or secured appliances, leading to unnecessary accidents. While no-one can know where roof sheeting is going to need replacement, it is quite foreseeable that gutters will need clearing, roof lights will need cleaning and the provision of proper and suitable walkways should be allowed for in the design and erection of the building by those involved.

12.4.4 In multi-storey structures the outside face of the building (or certainly the windows) will need cleaning at fairly regular intervals. The cleaning of windows can be carried out from ladders for buildings up to three storeys, but above this level either the windows must be so designed that both faces can be cleaned from inside or suitable permanent or temporary installation can be used. If permanent suspended scaffolds are to be employed, then these must be properly designed, installed, regularly inspected, examined and serviced. The access to the suspended cradles must be safe, and training must be given to those operating the cradles.

12.4.5 For window cleaning the provision of permanently fixed anchorages to which safety belt or harness can be attached also provides a safe place of work. Installation of these anchorages must be carried out by a competent person, and the anchorages must be in accordance with the provisions of the British Standard. The use of centrally pivoted windows may provide a hazard,

either by leaving too long a reach for the cleaner or, if a harness cannot be attached, may also provide a chute down which the cleaner could slide if not properly attached to a safe place. The installation of permanent anchorages requires thought if they are to provide safety at all times and encouragement for their use by their operatives. Alternatively, provision of a mobile anchor is possible, so that window cleaners may be kept safe even if the building is such that permanent anchorages cannot be fixed.

12.4.6 Temporary requirements

Where the provision of permanent access has not been included, there are a number of possible means by which temporary access can be provided, but each of these means can be dangerous if not properly installed or used.

12.4.7 Ladders

These are in use everywhere and contribute every year to the number of accidents and deaths from falls. The use of a ladder can constitute a safe place of work, but in order to do so, the ladder should be fixed at its head and be prevented from slipping down or along the wall. If fixing at the head is not possible, then the foot of the ladder must be fixed or held by another person. An alternative fixing by means of guys tensioned to the wall is shown in Figure 12.13. Although long ladders have been used safely it is recommended that the maximum height is 9 m. The provision of permanently installed fixing points to which ladders can be tied is recommended, together with a suitable lashing attached permanently to the ladder head.

12.4.8 Where a ladder is used to gain access to a platform, the head of the ladder should be at least 1 m above the level of the platform, unless an alternative handhold is provided. Ladders should not rest against eaves gutters, especially those made from plastics materials, but should be supported from the wall by the use of a properly made ladder stay. A lashing to the ladder head should still be used.

12.4.9 Step ladders and trestles should be regularly inspected to ensure that they are in good working order and that their ropes are sound and secure. This need for good working order applies to all equipment, as does the need to ensure that all ladders and step ladders are placed and used on sound bases, and that boxes, bricks and other such imperfect supports are not used. For work on

Figure 12.7 Crawling board up to a temporary roof platform

Figure 12.8 Permanent window cleaning ladder

staircases where the feet of a ladder may be at differing levels, proprietary devices can be fixed to the ladder feet to deal safely with this difference. The use of non-slip inserts at foot and head is advocated for all ladders.

12.4.10 Where access is required on sloping roofs, attention is drawn to the wording in clause 5.4.2 of BS 5534: Part 1: 1978 (*Slating and tiling*) where it is recommended that any roof of slating and tiling should be treated as fragile. Today, where the sizes of timber roof members have been reduced or spacing has been increased, there is no doubt that overstressing of, for example, tiling battens will be caused by the weight of a person walking in a haphazard manner over a roof slope. The use of a roof ladder is strongly recommended on any angle of slope regardless of the state of the roof covering. These roof ladders should be properly designed and constructed with correctly fitted ridge hooks. The use of timber blocks fitted to boards should not be permitted, as accidents have been caused by the blocks becoming detached from the board, by the blocks riding over the ridge tile or the ridge tile itself becoming detached from the roof.

12.4.11 Where access is required over fragile materials on the flat, or to gain access between roof trusses, it can be effected by the use of lightweight stagings. These can be obtained in timber or aluminium, and should be used with the guard rails fitted where there is likely to be a fall greater than 2 m. Stagings can be used on trestles up to a height of 4.5 m, provided the stagings cannot be displaced.

12.4.12 A number of firms supply access towers made from tubular section aluminium alloy, on hire or for sale. These towers vary in size on plan from 1.2 m × 0.90 m to 2.4 m × 1.5 m. They are supplied either to be static or, if required in a number of places of work, can be fitted with castors so that they may be easily moved. The height-to-base ratio should be limited, but they can be fitted with outriggers or stabilizers to extend the height of platform or they should be tied to the building at vertical intervals to maintain stability of the towers. Each manufacturer's product varies slightly, and they should be erected and stabilized strictly in accordance with the instructions. Moving of mobile towers should be carried out only by the application of force at or near the base, and persons should never remain on the platform while the tower is in motion. Access to the working platform is by using a ladder tied within the frame at one end, by the use of a ladder constructed integrally with the frame tubes, or by means of purpose-made steps forming part of the system.

12.4.13 For larger areas of required access, or for general access to building fascias, a scaffold tube and fitting system using aluminium alloy or steel tube can be used. There are also a number of proprietary scaffold systems in use relying on purpose-made jointing to connect the members and to maintain rigidity of the scaffold structure. Erection and use of all these scaffold structures and the necessary tying of the scaffold to the building should be carried out in accordance with BS 1139: 1987 and BS 5973: 1990.

12.4.14 In addition to the provision of permanent suspended scaffolds, it is possible to erect temporary cradles for work along and across the faces of buildings. It is necessary for outrigger poles to be erected on which the trolley track is suspended. These poles must be counterweighted so that there is sufficient resistance moment against overturning with a factor of safety of three times the overturning moment. This counterweighting is usually proved by the requisite number of steel weights, but it may be possible to have permanently built-in anchors. If these anchors are used, they should preferably be

Figure 12.9 Permanent sliding anchor for window cleaning

of non-ferrous material, but in any case should be tested and examined each time before being used.

12.4.15 In roofwork frame painting a slung scaffold is often used, particularly where the roof is supported by a truss. In such cases support for the suspension wires can be made in a number of places. Where the roof is supported by means of joists, consideration should be given to the provision of permanent suspension points to facilitate such erections. Suspension wires should always be made from tested wire-rope and never from soft wire-rope lashings. The point of suspension should be made with proper rope shackles, and any thin sections such as steel joist webs should be padded to prevent cutting of the wire.

12.4.16 The access to such items of plant as overhead travelling cranes and their attendant crane rails and electric cables must be given detailed consideration. The siting of the main cables can be critical, and access should preferably be of a permanent nature with the access barred to the casual user. Many ladders are constructed incorporating metal fitments and wire-rope

strengthening. It is suggested that maintenance ladders should be constructed without metal work if this is possible. Access to dangerous areas where there is moving machinery or live electric cables should not be attempted or permitted by the untrained or inexperienced.

12.4.17 In a manner similar to that permitted for access to factory machinery, access to any moving or live plant should be authorized only by the use of a 'permit to work' system. This should be made known to and used by all relevant personnel under strictest supervision.

12.4.18 Where there is difficulty in the provision of either permanent or temporary types of access previously referred to, it is possible to use various types of mechanical access. These include a small platform operated in a vertical direction only by means of a gas cylinder giving a rise of some 11 m, a larger platform size also operated in a vertical direction only by means of a scissor action or a hydraulic telescopic action, and hydraulically operated articulated arms which are capable of attaining 40 m above ground level in the large sizes, with a three-dimensional reach. Some of

Figure 12.10 Drilled in anchors used to tie the scaffolding to building

Figure 12.11 Use of a safety belt on a roof

these machines can be wheeled through normal doors or be driven into position, but their area of use may be restricted by door sizes or clear spaces.

12.4.19 Where there is likely to be a hazard which may lead to a fall, particularly in roof work, or where access has to be made over a roof, it is of the utmost importance that this access should only be made by trained, competent and fit people. Where access is over or adjacent to fragile materials such as fibre-cement sheets or plastic sheeting, then the provision of lightweight staging or roof-crawling ladders is an absolute necessity. If

anyone can pass adjacent to fragile materials then guard rails should be provided to the stagings, or the fragile areas covered so as to prevent people accidentally falling through these areas. On extensive areas of sloping roof maintenance barriers should be provided at the lower edge of the sloping roof, or guard rails should be placed at the edge of any flat roof.

12.4.20 In order to carry out inspection work, or where the special nature or circumstances of the work precludes the use of barriers or guard rails, it may be possible to maintain safe access by the use of safety belts or harnesses attached to anchorages by their lanyards. The use of longer fixed safety lines is also a possibility, when the use of a fall-arresting device will be necessary so that the length of fall is restricted. There are available a number of proprietary self-reeling inertia safety lines of varying length which also limit the fall of a person. The use of such devices is dependent on the provision of anchorages strong enough to support the shock load of a fall. A special roof anchorage bolt is available which can be temporarily fixed to the upstanding fibre-cement sheet hook bolts and enables them to sustain a shock loading without damage.

12.4.21 In some cases it is possible to rig industrial safety nets beneath hazardous work areas. The rigging and dismantling of these nets calls for the use of trained people who themselves may be put at risk in order to provide safety for others. In order to use safety nets, provision must be made for sufficient attachment points and the net properly rigged with an initial sag, with further clearance beneath the sag so that a falling

Figure 12.12 Suspended industrial safety net. Note that the sag is excessive and that the net is too close to the ground

body does not hit obstructions beneath the net. It is usual that these requirements will preclude the safety use of nets within an occupied building. Nets can sometimes be used at the edges of buildings, on tower structures connected to steel frames, or as complete enclosures to the face of a scaffold. Normally safety nets are composed of a 100 mm mesh, with debris nets of a smaller mesh being used additionally to prevent small tools and debris hitting persons below.

12.5 Operational methods

12.5.1 Planned maintenance is dealt with elsewhere in this book, but it is essential that for safe working, the safety policy should state the chain of command. The area of work and methods and means of access should be determined and adhered to, and, where necessary, the 'permit to work' system should be adopted. That routine inspections should be carried out in a systematic manner is obvious, but where a non-routine event occurs there is the possibility of a less than perfect approach leading to an untrained person achieving access in an incorrect manner, and working in a hazardous situation without the knowledge of anyone in authority which may result in death or serious injury. Further, should the person return

Figure 12.14 Access cradle. (Courtesy of GKN/Mills Ltd)

safely having completed the hazardous job, he or she may become used to improper working practices and may continue or train others in these incorrect ways.

12.5.2 It is the duty of each employer under the 1974 Act to prepare a written statement of policy with respect to the health and safety at work of employees. The organization and arrangements in force for carrying out that policy must be stated and brought to the notice of all employees. In order to satisfy this section of the Act such matters as the methods of safe working on and access to areas of hazardous work should be recorded, and employees made to follow the procedure laid down.

12.5.3 Within each organization it will be necessary to ensure that only trained and competent personnel carry out work involving hazards. Therefore it is essential that each person is properly supervised, and that no-one is able to approach a hazardous area alone or without the knowledge of other persons. It is vital that not only is each person trained in his or her work but

Figure 12.13 Example of a guyed ladder

Table 12.1 Injuries to employees reported to enforcement authorities 1986/87 – 1990/91

Standard industrial classification	Fatal					Non-fatal major					Over 3 days					All reported injuries				
	1986/87	1987/88	1988/89	1989/90	1990/91	1986/87	1987/88	1988/89	1989/90	1990/91	1986/87	1987/88	1988/89	1989/90	1990/91	1986/87	1987/88	1988/89	1989/90	1990/91
Agriculture, forestry and fishing	27	21	21	23	65	429	498	451	403	609	1043	1349	1473	1496	1389	1499	1868	1945	1922	2063
Energy and water supply industries	30	33	203ᵃ	31	30	1718	1397	1262	1140	1092	19621	15798	13728	11684	10103	21369	17228	15193	12855	11225
Extraction of minerals and ores other than fuels, manufacture of metals, mineral products and chemicals	42	42	34	34	33	1694	1544	1647	1524	1420	11100	11054	11122	11436	10319	12836	12640	12803	12994	11772
Metal goods, engineering and vehicles industry	38	35	29	46	29	2657	2647	2595	2712	2627	19393	18568	20029	21730	20238	22088	21250	22653	24488	22894
Other manufacturing industries	29	22	31	28	37	3027	3042	3138	3129	2884	23553	23112	24990	26840	25212	26609	26176	28159	29997	28133
Construction	99	103	101	100	129	2736	2767	2907	3180	3942	16468	16622	16597	17177	17929	19303	19492	19605	20457	22000
Distribution, Hotels and catering; Repairs	22	20	33	31	48	1872	2071	2061	2200	3462	10585	11870	12489	14624	15158	12479	13961	14583	16855	18668
Transport and Communication	34	48	45	44	50	1135	1184	1258	1312	1503	11493	11969	13224	14865	14755	12662	13201	14527	16221	16308
Banking, Finance, Insurance, Business Services and leasing	8	6	9	10	8	198	215	211	271	381	1058	1082	1174	1459	1472	1264	1303	1394	1740	1861
Other Services	16	22	22	23	109	4852	4466	4280	4406	12518	42822	44164	44381	43457	41255	47690	48652	48683	47886	53882

ᵃ Includes the 167 fatalities arising from the Piper Alpha disaster.

that he receives training so that he can appreciate the possibility of other hazards, and take steps to obviate them as far as he is able. The knowledge of safe working practices and the provision of the requisite equipment must be part of management's thinking and application.

12.5.4 Operatives themselves must ensure they know what is required, see that it is provided before commencing work, and check that their colleagues are familiar with these requirements. As from 1 October 1978, Safety Representatives have to be appointed to investigate potential hazards and complaints, carry out inspections and make representations to the employer on health and safety matters. Regular meetings of safety committees should be held at which the safety representatives attend.

12.6 Statistics

Each year a report is made on the accidents occurring in various processes. Tables 12.1 and 12.2 show the number of accidents, including deaths, for various industries from 1986/87 to 1990/91 with the fatal injuries in construction in the years 1986/87 to 1989/90.

Table 12.2 Fatal injuries to employees and self-employed people in construction, 1986/87 – 1989/90

Nature of accident	Employees	Self-employed	Total
Fall from over 2 m	183	93	276
Fall from less than 2 m	9	8	17
Trapped by object collapsing or overturning	60	21	81
Struck by moving vehicle	67	6	73
Struck by moving (including flying or falling) object	37	10	47
Contact with electricity or electrical discharge	16	10	26
Asphyxiation	11	2	13
Contact with moving machinery or material being machined	6	1	7
Other	8	4	12
Total	397	155	552

13
Maintenance policy, programming and information feedback
Douglas L. Warner

13.1 Introduction

13.1.1 The stock of buildings in which we reside, earn our living and relax has been amassed over many generations and represents to the nation a very large capital asset. Ideally this should be maintained in optimum condition so as to maximize both value in use and intrinsic worth. It is perhaps now slowly being appreciated that skills no less significant than those which planned and constructed the original building are also necessary to maintain it. Maintenance begins on the day the building is finished.

13.2 Maintenance policy

13.2.1 For some, a negative approach is made to maintenance; problems are dealt with only as they occur. Redecoration, servicing and other cyclical needs are left as long as possible, and the whole question of maintenance expenditure is seen as non-income producing and in those circumstances is usually one of the first items to be deferred when savings have to be made. A properly considered maintenance policy should be based upon forward planning and preventative action and will provide a building owner/user with value for the money that has been expended, will enable the building to be fully utilized and will protect its asset value and also the owner against breach of the mounting number of statutory and legal obligations.

13.2.2 Building maintenance is defined in BS 8210: 1986 as work other than daily and routine cleaning necessary to maintain the performance of the building fabric and its services.

13.2.3 The objective of maintenance policy should therefore be considered as retaining buildings in an optimum state by the most economic means. What is an 'optimum state'? Establishing the standard of care, and setting a level of expenditure necessary to achieve the aims but no more, is the substance of maintenance policy. The following paragraphs outline factors requiring consideration in setting a maintenance policy.

13.3 Use and standard of care

The use to which a building is put, the suitability of its design for that purpose, and the intensity and the way in which it is used by its occupants will directly affect maintenance requirements. The costs in terms of time, effort and materials will vary greatly – as, for example, between the requirements of a modern, sophisticated, service-filled office building to maintain it in an optimum condition and those of an ambient warehouse storing low-value goods. The future or anticipated use will also impinge. Buildings planned for redevelopment may be kept safe, wind- and watertight, to enable their continued use, but no improvements will be contemplated either to services or to structure and the degree of redecoration and repair will be minimal (Figure 13.1).

13.3.1 The standard to which a building should be maintained will be largely determined by use. The aim will be cost-effective preventative action, taken so as to eliminate avoidable breakdowns and failures. In some situations failure cannot be contemplated. The power supply to a hospital's operating theatre may be an example; here, monitoring, regular service and replacement of possibly defective items before failure will be necessary.

13.3.2 Bernard Speight, in the first edition of this book, sounded a warning against over-maintenance and used external painting as an example. He cited a tendency – usually made necessary by lease

195

Figure 13.1 Large property owners require a highly organized maintenance service in order to keep their buildings in peak condition. Shell UK Ltd have a full-scale maintenance department within the South Bank complex which is run by Trollope & Colls Ltd for the maintenance of the Shell Buildings. (Photograph by kind permission of the Shell UK Ltd)

covenants – to paint exteriors at specific intervals. The judgement that has to be made is whether a further year's delay would result in additional repair and preparation work greater than any savings that may result.

13.4 Design

13.4.1 Good design, proper detailing, well-supervised and sound construction, and the correct use of the appropriate materials all help to reduce maintenance costs without adding significantly to the overall cost of a new building. In a well-designed and properly detailed building, materials and components so used would normally last a full predictable service life, reducing the risk of unexpected failure and making the burden of budgeting easier upon those responsible for setting a planned maintenance regime (Chapter 1).

13.4.2 In ways that energy conservation now forms part of the design consideration, so should maintenance, with thought and consideration given to the elimination of as many maintenance problems as possible at the design stage. Developers selling on their completed buildings do not, of course, have the same priorities as to cost in use as those who will later own or occupy them. Perhaps some criteria or indices could be devised for new buildings to provide information for prospective purchasers as to how cost efficient a building might be in maintenance terms.

13.5 Accessibility

13.5.1 The ability to gain ready access to any part of the building or its services for maintenance purposes is an obvious design factor, and one which will much influence the costs of implementing a maintenance policy (Chapter 12).

13.5.2 All maintenance activities, including those of monitoring and inspection, require safe access and this is a question which should be addressed to an experienced maintenance manager at the design stage. Where possible, facilities such as fixed-track cradles, ladders and stairways, barriers, roof-edge protection and fixed wires or safety eye-bolts for use with lanyards and harnesses should be installed and access should ideally be made available to roof surfaces from the inside of buildings. It is also important that ducts serving plumbing, electrical, air conditioning and other services are properly sized and remain fully accessible. It is often impossible to remove plant once it has been installed.

13.5.3 Where necessary, a wide variety of temporary access equipment is available for maintenance work, ranging from ladders, hydraulic platforms or hoists to full-access scaffolds (Figures 13.2 and 13.3). If a local and isolated repair is necessary in an inaccessible location then the short-term hire of a hydraulic platform hoist will be more appropriate than the more costly provision of an access scaffold. Scaffolding will usually account for perhaps 20 per cent or 25 per cent of the costs of a normal external redecoration/repair contract in a mansion terrace

Figure 13.2 Scaffolding in course of erection at Chichester Cathedral for maintenance work. (Photograph by Keystone Press Agency Ltd for Cathedral Works Organization, Chichester)

in central London, and if such access is only provided on a 5-year cycle then full use must be made on each occasion with roof, rainwater goods, and repairs to masonry and cladding being undertaken at the same time.

13.6 Statutory considerations

13.6.1 The safety and wellbeing of workers on construction sites has traditionally been the

Figure 13.3 A full-access scaffold to a city-centre building necessary for external repair and redecoration

concern of building contractors; they and their sub-contractors have employed the labourers and craftsmen and contractually the onus and responsibility for site safety has been placed squarely on them. The emphasis is now changing, both from legislation such as the Health and Safety at Work Act and present and future directives emanating from Brussels. There are proposed directives on safety and health in construction, and proposals for safety upon temporary – i.e. construction – sites (Chapter 12).

13.6.2 There is a shared responsibility between the owner/occupier and contractor where maintenance work is undertaken within an occupied building, and in such cases the owner/occupier, who is in control of the premises, 'has a duty to take such steps as are reasonable in their position to ensure so far as is reasonably practicable that there is no risk to health and safety'. This requirement would relate not only to the occupied premises but also to connecting corridors, entrances and exits and also to machinery, plant and materials on site (Sections 4(1) and 4(2) of the Health and Safety at Work etc. Act 1974). There is a need therefore to ensure that plant and machinery is properly guarded and if access is necessary at roof level to maintain plant and equipment, then adequate perimeter guardrails or other suitable safety measures must be provided. The provision, testing and continued maintenance of these safety items must be included in any maintenance regime.

13.6.3 Future legislation, it seems, will also require those employing contractors to provide known information on hazards and likely dangers at an early stage, to enable financial provision to be made in any tender. There will also be a requirement to assess the competence of the contractor to undertake work in a manner not detrimental to health and safety. Perhaps this will mean that if, for example, a contractor intends to use ladders when proper access scaffolding is necessary to effect a repair, then the Employer may share responsibility with the contractor for any subsequent accident.

13.6.4 BS8210: 1986 draws particular attention to those involved in organizing maintenance operations to health and safety problems which can arise. Work with asbestos requires special attention, the use of flammable and toxic substances creates danger (cleaning materials, damp-proofing solutions, timber preservatives and adhesives are examples), temporary electrical installations and the fire risks associated with heat-producing processes are also areas where care is necessary. Safeguards and procedures which reasonably protect those working within and visiting premises must be part of any maintenance policy. The *Legionella* outbreaks of the 1980s have now focused attention upon the further need to maintain air conditioning and water services in an environmentally safe condition.

13.7 Prevention and priorities

13.7.1 In any well-run business there will be an expectation that outgoings over the field of the company's activities will conform to a pre-set budget, and expenditure upon maintenance and repair will form part of that expectation. No doubt even in buoyant times constraints upon spending will be imposed, and yet the Building Manager must attempt to maintain the building and its services in such condition that it enables its occupants to work and enjoy the environment created almost without comment. It is obvious that choices will have to be made and priorities set, but ideally these decisions will be taken in the light of an overall plan for maintenance repair and improvement.

13.7.2 To the unprepared, priorities will tend to set themselves, and the maintenance team will be able to do no more than follow lamely, making unscheduled expenditure, resulting in a need to defer works which might in themselves cause problems at a later time.

13.7.3 In a new or refurbished building, programming difficulties are less acute, and given that there will be initial teething defects, reasonably accurate forward cost planning of maintenance and servicing can be achieved. There is, however, a need to ensure that the building as handed over has been properly completed and fully commissioned.

13.7.4 In older buildings, perhaps those which in the past have been neglected, setting priorities is more difficult. There will be a need to deal with the backlog of general disrepair, long-term planning for major refurbishment will require consideration, normally recurring repairs and redecorations and the usual crop of emergency repairs will demand attention. Questions of safety may also be outstanding and, especially in buildings where there is public access, conflicting demands may only be capable of resolution by the agreement of a programme of necessary improvements with the statutory authorities concerned, so that work and expenditure can be phased over a suitable and mutually convenient period.

13.7.5 BS 3811: 1984 (Glossary of maintenance management terms in terotechnology – defined as a combination of management, financial, engineering, building and other practices applied to physical assets in pursuit of economic life-cycle costs) reproduces a maintenance matrix sub-divided into planned and unplanned maintenance. The unplanned is shown to be failure-led, requiring immediate attention to restore matters to a state of normal use, whereas the planned or preventative maintenance, although with a degree of corrective or emergency attention necessary, is indicated as work undertaken to a predetermined regime, and initiated as the result of knowledge of the condition of an item from routine or continuous monitoring.

13.7.6 The need for a detailed maintenance policy to be in place begins as soon as a building is finished. The use and abuse, from both human and natural sources, affects the degree of attention which may be required. Ideally, such matters would have been considered at the design stage and many of the more obvious problems eliminated.

13.7.7 Too strong an emphasis on prevention as a mainstay of any maintenance policy, wherever this is practicable, cannot be made.

13.7.8 A further important point is made by B. A. Speight in a booklet *The Care of Buildings* (Cluttons, 1983), and that is the need for timely attention to maintenance. The example he quotes is of a dry-rot outbreak, disruptive and costly to eradicate, and caused by a single leaking rainwater pipe of long standing, which in itself was remedied at a small fraction of the cost necessary to deal with the dry rot. General neglect of such routine or day-to-day 'housekeeping' matters leads inevitably to greater effort and expense necessary upon periodic maintenance, and in the long term to major repair, which might have been avoided or postponed.

13.8 Means of undertaking maintenance

13.8.1 Maintenance work, sub-divided into three main categories, routine or day to day, periodic and major repair or rehabilitation, will – dependent upon its nature and complexity – be undertaken either by directly employed labour or by outside contractors. These may be acting upon direct instructions, as term contractors or, in the case of major works, will perhaps have won the work in competition.

13.8.2 Directly employed labour

The use of maintenance staff on direct call and who are familiar with a building, its occupants and its past problems are strong arguments in favour of directly employed labour. In small establishments a handyman – one whose limitations are known – can adequately deal with minor day-to-day matters. Similarly, where the case demands, specialist tradesmen can be employed to undertake, for example, masonry repairs upon a cathedral or other large stone fabric structure.

13.8.3 A local authority may employ a larger direct works department for housing repair and maintenance, and in order to monitor cost performance it will be important that regular comparison is made with outside contractors on a work-sampling basis. Productivity levels can be maintained with work study-based incentive bonus schemes.

13.8.4 Contractors

Competitive quotations for day-to-day repair and routine maintenance can be obtained by the preparation of a detailed repair schedule or specification sub-divided into trades and priced by competing contractors as a schedule of rates. Such term contracts are generally let for a period of 2 or 3 years. Some work will be outside the scope of most directly employed labour departments – asphalting and specialist roofing works are examples. In order to obtain realistic quotations from such specialists, or indeed from any contractor, it is important that specifications and other documentation are clear and unambiguous, and detail the full scope of what is required.

13.8.5 The increased sophistication of building services now demands the employment of specialist mechanical service contractors employed under service agreements not only to carry out regular servicing but also to provide a breakdown service, frequently on a 24-hour basis.

13.8.6 The need is to select a suitable and competent contractor, as well as one who is cost competitive to undertake work upon the fabric and building services. It may be necessary to consider taking up references from past employers, inspecting work in progress on other sites, insisting upon membership of appropriate trade bodies, or seeking personal recommendation.

13.8.7 The task of selecting who is best suited to undertake such repair and servicing work is only one aspect of maintenance policy. In many cases the choice will be a simple one, and may even be self-selecting, but for larger and more complex buildings a more formal process will be necessary.

13.9 Programming

13.9.1 Simply put, sound maintenance is largely a question of doing the right thing at the right time.

13.9.2 At the day-to-day level, programming of maintenance work should ensure as far as is possible that there is no interference or any adverse effect upon the normal functioning of the building, and that abortive work is kept to a minimum.

13.9.3 An initial detailed condition survey and further regular inspections will enable an accurate anticipation of possible problems and allow a prioritized programme of work to be set. Emergency repairs will require first consideration, including those necessary to prevent water ingress, to stabilize possibly dangerous elements of the structure, and to eliminate fire and other safety risks. Care must be taken not to omit apparently non-critical items which could develop into major defects (Figures 13.4 and 13.5). Reference has

Figure 13.4 Stonework repairs to St Michael's Church, Southampton. (Photograph by Chichester Cathedral Works Organization)

Figure 13.5 A steel-framed, pitched roof structure erected over an original flat roof. A history of leaks and poor insulation brought about this change in design

been made in paragraph 13.7.8 to problems caused by a single leaking rainwater pipe.

13.9.4 Normal cyclical redecoration, monitored to keep pace with user requirements, and known items of non-critical repair can then be incorporated into a rolling programme covering, say, 5 to 7 years. The programme can anticipate such matters as the need to re-cover or undertake comprehensive repairs to roofs, or to repoint brickwork, or to renew mastics or other sealing compounds, and can also anticipate the necessary renewal of other components of known life expectancy.

13.9.5 Any maintenance programme should only be formulated following full consultation with the building owner/occupier, so as to take into account operational and financial constraints and any other relevant data.

13.9.6 To reiterate, planned maintenance is a question of doing the right thing at the right time, ensuring that all parts of a building and its services are inspected, and that appropriate maintenance is undertaken when necessary in an effective and cost-efficient manner.

13.10 Management of maintenance programmes

13.10.1 Setting an achievable maintenance programme upon any single or group of buildings is a major task. The programme may require 'fine

tuning' and updating from time to time, but, once in being, requires management and administration to ensure that the tasks set are undertaken in the proper sequence by the right people, and that information generated is used to the best advantage. One prime function of the management programme is to help contain and provide an evenly budgeted flow of maintenance costs. The maintenance programme inadequately supervised and assumed to be operating but in reality suffering neglect, is, in B. A. Speight's words, 'a travesty, and at worst a danger'.

13.10.2 There are now available a number of software packages specifically formulated to help in the routine administrative task and aid the costing of maintenance against budget. Information can be stored and retrieved in a number of open files, increasing flexibility and the maintenance manager's ability to generate *ad hoc* reports. The maintenance programme will create large quantities of information on the property, its occupants, past repairs, services' etc'. The data need not be limited to information retrievable but, with proper feedback, budgets and financial models can be constructed for 'cost-in-use' analysis.

13.11 Budgeting

13.11.1 A necessary prerequisite for sound budgeting is accurate knowledge not only of the condition of the building but also of constraints to programming, both physical and financial.

13.11.2 The initial detailed condition survey and later routine inspections will enable a programme of work to be formulated over a period of years, and for which funds will need to be set aside. Assessing realistic costs for the work as programmed is a further necessary step which, if wrong, will lead to either under- or overspending or variations in the amount of work that can be completed, to the general discredit of the maintenance programme and policy as a whole.

13.11.3 Wherever possible, budget information should be obtained from experience of past maintenance, repair, servicing or improvement work to the buildings in question, or those which are comparable. Until such time as a solid database has been created, there will be a need to utilize maintenance cost indices and other reliable forms of outside information. These should, however, only be used with care, for there will be many varying factors acting to influence costs, and such factors are not always readily apparent to the casual observer.

13.12 Information feedback

13.12.1 An efficient data-retrieval system will be able to provide information upon such matters as design. It may be found that high or repeated maintenance costs are associated with a particular element or design feature and that the correct solution is not repair but replacement, or some other more major form of corrective maintenance. Ideally the feedback or retrieval system should then ensure that the need for undertaking such action is notified to those responsible so as to avoid the defective items being incorporated into future designs. In the case of public sector housing or other repetitive long-term schemes such procedure would enable features creating high levels of maintenance cost to be progressively built out, it is hoped, for obvious cost in use savings.

13.12.2 Innovation and technical change continues apace, and inevitably certain new materials or building components will fail. Long-term maintenance and cost-in-use considerations are not usually in the forefront of the designer's mind at the conceptual stage of any project, but with pressure from investors and occupiers the use of collateral warranties is becoming more widespread, extending the cloak of responsibility, and the question of professional liability is also one of which prudent architects, engineers and surveyors are constantly aware.

13.12.3 As briefly mentioned earlier, there is much to commend the suggestion that building surveyors or maintenance managers who deal day to day with building maintenance should be co-opted into the design team, where the benefit of their experience might prevent mistakes being repeated.

13.13 Conclusions

13.13.1 The principles underlying maintenance policy are simple, straightforward and largely common sense, but the creation and implementation of a sound maintenance policy requires – in addition to the numerous skills and expertise involved – a desire and commitment by all concerned to ensure that it succeeds.

13.13.2 It has been explained that there is a direct relationship between present use, future plans and the required standard of maintenance care for any building, and that the degree of care necessary will depend largely on how well the building has been designed and constructed to meet the twin demands of use and the effects of the weather.

13.13.3 Questions of safety have been raised and now there is an indication that the onus of responsibility may have shifted slightly toward the building owner/occupier.

13.13.4 From the initial careful survey through the various stages outlined in this chapter to the implementation of an agreed maintenance policy, the goal will have been to provide an adequate level of care for the building in the short, medium and long terms, against a sustainable budget that allows an even spread of maintenance costs.

References

BS 3811: 1984, Glossary of maintenance management terms in terotechnology
BS 8210: 1986, Guide to building maintenance management
Speight, B. A., *Building Maintenance and Preservation*, 1st edn, Chapter 12, Butterworth-Heinemann, Oxford (1980)
Speight, B. A., *The Care of Buildings*, Cluttons (1983)

14
Fire safety and means of escape
Margaret Law

14.1 Introduction

14.1.1 It is generally considered by fire safety engineers that well-maintained buildings present lower fire risks and that 'good housekeeping' is an important part of fire-safety management. The reason for this is not only that the various fire-protection systems are likely to be in good order but also that many of the potential sources of fire and fire spread – rubbish, faulty wiring, etc. – are likely to be removed efficiently. It is immediately clear that maintenance has an important effect on the two main approaches to achieving fire safety:

1. *Fire prevention* – designed to reduce the chance of a fire starting.
2. *Fire protection* – designed to mitigate the effects of a fire should it nevertheless occur.

14.1.2 Some maintenance operations can, of course, themselves increase the fire risk, either directly (for example, welding) or indirectly (for example, when the water supply for sprinklers is temporarily turned off). It is not the intention here to describe in detail the fire-safety measures needed while carrying out various maintenance operations; these are normally covered by the relevant codes of practice and legal requirements. It is also not possible, in the space available, to describe the planning of maintenance procedures, where fire safety is concerned; particularly since each building is operated differently, according to its use. However, these maintenance operations and procedures are only a means to an end: the maintenance of the overall fire safety of the building, for both the people and the property it contains.

14.1.3 Once we begin to consider the overall fire safety of the building two important points become clear. First, that since the reduction of the probability of fire damage in the building must be achieved by a number of measures, the building designer must take into account the impact of

maintenance, as well as other operations, on the effectiveness of these measures once the building is in use and then plan accordingly. Second, many changes will occur during the life of any building and, although the maintenance team may be diligent, much of their effort can be wasted and hazards introduced if the impact of these changes on the fire safety design is not understood or recognized by the building owner and occupier.

Figure 14.1 Black smoke rises from the Aquadrome and Summerland holiday complex in 1973. (Copyright: Press Association)

14.1.4 This section is therefore devoted to a general description of fire-safety measures, intended to give the designer and the owner an understanding of the principles of maintaining fire safety in a building.

14.2 Fire behaviour

14.2.1 It is convenient to divide fire behaviour into three stages:

202

1. *Growth* It is at this stage that fire detectors and alarms are designed to operate and people in the vicinity of the outbreak have the opportunity to escape.
2. *Full development* When the room or compartment becomes fully involved in fire, the oxygen content is low and any people who have not already escaped are unlikely to survive.
3. *Decay* This occurs when the fire is either brought under control or burnt out.

The maximum rates of burning, the maximum structural damage and the greatest risk of fire spread occur during the second stage.

14.2.2 Fires produce heat, toxic and corrosive compounds, and smoke particles. Deaths and injuries are caused by burns, on the skin and in the lungs, by hyperthermia and by inhalation of toxic gases. Smoke particles and irritant gases reduce visibility and, while not directly life-threatening, they can delay escape so that people are overtaken by the fire. Smoke damage and corrosion of surface finishes and contents can be a major loss. The heat generated by the fire causes damage to structure and contents and can spread fire from room to room or building to building by convection, conduction and radiation.

14.3 Fire prevention

14.3.1 Stated simply, fire can occur when a source of heat, combustible material and oxygen are

Figure 14.3 Smoke in fire situation can often cause more deaths than the fire itself. The BRE use a model of a shopping mall to study the travel of smoke in the case of fire. (Photograph: Fire Research Station, Borehamwood, Herts)

brought together. Since oxygen is normally always present, fire prevention can be described as trying to make sure that combustible materials are kept separated from sources of heat. If fire prevention were perfect no other fire-safety measures would be needed.

14.3.2 Energy systems are potential ignition sources and must be designed so that any heat generated is allowed to dissipate without the risk of ignition of adjacent combustible materials. Careful control and installation of service systems is necessary, particularly with the physical separation of fuel and energy. Once the systems are in use, regular maintenance is needed to avoid, first, an accidental increase in heat supply at, for example, local hot spots in moving parts, series faults, or arcing in damaged electrical wiring, and second, accidental leakage of fuel supplies from gas or oil systems. If rubbish is allowed to accumulate it can prevent the safe heat dissipation from machinery or heating systems and provide fuel for a potential fire.

14.3.3 As time goes on, new machinery and equipment will be installed, furniture and fittings will be moved or altered and thus potential fire hazards may be introduced which were not catered for when the building was first designed. Fire-prevention measures should therefore be regularly reviewed.

Figure 14.2 An example of bad housekeeping with a fire exit blocked by stacked goods. (Photograph: Surrey Fire Brigade, Reigate)

14.4 Fire protection – active

14.4.1 Fire-protection measures are generally considered under two headings – active and passive. Active measures are those which only operate once a fire has occurred; passive measures are inherent in the design, layout and materials of the building and, in principle, are present at all times. It is often stated that active measures cannot be relied on to the same extent as passive measures which are 'always there'. However, both types have their advantages and disadvantages and both are of value in the overall fire-safety design.

14.4.2 Firefighting

The oldest, and in many buildings the only, active measures employed is firefighting. This includes action by the occupants, using first-aid appliances such as buckets of sand and hand extinguishers, and action by the fire brigade. A large number of fires are controlled or extinguished by the occupants before the fire brigade arrives and many others are successfully tackled without the fire brigade being called. It therefore makes sense to have regular maintenance of first-aid appliances and to make sure that they are always accessible and in the right place. An extinguisher should not be used to hold open a door, it should be at the fire point (Figure 14.5a).

Figure 14.4 Precast concrete grids which allow grass to grow through to provide hard access for fire brigade appliances with minimum interference with soft landscape. (Cement & Concrete Association)

Facilities provided for the fire brigade use include water supplies, wet and dry risers, firefighting staircases and lifts, and controls to override the building systems controls. All need to be maintained to the standard required by the brigade. In normal conditions the firefighting lifts can be used for passengers but must be kept clear of goods. Pressurization of the firefighting stairways in high-rise buildings and buildings with deep basements may be used instead of ventilation ducts and openings in order to assist the fire fighters. (See paragraph 14.4.7).

14.4.3 Detection and alarm

Detectors are used to raise the alarm and alert the occupants. They can also locate the fire and be linked directly to the fire brigade. Early detection and subsequently rapid fire brigade attendance has a major effect in preventing a fire becoming large. Many fires which have caused major losses have developed unseen at night time. It has been estimated that installation of automatic detection systems reduces the chance of such large fires by a factor of at least two. Automatic detection systems were originally used mainly for property protection but are increasingly being installed for life safety as well, when there are large numbers of people at risk, as in a covered shopping centre, or when there is a 'sleeping risk', as in hotels, hostels and old peoples' homes.

14.4.4 There are various types of detector – smoke, fixed temperature or rate of temperature rise – and the type used depends on the potential fire hazard, its situation and the general environment. The detector is positioned to be near the plume of smoke, flames or hot gases which could arise from a potential fire. If the detector is subsequently obscured, by high racking for example, or the potential fire hazard changes in nature, then the effectiveness of the detection system can be impaired. It is also important to check that audible alarms remain audible when there are major changes in layout or machinery.

14.4.5 While detection in itself does not directly affect the fire behaviour, the value of detectors can be increased if they are used to operate other protection systems – automatic extinction, smoke control, door closers, etc.

14.4.6 Automatic extinction

The automatic sprinkler system in its simplest form uses a head incorporating a detector which

(a)

(b)

Figure 14.5 (a) Fire extinguisher used to hold open a door when it should be situated at the fire point, (b) An automatic sprinkler head at the point of operation, showing the release of water when heated by a plume of gases rising from the fire below. (Matthew Hall Mechanical Services Ltd)

releases the water when heated by a plume of gases rising from a fire below. An alarm is automatically sounded. Sprinkler systems are designed to control or extinguish fires while they are small. Any obstruction, which obscures a sprinkler or deflects the heated plume so that sprinklers remote from the fire operate, may mean either that the fire grows until it is too large to be controlled or that even if the correct sprinkler operates, the water may be deflected from the fire, which once again can grow large.

Sprinklers, because they limit the amount of heat and toxic products generated, can contribute to life safety. When this is their primary purpose there are special requirements. In particular, only one zone at a time can be shut down for maintenance and the fire authority needs to be informed. When only certain areas are sprinkler protected for life safety, then it may be necessary to shield them from any fire in adjacent unsprinklered areas by using fire-resistant barriers.

14.4.7 Smoke control

The term 'smoke control' is used to describe the control of the movement of both smoke and toxic gases which, of course, will be generated simultaneously. Automatic smoke control can be used to close air-conditioning systems, to keep escape routes clear of smoke and to vent smoke so that the fire brigade can enter the building to tackle the fire. The design of the detector system needs to be reviewed to check that it is still appropriate to the type and location of the potential smoke hazard.

Pressurization systems are designed to stop smoke entering protected routes, such as stairways, by promoting a favourable air flow across the entrance doors. They may be two-stage, running at a low level during normal conditions and being boosted on detection of smoke, or one-stage, only operating when smoke is detected in the building. Regular testing is advised. In order for pressurization to be effective in an emergency, only a limited number of doors should be open at any one time on each protected route.

In buildings where it may be necessary to travel a relatively long distance to reach the exit – for example, in a covered shopping centre – it is normal practice to provide smoke control for the malls or walkways. An extract system, natural or mechanical, may rely on drop-down screens to channel the smoke or to form reservoirs, and regular testing of both the screens and the vents is advised.

14.4.8 Door controls

Certain doors and shutters are designed as part of the fire-protection system to restrict the passage of fire and smoke. When doors need to be kept open for long periods during the normal operation of the building they can be fitted with automatic closing devices operated by fire detectors, but it is obvious that the area must be kept clear so that the doors can close freely when necessary.

14.5 Fire protection – passive

14.5.1 Means of escape

Escape routes are designed on the principle that a person confronted by a fire should be able to turn away and proceed by his or her own unaided efforts to an exit leading either directly or via a protected route to a place of safety. Thus there should normally be a choice of escape routes leading in substantially different directions. The place of safety is usually the open air at ground level but can be a separate part of the building which is completely 'compartmented' from the portion containing the fire. It is implicit in the design that the protected route, usually a stairway, enclosed in fire-resisting construction, will be kept free of combustible material and of obstructions. In addition, doors will be provided to reduce the likelihood of smoke entering the route and ventilation (by an openable window, for example) will be needed. The maintenance of safe escape routes is important. If there are security problems (at the final exit door, for example) suitable locks can be installed. If people insist on keeping the smoke doors open during the normal operation of the building then automatic closers operated by fire detectors can be used. An alternative approach (in hotels, for example) is for evening patrols to close the smoke doors once people have retired to bed.

14.5.2 No rubbish or storage should be allowed to accumulate on the routes. The emergency lighting and signposting should not be obscured. The people who use the building should understand the escape route system.

14.5.3 Compartmentation

Compartmentation is designed to restrict fire spread by containing the fire within a certain area bounded by fire-resistant walls, floors, ceilings. The major points of weakness in the compartment are doors, service ducts and other voids. As explained earlier, if fire doors are continually being left open then a system of automatic closers or supplementary automatic shutters may be needed. An important point to note is that while fire doors are effective barriers to flame and hot gases they are normally poorly insulated. This lack of insulation is not of consequence while the door is in regular use, since no combustible materials would be expected to be nearby, but should the circulation area change, and the door be kept permanently shut, it must be recognized that the door could transmit enough heat to spread a fire to

Figure 14.6 Collapsed pre-formed concrete wall panels together with supporting columns which sheared off at the base. (Photograph: Wiltshire Fire Brigade)

combustible goods stacked against it on the other side.

14.5.4 If fire is able to enter a service duct or other void then the compartmentation may be breached. To guard against this, either the duct must be enclosed in fire-resisting construction or there must be a fire-resistant barrier wherever a compartment wall or floor is penetrated. It is most important that during maintenance operations this principle is understood.

14.5.5 Structural fire protection

Building fires attain temperatures of the order of 1000°C and all structural materials are affected adversely when heated by building fires. Depending on the materials, they may expand, shrink, spall, change their nature, or burn, and they all lose strength which may or may not be regained on cooling. 'Structural fire protection' is the term used to describe the detailing and cladding of elements of structure (which may be made of one or more materials), so that they can perform satisfactorily during a fire. The requirements for these elements are the provision, for a certain time, of one or more of the following:

1. *Stability*: resistance to collapse or excessive deflection.
2. *Integrity*: resistance to penetration of flame and hot gases.
3. *Insulation*: resistance to excessive temperature rise on the unheated face.

14.5.6 The detailing of joints and fixings is an important feature which, if neglected, can cause

Figure 14.7 Unprotected structural steelwork after a major fire. (Photograph: Clifford Ashton)

premature failure. It is necessary therefore, when carrying out maintenance operations to pay attention to these details and where proprietary systems are concerned the manufacturer should be consulted if at all possible.

14.5.7 The choice of structural material and the cladding, if any, will depend on many things, including the environment – wet, dry, corrosive, etc. – the likelihood of mechanical damage and the ease or otherwise of maintenance. Some features of the main structural materials are summarized below.

14.5.8 Concrete loses strength significantly once its temperature exceeds 300°C. Fortunately, the diffusion of heat into the concrete is fairly slow so that while the exposed surface may become very hot, the main body of the element can remain cool for some time. Fire-resistant design takes this into account by increasing the thickness of the concrete to provide the 'expendable' portion. Since reinforcing or prestressing steel will also lose strength when heated it must have adequate concrete cover to insulate it from any fire exposure. It is clear that concrete which has spalled or is showing cracks or signs of mechanical damage may suffer premature failure should a fire occur.

14.5.9 While unprotected steel can be used successfully in doors, shutters and non-load-bearing partitions, it must be protected from overheating if it is to perform a load-bearing function. Unless the fire exposure will be low, structural steelwork is normally encased in fire-protective materials. There are many such materials – wet, dry, spray and board applications. Some are more susceptible than others to damage

during maintenance operations, and all should be inspected for signs of damage or deterioration. Particular attention should be paid to the integrity of suspended ceilings which are used to protect steel beams. Intumescent paint is sometimes used to protect steel elements; this paint expands when heated to form an insulating crust or meringue. If it needs renewal the manufacturer should be consulted on the method of application. It is also important to ensure that there is enough space available for the paint to expand.

14.5.10 Although timber burns, the charred layers on the exposed surfaces act as insulation to the unburnt portion beneath, and timber structural elements are designed with an extra layer which can be 'sacrificed' if exposed to a fire. Joints which can open up and metal fixings that conduct heat to the interior will impair the performance of the timber element. Fire-retardant treatments are designed to reduce the risk of ignition and flame spread, but do not reduce the rate of charring once the fully developed stage of the fire is reached, so that the treatments do not improve the structural performance.

14.5.11 Surface finishes and materials

Surface finishes should be selected to reduce the chance of ignition and spread of flame. They must be of a high standard, or non-combustible, particularly in circulation areas. The performance of the finish can be affected by the base on which it is supported. For example, a substrate which is a good absorber of heat will keep the finish cool and flame is less likely to spread. Many combustible finishes can give improved performance if they have a fire-retardant treatment, but these treatments may be affected

Figure 14.8 Fire damage to a timber-framed building showing the charred timber remaining after the cladding has been destroyed. (Photograph: Fife Fire Brigade, Scotland)

adversely by conditions in use and the manufacturer's instructions should be noted carefully. Surfaces which are damaged or where layers are beginning to peel off may ignite easily and burn rapidly.

14.5.12 Fire-retardant treatments are often used for curtains and other fabrics and there may be special laundering or dry-cleaning requirements if the treatment is to be preserved. Unfortunately, fire-retardant treatments tend to increase the production of smoke. If this is not acceptable, a non-combustible material may be the only choice.

14.6 Conclusions

14.6.1 When a new building is handed over, the purpose of the fire safety measures, and the maintenance implications, should be recorded. The effects of changes, as far as fire safety is concerned, will then be more readily understood. Once the building is in use, maintenance for fire safety can be considered under three broad headings:

1. Maintenance of the fire-prevention and fire-protection measures.
2. Safe procedures for all maintenance operations.
3. Regular review of the overall fire safety of the building.

14.6.2 These subjects are of equal importance, yet the third one, the review, tends to be neglected, if only because there is no appropriate Code of Practice. Such a review would always begin with a statement, or a reassessment, of the basic objectives. These would be safeguarding the people using the building, safeguarding nearby buildings, protecting the structure and contents of the building, and providing safe access for the fire brigade. The importance of these objectives and the balance of measures designed to meet them depends on the use of the building. The life risk in a school, for example, is high, while in a warehouse it will be low. On the other hand, the property risk in a warehouse is high. Legal requirements are primarily concerned with life safety and the prevention of conflagrations. Insurance companies are primarily concerned with property protection. The measures adopted to meet legal and insurance requirements tend to support each other, but it must be recognized that the degree of protection for legal purposes may be less, where property is concerned, than for insurance purposes.

14.6.3 The detailed maintenance requirements may vary not only with the local authority involved but also with the insurance company. Thus although the basic objectives will be the same, the detailed design of the measures may be different. It is important, therefore, to be familiar with the legal and other obligations for the building under consideration.

15
Rehabilitation and re-use of existing buildings
Alan Johnson

15.1 Introduction

15.1.1 Before widespread industrialization, the erection of highly specialized buildings was rare and the normal practice was to adapt buildings for different purposes over their lives. Evidence of the logical development of this policy can be found in early industrial buildings. At least one small nineteenth-century hat factory in Denton, Manchester, UK, was built with certain architectural features included so that it might easily be converted into a terrace of houses should there be a downturn in trade. However, the general adoption of the policy of specialization engendered by the Industrial Revolution caused the demolition and rebuilding of specialized building types to be the norm, in contrast to the old principle of adaptation and extension. It is only in the last 15 to 20 years that there has developed in the Western world the idea that redundant industrial and commercial buildings can be conserved and put to new uses.

15.1.2 Advances in industry and commerce, including the growth of industrial and office automation, and user demands for more comfortable environments for work and leisure have led to large numbers of buildings becoming obsolete or redundant and these changes have provided an abundance of buildings suitable for rehabilitation and re-use. Obvious examples include many textile mills, factories and warehouses in industrial areas too crowded for easy access from new highways, institutional buildings such as schools and hospitals made obsolete by changing educational and medical policies, older office buildings and much nineteenth-century, early twentieth-century and inter-war housing. In the UK in particular, another major source of buildings for rehabilitation is the large number of redundant church buildings. Many Anglican churches, as

well as those of other denominations, are becoming redundant because of changing population patterns and a general decline in religious observance. A considerable number of redundant church buildings have been successfully rehabilitated to provide residential, recreational or office accommodation.

15.1.3 In addition to their widespread availability, a further factor in favour of their rehabilitation and re-use is that many of these buildings were soundly built and remain structurally secure. Although on the face of it 'unfit for modern use', despite obsolescence and neglect, the traditional methods of construction used to build them have left potential developers with a legacy of stable, durable structures which can provide an ideal basis for improvement and re-use.

15.2 Economic advantages of rehabilitation

15.2.1 Another principal advantage of rehabilitation rather than redevelopment is that, in the majority of cases, the 'new' accommodation can be created more quickly. The physical work required to rehabilitate an existing building will normally take considerably less time than demolition, site clearance and the construction of a new one unless comprehensive structural alterations or repairs are necessary. In addition to the time saved during the construction phase, time is also saved during the pre-contract design and official permissions phases, which normally take much longer for new development than for rehabilitation, even where a change of use of the existing building is proposed. This saving of time during the pre-contract design, planning and construction phases often means that rehabilitation will provide the new accommodation in half to three-quarters of the time needed for demolition and new construction, granting the following economic advantages:

1. A shorter development period reduces both the cost of financing the project and the effect of inflation on building costs.
2. The client obtains the building more quickly and therefore begins to earn revenue from it at an earlier date.

15.2.2 The cost of converting a building is generally much less than that of new construction because many of the building elements already exist. However, the nature of the existing construction and its condition impose an important influence on the conversion costs. Adjustments of unacceptable planning or constructional arrangements may prove to be very expensive; combustible elements of structure may require upgrading to comply with current standards of fire resistance; new fire-escape stairs and enclosures are likely to be required, all of which will add to the cost of conversion. If the building is in poor physical condition because of deterioration, the conversion will also involve the expense of repair and renovation work which may have a significant effect on overall costs. It is essential that any building being considered for rehabilitation is subjected to a detailed survey to confirm its structural/constructional quality. This process must include a quantification of the likely cost of any structural repairs, and this information will condition the feasibility of proceeding with any proposed rehabilitation project.

15.3 Environmental advantages

15.3.1 Few will be persuaded of the advantages of rehabilitating and re-using old buildings if the costs are likely to be greater than those of new construction. The exception to this general rule applies where there are overriding environmental benefits from rehabilitation, as occurs where buildings of architectural or historic interest should be retained for their contribution to visual amenity, to culture, or to the interpretation of history. Conservation of an attractive environment combined with rehabilitation of old buildings to provide modern accommodation can often be translated into a financial advantage. Many older buildings are more ornate and display more 'character' than more recent structures, having been constructed by highly skilled craftsmen using high-quality materials. Such buildings may be attractive to certain users, including banks and insurance companies, who like to project an image of solidity, prestige and prosperity to their customers.

15.3.2 Important to the monetary value of a rehabilitated building is the physical context. If a building stands in close proximity to other architecturally attractive old buildings, its appeal and value is increased provided the rehabilitation work which has been carried out maintains and, ideally, enhances its character and architectural integrity.

15.3.3 In the broader context of care for the global natural and artificial environment and the increasing attention to issues of energy conservation which is leading architects and building owners to view existing buildings as a resource for potential re-use rather than replacement, it is not only the architecturally ornate structures which are considered as suitable 'raw material' for rehabilitation. Plainer, more utilitarian buildings are increasingly subjects of this treatment. West Germany offers the example of a large water tower which has been converted into a luxury hotel, while in southern France the surface structures of a former lead mine have been converted into dwellings. The conversion of a former hosiery factory in an English Midlands city into offices which is illustrated in this chapter is a notable success in this field.

15.4 Social advantages

15.4.1 One of the most disruptive aspects of the policy of 'comprehensive redevelopment' practised widely in the UK in the 1960s and 1970s was that established communities that had existed for generations were broken up for ever. Since that time it has been realized that the creation of communities is a complex process which is not achieved merely by architects and town planners wishing it into being. In addition to the economic advantages, the sociological advantages of preserving established, stable communities by adopting a policy of housing rehabilitation rather than wholesale clearance and rebuilding have been recognized and this policy now dominates the activity of provision of 'social housing' by both public authorities and the voluntary sector (i.e. cooperatives and housing associations).

15.5 Use of existing infrastructure

15.5.1 Medium- and large-scale rehabilitation of obsolete housing by private developers, public authorities and the voluntary sector achieves significant cost savings by retaining and re-using not only the dwellings but also the existing infrastructure. Where new housing is erected, not only is additional land required for vehicle and pedestrian access and underground services but finance must be found to pay for the new roads, street lighting, sewerage, gas, electricity and water supplies, telecommunications etc., all of which add considerably to the cost. Where the rehabilitation option is chosen, all or most of this infrastructure already exists and the planned expenditure may be largely devoted to upgrading the dwellings. The provision of a new infrastructure for the housing being obviated, the development period is also shortened, offering further cost savings.

15.6 Controlling laws and codes

15.6.1 In most modern democratic countries there exists a structure of law to control development of land, to enforce space standards within new buildings and standards of construction which will ensure public safety, structural stability etc. 'Planning permission' or control of land use is a form of control naturally most relevant to the erection of new buildings, although the legal definition of 'development' may draw alterations to existing buildings into the net. In the UK this issue is dealt with by the definition that the carrying out of works for the maintenance, improvement or alteration of a building, affecting only the interior of the building and not materially affecting its external appearance, does not constitute development and therefore does not require Planning Permission. Avoidance of the need to obtain Planning Permission grants a further economic advantage to rehabilitation over new-build as it shortens the development period resulting in a corresponding cost saving. Under English law, even if the exterior appearance of a building is not affected, Planning Permission will still be required for a 'material change of use'. A Use Classes Order (1987) defines several different use classes and any proposed change from one to another requires Planning Permission.

15.6.2 There are many examples of rehabilitation projects which do not require Planning Permission according to the scale of alteration contemplated. Rehabilitation ranges from interior redecoration to near-total reconstruction of a building with many intermediate conditions. A classic example of a scheme not requiring planning permission is the interior upgrading of an outdated office building to provide modern facilities – the rehabilitated building remains in the same 'Use Class' allowing extensive internal alterations to be effected without the need for Planning Permission. This exercise can extend to complete gutting of the building and provision of a new internal structure, providing the external appearance remains the same.

15.6.3 Control of land use is only one aspect of legal control of development. Public health and safety concerns are largely dealt with in England by The National Building Regulations (1991). These relate to structural stability, provision of adequate and soundly constructed waste and stormwater drainage, thermal and sound insulation, means of escape in case of fire, fire protection of structural elements and provision of access for disabled people. Most of these considerations are likely to apply to any building rehabilitation project, necessitating application for 'Building Regulations Permission' which is administered in England and Wales by local authority Building Control Officers.

15.6.4 The third area of official control of development relevant to re-use of existing buildings is protection accorded to historic buildings. As an arm of town planning controls, each UK local authority maintains a list of buildings of architectural or historic interest. Very old buildings are almost invariably protected as Scheduled Ancient Monuments; most buildings erected between 1700 and 1840 qualify for inclusion in the list. For buildings after 1840, selection operates; only buildings of definite quality are on the list – a standard which applies even more strictly to post-1914 buildings. However, it is apparent that many buildings being considered for upgrading or conversion to new uses will have atttained 'listed' status, necessitating application for the official permission approving proposals for alterations; Listed Building Consent. This is applied for in the same way as Planning Permission – by completing and submitting a standard pro forma together with drawings fully illustrating the proposals. Acceptance of a proposal depends not only on their architectural quality and impact upon the existing building but also on the grade of listed building for which the scheme has been formulated. A Grade I listed building (of which there are around 6000 examples in the UK) is considered to be of exceptional importance and the opportunities for making alterations to it are likely to be very small. However, the majority of listed buildings (more than 400 000 structures) are of Grade II, a status which tends to result from a noteworthy exterior or distinctive front elevation, allowing radical internal alterations to be made without adverse effect on the feature which caused the building to be listed (see Chapter 11).

15.6.5 A further constraint on freedom to remodel an existing building for greater efficiency or new uses may be its location in an historic area where its contribution to the aesthetic value of a group of buildings is significant. In the UK such historic areas are designated Conservation Areas and any application for 'conservation area consent' to approve significant alterations or partial demolition of an unlisted building may require the applicant to demonstrate that the proposal will 'preserve or enhance' the character of the Conservation Area.

15.7 Scales of rehabilitation

15.7.1 Building rehabilitation ranges from interior redecoration to near-total reconstruction with a wide range of intermediate prescriptions for upgrading, remodelling and renovation. An area of great activity in recent years has been 'retrofit' – the installation of new services and fittings in existing commercial buildings, many of them not more than 25 years old. This form of upgrading of facilities has been stimulated chiefly by computerization and other advances in information technology which have increased requirements for electrical power and space for accommodating the resultant plethora of cables. Another activity generated by the often poor performance of quite modern multi-storey buildings – in this case, the shortcomings of large-panel cladding systems – is overcladding; the addition of a rainscreening and thermally insulating 'overcoat' to those panel-clad residential blocks erected in the 1960s and 1970s which have not proved to be weathertight or can offer only poor thermal performance. This principle also offers opportunities for the refurbishment and upgrading of commercial buildings because the addition of an additional external 'screen' bracketed off the original shell by a new lightweight structure creates a void which may be used to accommodate air-conditioning ductwork as well as the many cables needed for modern telecommunications, networking of computers, etc. There are further reasons for applying overcladding; it can be cosmetic – to increase the appeal of an existing building in competition with newer accommodation – and, if not required to resolve a basic lack of weathertightness in the original external envelope, it may be applied as a form of upgrading the appearance of elevations by concealing unsightly staining. Although a potentially successful treatment for all these conditions, it is necessary to take extreme care in the detailed design, fabrication and installation of overcladding. Gaining an understanding of the extent and underlying causes of any deficiencies in the original construction is essential before proceeding with this treatment. In the absence of this knowledge, the underlying problems may continue or even be intensified by the addition of overcladding. The British Building Research Establishment strongly recommends that prior to any covering up of concrete construction by

cladding, any defective areas should be carefully repaired. See also chapter 11.

15.7.2 More radical than this fairly superficial work is partial restructuring of a building, the most modest version of which is strengthening of timber floors in traditional short-span timber and load-bearing masonry structures. Such strengthening is often carried out where old buildings are converted to office use, existing timber floors being incapable of carrying the excessive localized loads imposed by modern office equipment. The most radical means of satisfying this requirement is the replacement of existing timber floor beams with timber or steel members capable of supporting the increased loads. This solution involves major disruption to existing structure, as it demands removal of floordecks, ceilings, etc., and considerable expense and inconvenience results. An alternative solution is to strengthen existing timber beams by adding steel channel members to both sides of each beam to improve load-bearing capacity. This option is hardly less disruptive than the first method, so a third principle which has less impact upon local building fabric and which involves stiffening of existing floor beams with steel or additional timber fixed to their top or bottom surfaces is more frequently employed. Stiffening of timber floor beams may also be achieved by adding steel plates to both sides of the existing members. Under the general category of 'restructuring', the techniques for insertion of new structural elements – columns and beams of concrete or steel and load-bearing masonry – into existing buildings hardly differ from those employed in new construction, but it is important to add some information on what is perhaps the most marginal example of building re-use before total redevelopment, yet which involves the greatest effort by contractors in this area of work – *facade retention*.

15.7.3 Facade retention is a technique in which most of the existing building is demolished, only main elevations being retained, usually for the important contribution they make to the streetscape. This policy has become popular in the UK because it allows older commercial buildings enjoying 'listed' status to be effectively redeveloped for more efficient use of their sites while retaining the features of the buildings – usually the front elevation, or street facade – which caused them to be listed in the first place. Clearly, the main concern in this sort of work is the careful design of often sophisticated temporary works. For example, the raking and flying shores which must be kept in place until the retained facades can be reunited with the new building structure being erected behind, although close attention must also be given to potential differential movement in the design and detailing of junctions in the permanent construction between new and retained building fabric.

15.7.4 In terms of the financial viability of any option within this spectrum of rehabilitation it should be noted that retention of the whole of the external envelope (including the roof), combined with internal remodelling ranging from redecoration and minor upgrading of existing building services to major internal structural alterations including new lifts, stairs and sanitary accommodation, proves to be the most cost-effective approach. Demolition of the mass of a building, retaining only the main facade for re-use, often costs more than redevelopment. However, the 'facade retention' option is often viewed as inescapable by commercial developers seeking the best possible financial return from an existing property while being legally obliged to preserve a 'listed' street elevation.

15.8 Housing rehabilitation

15.8.1 The most significant factor for the activity of rehabilitation of housing is the poor condition of the existing housing stock, a circumstance which is particularly marked in the UK as a result of the legacy of much poor-quality housing inherited from nineteenth-century development. Many dwellings which were originally of good quality deteriorate because of neglect and decay over time. Periodic investigations into the condition of all types of permanent dwellings have produced many alarming statistics. On examining these data it is clear that there is an urgent need to inject large sums of money into upgrading the existing housing stock, both now and for the foreseeable future. In the UK the task is so great that rehabilitation of housing has become, and will remain, a permanent component of construction industry activity.

15.8.2 In 1981 there were 18.1 million dwellings in England, of which 1.1 million were found to be unfit for human habitation. At least 2 million dwellings lacked one or more of the five basic amenities such as a fixed bath or shower or exclusive use of an inside water closet. In terms of building repair, the 1981 survey found that just over 1 million dwellings needed repairs costing more than £7000 to bring them up to a satisfactory condition. Perhaps the most significant finding of the survey was that only 76 per cent of all dwellings in England were in a satisfactory condition and fit for habitation with all amenities present, the cost of all necessary repairs not

exceeding £2500 per dwelling. Almost a quarter of all dwellings suffered some shortcomings and were in need of quite extensive repairs. It has been estimated that the total cost of rectifying these problems and bringing the whole of the housing stock up to the minimum officially defined standards would be more than £30 billion. The position is so critical that even with the increased sums that have been devoted to housing improvement since the start of the 1980s, it is unlikely that a permanent solution will ever be achieved in the absence of a completely different scale of funding.

15.8.3 The extent of work required to improve existing dwellings to meet current standards naturally varies considerably according to their condition. Some houses which are in generally good condition may only require minor work, such as the renewal of a damp-proof course.

15.8.4 Many dwellings continue to be in need of major improvement. One of the most common deficiencies demanding correction, particularly in UK houses erected before 1919, is the absence of an inside water closet and/or fixed bath. In England in 1981 it was found that in 402 000 cases the only water closet was outside the dwelling. A third of a million dwellings were without a fixed bath. To correct this deficiency it is normally necessary to provide a new bathroom within the existing building or in a purpose-built extension. Another common shortcoming of older dwellings is inadequate kitchen accommodation. Correction of this deficiency, similarly, may entail erection of an extension to the existing house to enlarge or replace an undersize kitchen (Figure 15.1). In the upgrading of larger houses, where more internal space is available, it is naturally more cost-effective to provide the new facilities within the existing interior. There may be several feasible solutions for installing new amenities in most dwellings.

15.8.5 After two or three decades of great activity and consequent improvements in procedures in this area of work, it is widely recognized that older properties which are structurally sound can be economically rehabilitated to standards very close to those of new construction. The provision of modern bathrooms and kitchens usually heads the list of priorities and evidence of the addition of these amenities is often the most visible sign of a rehabilitation project. However, improvement of sub-standard housing embraces a wide range of other operations which aim to solve problems of unfitness and disrepair. These operations may be effected to upgrade dwellings to, or beyond, official minimum standards. Examples of such work include the achievement of a very high

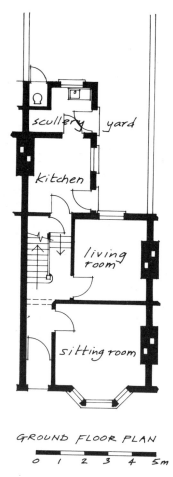
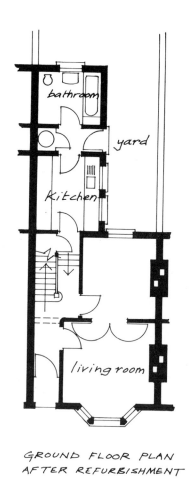

GROUND FLOOR PLAN

0 1 2 3 4 5m

GROUND FLOOR PLAN
AFTER REFURBISHMENT

Figure 15.1 The ground-floor plan of a two-storey terraced house, originally lacking adequate sanitary and kitchen facilities, before and after rehabilitation/extension

standard of thermal insulation and improved fire resistance for structural elements. The range of operations needed to upgrade sub-standard dwellings at least to official standards commonly includes:

1. The incorporation of new damp-proof courses.
2. Renewal of coverings to pitched and flat roofs.
3. Upgrading thermal insulation of roofs.
4. Preventing condensation by means of improved thermal insulation, heating and ventilation and the inclusion of vapour barriers.
5. Upgrading thermal insulation of external walls.
6. Preventing rainwater penetration through solid external walls.
7. Eradicating fungal and insect attack, including replacement or repair of affected timbers.
8. Strengthening timber floors weakened by decay or alteration or where required to support heavier loads.

15.8.6 In addition to the continuing challenge of upgrading large numbers of constructionally sub-standard dwellings to meet modern requirements, there is the equally important problem of the worsening mismatch between the range and mix of accommodation provided by the existing housing stock and the changing constitution of society. With the general reduction in the size of families in Western countries and the rapid growth in the number of one-person households (economic autonomy being an option open to an increasing number of adults) the 'anatomies' of many dwelling types no longer match the demands of populations. Hence much accommodation is used inefficiently and the provision of a mix of dwelling types which better fulfils the demands of contemporary society is an urgent need. Even accounting for the essential role which must continue to be played by new construction, it is clear that there will have to be increased activity for the foreseeable future to create a housing stock better suited to the changing composition of society not only by conversion of many non-domestic buildings into dwellings but also by the conversion of large dwellings into multiple occupation for use by today's smaller families and single people (Figures 15.2 and 15.3).

Figure 15.2 Even comparatively modern educational buildings can become redundant in consequence of changing priorities. This view of Tilehouse Primary School, Denham, Buckinghamshire, UK, which was taken in 1983 might suggest that the scope for its re-use was limited. (Photograph: John R. Pantlin)

Figure 15.3 A similar shot of the school taken in 1987 shows how the building has been successfully converted into sheltered housing for elderly people. (Photograph: Alan Johnson)

15.9 Rehabilitation and re-use of non-domestic buildings

15.9.1 Unlike rehabilitation of housing, where the most significant stimulus for the activity is the poor condition of the existing stock, the main reason for the rehabilitation of other building types is likely to be obsolescence or redundancy, often combined with poor physical condition.

15.9.2 In relation to obsolescence, the majority of industrial buildings are designed to manufacture specific products using particular production techniques. Since most older buildings were constructed, the products of industry and the methods used for their manufacture have changed considerably, rendering many factories obsolete. This forces manufacturers to face the choice of erecting new buildings better fitted to the current

Figure 15.4 The internal form of many churches poses problems for re-use of this building type for dwellings. In many instances in UK practice, some of these problems have been overcome by judicious use of proprietary roof windows, as is shown in this view of work-in-progress on a church conversion in south London. (Photograph: Alan Johnson)

means of production or rehabilitating the obsolete premises to meet the new requirements. Often, distance from modern high-speed motorways prevents continued use of these buildings as factories, but as they are usually close to town and city centres, a common new use for such structures is as office accommodation. This was precisely the formula adopted in the rehabilitation of a former hosiery factory in Lower Brown Street, Leicester, UK, the main features of which are described and illustrated in this chapter.

15.9.3 Redundancy of buildings should not be confused with obsolescence. For instance, an existing church may be in excellent structural condition and continue to offer suitable accommodation for communal worship, but if it is no longer used efficiently because of a decreasing congregation, it becomes superfluous or redundant (Figures 15.4 and 15.5). It will be appreciated that any building which becomes surplus to requirements for a variety of reasons, despite

being in good physical condition and remaining suitable for the purpose for which it was designed, is redundant. There are large numbers of redundant factories, schools, churches, agricultural buildings, railway stations, mills, warehouses and many other building types that are ready for rehabilitation and re-use.

15.9.4 Generally, occupied buildings tend to be well maintained and poor physical condition is more commonly encountered in those which have been vacated because of obsolescence or redundancy. If a building stands empty for a long period, neglect, vandalism and the effects of the elements allied with lack of maintenance can lead to rapid deterioration of the fabric. The accelerating pace of technological change is causing considerable growth in the number of obsolete and redundant buildings, yet there is, to date, no central register, either national or international, which lists these structures. Unfortunately, the effort that is put into producing regular, detailed statistics on the condition of housing is not extended to analysis of the performance of other

Figure 15.5 In this view of the completed south London church conversion it is clear how the five-storey living accommodation which has been created in a formerly unitary space is lit either by existing windows, new roof windows, or new French windows carefully inserted below existing apertures to serve the lowest storey. (Photograph: Alan Johnson)

building types. One result of this lack of information is that proposals for re-use of buildings tend to emerge from knowledge of the availability of a particular building and its susceptibility to re-use, rather than perception of a need followed by selection of a particular building to fulfil it. The main deficiency of this process is that redundant or obsolete buildings are often in a very poor state of repair and hence more expensive to rehabilitate by the time suitable new uses are found for them.

15.10 Susceptibility to change of use by constructional form

15.10.1 Certain types of buildings are more readily converted to new uses than others. The modern large building usually incorporates a strong structural 'skeleton' over which a lightweight 'skin' is applied to exclude the weather. In contrast, most buildings of traditional construction do not include such a discrete structure. In these cases, the external and internal walls fulfil a major load-bearing and stabilizing function. Where a discrete structural skeleton does not exist, it is likely that all elements of the building are acting together to prevent collapse. Most traditional buildings rely on gravity and friction to hold themselves together and have little tensile capacity. The scope for altering such structures significantly is slight unless major structural reinforcement or reconfiguration is carried out, whereas buildings of framed construction can often tolerate large-scale internal rearrangement and recladding without adverse effect on capacity to resist load or structural integrity. Although framed structures were heavily outnumbered – even among large buildings – until

Figure 15.6 The unconverted interior of the top storey of a disused hosiery factory in Leicester, UK, which became 'Mill House' following its conversion to offices by local architects Jonathan Smith & Partners. (Photograph: Jonathan Smith)

the twentieth century by buildings of load-bearing masonry, a growing number of at least partly framed buildings began to be erected from the end of the eighteenth century. The most numerous examples of this type were the multi-storey mills and factories erected for flour, yarn, textile and garment manufacture, first in the UK and then in innumerable industrial towns throughout the world.

15.10.2 The unconverted interior of Mill House, Lower Brown Street, Leicester, shows constructional arrangements typical of this common building type, which often proves to be a constructional form readily converted to suit a wide range of new uses (Figure 15.6).

15.11 Technical aspects of building rehabilitation and re-use (common staple defects of existing buildings)

15.11.1 Often, much more care was taken in the construction of large industrial and commercial buildings than in the erection of those of domestic scale because large and experienced companies were engaged to build them. Improvements in civil engineering practice initiated in France from the mid-eighteenth century eventually spread into building construction, so that by the final decades of the nineteenth century all elements of large buildings, from the foundations upwards, were likely to be carefully considered and well constructed.

15.11.2 Architects, engineers and surveyors engaged in projects involving the alteration of old buildings should take comfort from the fact that traditional builders tended to be conservative in their assessment of the load-bearing capacity of masonry or cast and wrought iron, but this confidence in the solidity of superstructures must be qualified by the recognition that they were either highly optimistic or apparently less concerned about the bearing capacity of foundations, and settlement often took place after construction. The main options in foundation upgrading are widening, to reduce bearing pressures and/or deepening, to found the building on stronger ground or ground less affected by loads from adjacent buildings and seasonal movements. Many of the related techniques can be executed only by specialists, although much foundation strengthening is still carried out by underpinning in the traditional way.

15.11.3 It is important to recognize at the outset of any assessment of a building for re-use that few older buildings retain the arrangement which was originally built. Many well-conceived and competently built structures suffer from poor-quality alterations undertaken by small-scale contractors much more ignorant of structural principles than the original builder. Structural problems are thereby introduced which the alterations succeed in concealing through replanning and redecoration.

15.11.4 Defects in existing buildings are often better understood if they can be monitored over a period of time. It is important to understand how any defects arose, whether they have reached a stable state, and how they would be affected by the conversion work.

15.12 Treatment of defects in existing buildings

15.12.1 Adding tensile capacity

Although there are some advantageous features which result from the way older buildings were constructed – for example, the ability of brick or stone masonry bedded in lime mortar to absorb differential settlement more successfully than cement–mortar construction – a lack of 'togetherness' often characterizes old buildings of load-bearing masonry. The need to supply some tensile capacity to help hold a building together can often be met by simple means such as anchoring the floor structures more firmly to the walls with local ties. Stone masonry and brickwork are often weak and new metal ties are very strong, so it is preferable to provide wall restraint in small 'doses' rather than in concentrated areas.

15.12.2 Preventing damp penetration

In northern Europe, the most common cause of building deterioration is dampness and it is believed that more than 1.5 million dwellings in the UK alone are seriously affected by this problem. The principal sources of dampness are rainwater penetration through roofs and external walls, rising damp through walls and solid floors and condensation.

15.12.3 Rainwater penetration through roofs

Rainwater penetration through pitched roofs – particularly those covered with overlapping slates or tiles which lack an internal lining of impervious sarking felt (not widely used in the UK before the end of the 1930s) – is due to the incursion, in extreme conditions, of wind-blown rain or snow

through the lap joints, or voids caused by missing or damaged slates or tiles. Pitched and flat roofs with sheet-metal coverings are susceptible to the same problem because the metal sheeting corrodes and changes shape in its long-term chemical reactions with air and rainwater, introducing fissures in the formerly impervious surface which readily admit moisture. For tiled or slated roofs, the proper treatment is complete removal of the roof covering and its reinstatement (in new or salvaged material) over a continuous sarking felt lining. Isolated defects in sheet-metal roofs may be rectified by local repairs (for example, lead-burning patches on sheet lead) or replacement of the roof covering only within the bays affected. More widespread erosion of the weathering surface will necessitate replacement of the entire roof covering.

15.12.4 Rainwater penetration through walls

The majority of older buildings have solid stone or brick external walls which are permeable and vulnerable to rainwater penetration. A common result of this condition of permanent dampness is deterioration of internal plasterwork and other interior finishes. This problem can be combated by applying treatments to either the inside or the outside surfaces of the masonry. Internally, it is possible to install damp-proof sheet linings which are then clad with *in-situ* render to achieve a surface suitable for decoration, or proprietary damp-resisting coatings applied by brush or spray. These liquid treatments are less certain of success in resisting damp penetration in the long term than the 'added membrane' principle effected by adding a sheet lining, so it is particularly important to ensure that the following internal *in-situ* plaster/render finish is of a more damp-resisting composition than conventional lightweight gypsum wall-finish plasters. External treatments range – with increasing likelihood of success – from the application of silicone and silicone-based water-repellent liquids, through painting with water-resistant masonry paints, to coating the exposed brickwork or stone masonry with *in-situ* sand/cement render in its traditional forms – stucco, roughcast or pebbledash.

15.12.5 Penetration of ground moisture

The penetration of ground moisture, in the form of rising dampness in walls, is a common problem in many old buildings and often results from the absence of a damp-proof course or the breakdown or bridging of the original damp-proof course caused by its age or alterations to the level of adjoining ground or external wall surface finishes (for example, addition of rendering or extensions of original rendering causing bridging of a damp-proof course). Although the installation of damp-proof courses in new UK construction was made compulsory by the Public Health Act 1875, this statute was not applied universally so that many pre-1900 buildings lack damp-proof courses and consequently suffer from severe rising dampness. The solution for this problem is the installation of a new damp-proof course to prevent further ground moisture from entering the building. Options available to achieve this end are insertion of a damp-proof course by piecemeal removal and re-erection of masonry, incorporating a new damp-proof course; sawing of a suitable low-level masonry bed joint and insertion of a new damp-proof course; or provision of a chemical damp-proof course by pressure-injection into the wall of a water-repellent fluid to create a band of masonry which will resist rising dampness.

15.12.6 In pre-1939 construction solid ground floors were not normally provided with damp-proof membranes and ground floors of concrete, stone slabs or fireclay tiles in older buildings are often found to be suffering from rising dampness. Where such surfaces are to remain exposed it is likely that much dampness will evaporate but if floor finishes are to be installed, a damp-proof membrane must be incorporated and the most effective means of providing it is the application of a two-coat mastic asphalt layer on sheathing felt, well bonded to the damp-proof course in the walls at its perimeter.

15.12.7 Preventing condensation

Water vapour in varying quantities is always present in the air and, the warmer the air, the greater its vapour-carrying capacity. Condensation occurs in a building when warm air, containing water vapour, comes into contact with a cold surface, reducing its temperature and therefore its vapour-carrying capacity. Excess water vapour that the air is incapable of carrying because of its reduced temperature is deposited as condensation on the cold surface. Condensation can also arise within a permeable building element where the dewpoint temperature at which the water vapour condenses occurs at some point within its thickness. This interstitial condensation is often more harmful than surface condensation because it is invisible and the consequent dampness may remain undetected until substantial damage and decay have resulted. Condensation is caused by a combination of different factors, but it is most likely to occur in buildings in which large

quantities of water vapour are produced, such as those accommodating certain industrial processes, and in housing where modern living habits create conditions which generate condensation. The risk of condensation occuring in rehabilitated buildings can be reduced by careful consideration of heating, ventilation and thermal insulation combined with vapour barriers.

15.12.8 Adequate heating reduces the risk of condensation in two ways; first, it warms room surfaces, preventing surface condensation. Second, it increases the moisture-carrying capacity of the air. Different heating methods vary considerably in the efficiency with which they combat condensation. Continuous background heating of the whole building is the most effective means of controlling condensation. Ventilation helps to remove air containing water vapour from a building. Ventilation by extractor fans is advisable where a large amount of water vapour is present. Otherwise, allowing natural ventilation via use of openable windows or ventilators is normally adequate.

Thermal insulation of walls, floors and roofs helps to reduce the risk of surface condensation by ensuring that the internal surfaces of these elements are kept above dewpoint temperature. However, inclusion of thermal insulation can increase the risk of interstitial condensation within building elements because the position of the dewpoint temperature is moved from the element's surface to a point within its thickness. To prevent interstitial condensation, the water vapour must be prevented from diffusing into the insulation, and this is achieved by providing a vapour barrier on the warm side of the insulation. In building rehabilitation projects, vapour barriers are often installed pre-bonded to other materials, familiar products being 'vapour check' plasterboard and 'vapour check thermal board', which are, respectively, gypsum wallboard laminated with a vapour barrier or a vapour barrier and expanded polystyrene/polyurethane thermal insulation. These products are used extensively to line formerly uninsulated solid external walls.

15.12.9 Eradicating timber decay

The majority of older buildings contain many more timber components than modern ones and timber decay is frequently encountered in rehabilitation projects, particularly where a building has suffered from neglect and lack of maintenance, including blocked or broken rainwater drains, displaced roof tiles and peeling paint finishes, all of which can lead to timber decay. Because the existence of timber decay is likely to be connected with other shortcomings in the condition of the building, these faults will require correction in addition to treatment of the timber. Problems may have been built into the original design through inadequate protection of timber from ground moisture or wind-blown rain penetrating thin skins of adjacent masonry, lack of ventilation, or roofslopes which were insufficiently steep to drain away all rainwater. Equally, defects may be due to earlier modification of the original construction and these changes need to be identified and corrected. Rot or insect attack are the main causes of deterioration of timber construction and when an attack is found it is advisable to obtain the advice of a specialist who can identify the type of decay and prescribe treatments for the timber, including specifying any parts which must be removed. It aids understanding of these phenomena to recognize that the function of all timber pests is to break down and convert dead trees into soil, allowing new trees unimpeded growth. The fact that the dead wood has been shaped by people for use as a building material is of no consequence to the agents of timber decay.

15.12.10 Rot in timber

Timber decay which is not due to insect action is normally the result of fungal attack. All fungi require a minimum of about 20 per cent moisture content in the host wood to germinate so that in a sound, dry building there is no need to fear fungal attack. The most serious and devastating wood fungus and the species most frequently found in old buildings in the UK is dry rot or *Serpula lachrymans*. Its airborne spores having alighted on damp timber and established themselves there, it has the unique ability to produce the moisture it needs for further growth, even when the moisture content of the timber falls below the level needed to sustain fungal growth. The fibres of the fungus seek out and feed off the cellulose in the wood to leave a dry and fragile shell of wood-fibre which cannot continue to perform any structural role required of the timber. The most vulnerable parts of a building include timber ground floors, joists built into solid external walls, and roof timbers. All timber is vulnerable in the face of a severe attack. Once rooted in moist timber, the fungus will not only readily extend on to dry timber but will also penetrate masonry and plasterwork to reach and infect distant sound timber.

15.12.11 Because dry rot is so prolific, its eradication involves not only the replacement of all affected wood with new timber pressure-impregnated with preservatives

containing fungicides but also the treatment of adjoining timbers, brickwork, plasterwork and adjacent areas well away from the site of the decay.

15.12.12 Wet rot is the other common wood-destroying fungus. It requires much wetter timber than dry rot in order to develop. Therefore it is more common in exterior joinery exposed to rain such as window sills and timber cladding. Outbreaks are invariably confined to the wet timber and wood immediately adjacent to it. Treatment and eradication of these quite localized outbreaks is therefore much simpler than action on dry-rot outbreaks and involves replacement of the affected timber with new preservative-impregnated wood and the elimination of the cause of the wet rot attack.

15.12.13 Insect action on timber

The destructive action of beetles on timber incorporated into buildings of temperate-zone countries is rarely as serious as fungal attack because an infected piece of wood is only weakened by the combined cross-sectional area of the tunnels which the insects have bored. The strength of surrounding wood remains unimpaired. In contrast, termite action on timber buildings of tropical countries is often disastrous as these insects are voracious consumers of most organically based building materials. Beetle attack which has ceased requires little attention except for any necessary repairs to the damaged timbers. However, an active attack should not be left unattended and can be identified by the clean appearance of the 'flight holes' and the tiny deposits of fresh bore dust which fall from them (Figure 15.7).

Figure 15.7 Insect action on timber. The characteristic 'flight holes' of the common furniture beetle (or 'woodworm') are clearly visible in this section of pine floorboard removed from a pre-World War I building undergoing renovation. (Photograph: Alan Johnson)

In exactly the way fungal attack is addressed, the eradication of insect pests necessitates identification of the unhealthy conditions which have invited the attack and cure of these conditions at source. Dampness and lack of ventilation are as welcoming to insects as they are to wood fungi and once these conditions are corrected, the beetles will feel less inclined to stay. All infested wood must be cleared away and burnt and related construction cavities must be cleaned out carefully. When the remaining sound timber is clean and readily accessible, it can be treated with insecticide.

Figure 15.8 An unconverted interior of a nineteenth-century multi-storey industrial building showing the characteristic 'semi-framed' arrangement of floor joists bearing on timber beams which are supported on cast-iron columns. (Photograph: Alan Johnson)

15.13 Improving fire resistance

15.13.1 A large proportion of older buildings which are suitable for rehabilitation to new uses contain elements of structure that fall far below current standards for fire resistance. Such elements either need to be replaced with new construction of acceptable fire resistance or, more usually, upgraded by one or more of a range of treatments to ensure that they comply with regulations. The suspended timber floor is a building element which almost invariably requires attention in building re-use projects, particularly those involving the multi-storey factory and warehouse buildings of traditional construction that were erected in the second half of the nineteenth and the early years of the twentieth centuries. In the majority of these buildings the undersides of floors did not receive a ceiling finish, comprising only floorboards on timber joists bearing on timber beams, all of which were left exposed to view from below. This was the condition which applied at Mill House, Leicester (Figure 15.8). Clearly, this construction offers a standard of fire resistance much lower than that demanded by current regulations. Upgrading of timber floors for improved performance in a fire is also necessary in the conversion of single dwellings of traditional construction (for example, large nineteenth- and early twentieth-century houses) into a number of self-contained flats. In this case, the focus of attention is the floors which will separate the newly created dwellings. Such new 'compartment' floors formed from existing construction in formerly unitary dwellings invariably have some form of ceiling finish which offers a degree of fire resistance, and so the upgrading treatment need not be as comprehensive as the protection which must be added to 'industrial' floors framed in exposed timber joists and beams.

15.13.2 Options for improving fire performance of timber floors include adding a fire-resisting layer beneath the existing floor members or ceiling or filling voids between the existing floor surface or ceiling. Where the former policy is adopted, suitable materials for uprating fire resistance include *in-situ* gypsum plaster on expanded metal mesh, plasterboard in one or more thicknesses with an *in-situ* plaster skim coat on its underside (as installed at Mill House, Leicester – Figure 15.9) and mineral-fibre fire-resisting boards. Where it is not practicable to replace an existing ceiling with a new construction of superior fire resistance – for example, where ceilings of ornate plasterwork must be retained in a building listed as of architectural or historic importance – fire resistance can be improved by filling the void between the floor surface and the retained ceiling with a fire-resisting material such as foamed perlite.

15.13.3 Re-use of older buildings often also calls for fire protection to be added to metal structural elements such as exposed iron columns or beams. This requirement is met most simply by applying thin-coat intumescent paints to their surfaces. When exposed to fire, this material foams to form a thick sponge-like layer, insulating the metal beneath from the intense heat. Recent general official acceptance of the use of intumescent paints for fire protection of metal structural elements has revolutionized the process of upgrading fire performance of existing buildings by supplementing the officially approved techniques of cladding with rigid boards, thick-coating sprayed protection and encasement in concrete with a method that is much more simply executed and consequently far less disruptive to existing fabric.

Figure 15.9 The interior of an upper-storey floor of Mill House, Leicester, UK, following its conversion to the offices of architects Jonathan Smith & Partners. Adequate fire protection of the formerly unprotected joisted timber floors of each storey has been achieved by installing new ceilings of skim-coated plasterboard between the original heavy timber beams. (Photograph: Alan Johnson)

Figure 15.10 Mill House, Leicester, UK; the goods entrance at the centre of the front elevation of the former factory has become the entrance to the development's integral car parking. (Photograph: Alan Johnson)

Figure 15.11 Mill House, Leicester, UK; a meeting room for the architects' office has been created in the block linking the two main wings of the former hosiery factory. (Photograph: Alan Johnson)

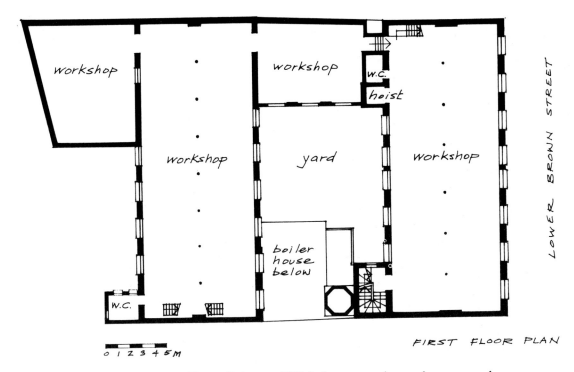

Figure 15.12 The first-floor plan of Mill House, Leicester, UK, before conversion work commenced

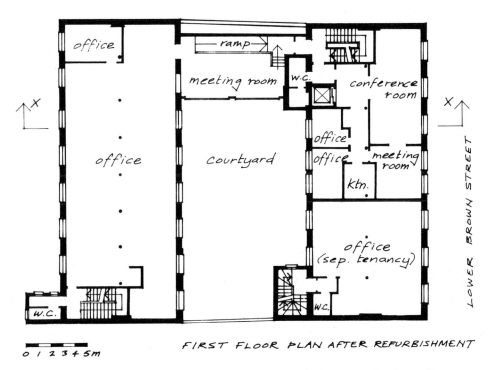

Figure 15.13 The first-floor plan of Mill House, Leicester, UK, following its conversion into offices

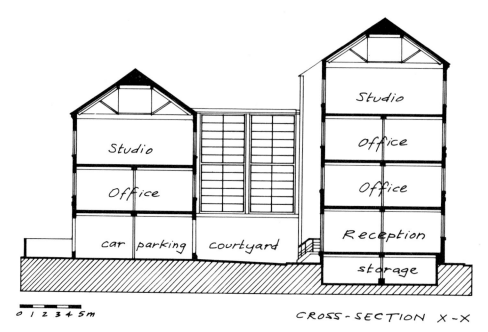

Figure 15.14 A cross-section through Mill House, Leicester, UK, showing the disposition of studio and office space and the new integral car parking

15.14 Upgrading thermal performance

15.14.1 Most rehabilitation projects include the exercise of upgrading thermal insulation of existing external walls. This improvement is usually achieved either by applying a layer of insulating material to the inside or outside face of the wall, or, in the case of cavity walls, by injecting an insulating fill into the cavity.

Internally applied insulation is effective when the building is heated intermittently because it prevents internally generated heat being lost into the walls. However, the addition of internal insulation makes the existing wall construction colder, increasing the risk of condensation. Hence provision of an efficient vapour barrier on the warm side of the added insulation is essential. The application of thermal insulation to the internal surfaces of external walls is often combined with the upgrading of surface finishes and this result is usually achieved by installing dry linings comprising plasterboard laminated with rigid insulation and a vapour barrier, as described earlier.

15.14.2 Externally applied insulation is most effective where the building has thick walls of high thermal capacity and is heated continuously. Despite its advantages in achieving a thermally stable building, the scope for applying thermal insulation externally will, in most cases, be constrained by the condition, configuration and appearance of the existing external wall surfaces. For example, it is unlikely to be acceptable to clad facades of ornate masonry with externally applied insulation. Although the use of external insulation obviates disruption of interiors, even where it is acceptable to change the external appearance in this way, certain modifications are obligatory to accommodate the increase in wall thickness, including modification of window frames at the least by extension of sills, and removal and repositioning of rainwater and wastewater pipes.

15.14.3 Various options exist for the addition of external insulation, including installation of thermal insulation behind conventional claddings where walls are finished in tile-hanging or boarding, application of lightweight insulating renders, adding rigid insulation boards with a render finish as well as flexible insulation with a render finish.

15.14.4 Pitched roofs may have their thermal performance upgraded by inserting an insulating layer at either ceiling or rafter level, immediately below the roof covering. Common forms of insulation for this purpose include quilts of glass fibre or mineral wool, rigid insulation boards of expanded polystyrene or polyurethane, or loose-fill materials such as expanded polystyrene beads or blown mineral wool, all of which can be installed to a thickness to suit the degree of thermal upgrading required. An essential provision in conjunction with the installation of thermal insulation in any pitched roof is adequate ventilation of the roofspace so that the risk of condensation is minimized.

15.14.5 Thermal upgrading of flat roofs depends very much on the anatomy of the existing construction. Concrete flat roofs can only be upgraded by adding insulation either beneath or on top of the slab. In the case of flat roofs framed up in timber joists and decking, a third option exists of inserting the insulation within the void between the ceiling and the decking. All the above-mentioned materials may be used to improve thermal performance of flat roofs, but great care must be taken with 'cold' roofs (i.e. where insulation is installed at ceiling level) to incorporate vapour barriers and effective insulation of voids sandwiched between the ceiling and the decking.

15.14.6 Heat loss through the floor of a building is slight in comparison with leakage through the external walls and roof but upgrading the thermal insulation of floors may improve comfort and reduce running costs. Ground-bearing slabs of concrete and suspended timber floors can both be upgraded fairly easily by the addition of proprietary panels comprising rigid insulation boards laminated with a conventional floordecking material such as chipboard, although the additional thickness of floor construction which results naturally demands modifications to existing door openings, their linings and other items of fitted joinery.

15.15 Upgrading acoustic performance

15.15.1 The need to upgrade the acoustic performance of elements of a building relates mainly to its proposed use. In rehabilitation work, acoustic upgrading is needed most frequently in schemes involving the conversion of existing buildings into a number of dwellings. In these cases it is always desirable – and often mandatory – to upgrade the sound insulation of walls and floors separating different occupancies in order to prevent or reduce transfer of noise between dwellings. Floors and walls may need to be resistant to the passage of both airborne sound (i.e. sound waves in the air striking walls or floors) and impact sound (sound waves set up in building elements as a result of impact).

15.15.2 Achieving adequate sound insulation in party walls between dwellings presents one of the main challenges in building rehabilitation projects. Existing walls are often difficult to upgrade because they interface with other elements, particularly floors, allowing unimpeded transmission of impact sound. However, the sound insulation of a wall can be improved by adding a separate leaf on one or, ideally, both sides of the wall, trapping a minimally 100 mm wide air gap. This may be achieved in lightweight construction such as timber studding and plasterwork, or by adding skins of brickwork or blockwork. Plainly, the overall thickness of the wall is increased considerably and the usable floor area is reduced. New party walls much more easily achieve the desired acoustic performance by careful detailing to isolate building elements and correct selection of construction materials (for example, use of 'party wall blocks' which offer good acoustic performance).

15.15.3 The other sensitive element is the party floor between occupancies and the problem is particularly acute where a dwelling or non-domestic use is sited over another dwelling. In this circumstance, the party floor is required by regulations to have adequate resistance to both airborne and impact sound. At the most modest level, installation of carpets with underlay makes a significant improvement to insulation against impact sound. A more effective measure is the provision of 'pugging' – sound-deadening materials installed within the thickness of floors formed from decking over joists or joists and beams. Loose sand to a density of 80 kg/m² is often used for this purpose and as this additional load is usually too heavy to be supported by existing ceilings, it is carried on 'pugging boards' and battens fixed between the joists. More effective still is the alteration of the existing construction to create a 'floating' floor, a principle which can be applied to either joisted timber floors or concrete slabs. In the case of suspended timber floors, the existing floorboards are lifted and relaid on battens located between the joists after an interleaving layer of minimally 25 mm thick glass-fibre or mineral-wool quilt has been laid over the joists. With the crushing action of the reinstated boards on the quilt, this modification raises the top surface of the floor by only approximately 10 mm.

15.15.4 Although concrete generally provides good sound insulation in its own right, in cases where it is desirable to improve acoustic performance, a floating 'raft' of floorboards or timber decking may be added, comprising decking on battens laid on the concrete floor over a minimally 25 mm thick layer of resilient quilt. If flanking transmission of sound is to be avoided, it is essential that the resilient layer is turned up at its edges to isolate the raft from perimeter walls, and that a gap is left between the underside of skirting boards and the raft.

These measures all involve modifications to the existing floor surface or the floor void. An alternative means of upgrading the acoustic performance of a party floor is to add a new ceiling beneath it, carried on its own structure and spaced as far below the existing ceiling as possible, an acoustically absorbent quilt being inserted between the new and existing ceilings.

15.16 Access for the disabled

15.16.1 Building Regulations now recognize that impaired mobility is the permanent condition of a significant part of the population, as well as a disability which will afflict most people at some time in their lives, by requiring easy access for disabled users to most areas of all new buildings. Naturally, Building Control Officers have less jurisdiction over existing buildings, but comprehensive alterations to an existing building may make it possible to enforce the level of provision required in new buildings in cases where the developer has not recognized the deficiency from the outset and advanced a proposal to cure it. A prime problem arises where the ground floor of the building is raised substantially above external ground level, as occurs very often in old buildings. Sometimes, space is not available for installation of wheelchair ramps to the maximum gradient of 1 in 12, or such features cannot be accommodated without compromising the appearance of an imposing entrance elevation, etc. Where it is impractical to add a ramp and it is not possible to insert an additional entrance into existing fabric at pavement level, an externally sited scissor-action platform hoist allowing access for wheelchair-bound building users may provide a solution.

15.16.2 Access to all main areas of a building interior is best effected by incorporating a conventional passenger lift within a purpose-built shaft but this option is not always viable or practicable in low-rise or structurally complex buildings where it is not intended to undertake 'major surgery' to remove small distinctions between floor levels or to disentangle existing structure/construction to provide a continuous vertical zone for a lift well. Several lift manufacturers have recognized that there is a growing market for easily installed devices for improving access for disabled people to upper storeys of dwellings and a wide range of 'stair lifts' is now available, some of which will move wheelchair-bound people as well as the more familiar type which obviates stair climbing by the ambulant disabled. Increasingly ingenious versions of this 'staircase funicular' principle are appearing, allowing addition of stair lifts not only to the straight-flight staircases for which they were first developed but also to those incorporating right-angle turns and winding treads.

15.17 Project administration, site management and forms of contract

15.17.1 Developing policies for the preparation of documents to specify and control merely the activity of rehabilitation of domestic buildings, ignoring other building types, presents a considerable challenge. There is a great variety of domestic buildings by size, plan-form, construction and condition and countless ways of repairing and converting them. The danger of attempting to develop a standard or model specification for such work which is comprehensive is the production of a document that is effectively unmanageable in use because it presents a virtually endless list of rarely used clauses. Some architectural practices, finding that they were constantly being asked to undertake, for example, the rehabilitation of low-rise nineteenth-century houses in the Greater London area, sensibly conducted an analysis of the anatomy of these buildings and accordingly produced a document: *Specification Clauses for Rehabilitation and Conservation Work*. Moves have been made to develop similar documents describing the operations common to the conversion of other building types (for example, the conversion of nineteenth-century factories, mills and warehouses to office use) and it is the emergence of documents on this 'generic type' basis which offers most hope for the compilation of a library of manuals that will guide specifiers to appropriate prescriptions for the treatment of the very wide range of structures which feature in building conversion work.

15.17.2 On the related subject of suitable forms of building contract to administer and control rehabilitation work executed by commercial building companies, although it is appreciated that, in a highly competitive business environment, best value for money is the main aim and close cost control of building procurement is demanded in consequence, work to existing buildings provides many more problems in this regard than does new-build because of the range of often unforeseeable difficulties that come to light when existing construction is opened up. The standard forms for building contracts let on a lump-sum basis have been developed primarily for use in new construction and rarely include provisions guaranteeing continued tight cost control when serious and unforeseeable difficulties arise. Equally, it is often unacceptable or hazardous to modify contract terms to try to cope with such circumstances, and clients are rarely persuaded of the merits of reducing the opportunities to achieve the converted building at the keenest possible cost by including a large contingency sum at the outset. For these reasons, conventional firm-price contracts arrived at as the result of a competitive tender procedure are often an unsatisfactory means of procuring a remodelled or renovated building through rehabilitation work and, despite the loss of the 'efficiency' spur built into firm-price contracts, to ensure continuously satisfactory client/contractor/consultant relations and the desired quality in the product, it is often better to administer the work on the basis of a 'cost reimbursement' contract in which the builder's fee for managing work of known value (excluding unforeseeable conditions which are costed separately) is pre-agreed in some way – usually at a fixed percentage of the contract cost.

15.17.3 In terms of site techniques, it should be clear that the virtually limitless complexities posed by the large range of rehabilitation projects often call for great ingenuity from builders. It is not surprising that particular contractors or specialist divisions of many of the larger companies offer greater expertise in this field than those builders whose main activity is new construction. With the growth of the rehabilitation market over the last two decades and the more recent reduction in new-build, many contractors now claim expertise in rehabilitation work. Such ingenuity frequently encompasses matters such as sophisticated site organization, careful and detailed sequencing and integration of building operations and managing continuity of building services shared with adjoining buildings or for the purpose of continued partial occupation of the subject building. Often these concerns influence the methods for maintaining structural stability of retained fabric during the work and design and planning of the installation of sometimes complex and sophisticated 'temporary works' in the form of raking, flying and dead shores to secure facades and other elements of structure may be as time-consuming as the effort expended in preparing the design and specification of the permanent works which shape the remodelled building.

15.18 Re-use for combined uses

15.18.1 The enlarging knowledge and expertise of professionals working in the field of building rehabilitation and re-use as well as greater interest in appraising the potential of redundant buildings for re-use naturally causes architects and engineers to tackle increasingly difficult challenges. An area of growing activity is the conversion of large

buildings to suit more than one new use – for example, the combination of retail and residential accommodation. Sometimes, a large redundant building cannot be re-used satisfactorily for one purpose and is only 'recyclable' by turning it over to a range of mutually tolerant new uses. Although massive load-bearing masonry structures such as nineteenth-century mills and factories are often first thought of in this connection, many twentieth-century buildings, and particularly those of the post-war period, are now under consideration for such treatment. It can be seen that if the current decline in demand for purpose-built offices becomes permanent in consequence of advances in information technology, this activity is likely to increase. It is a component of building rehabilitation work best illustrated by reference to a specific project – the conversion of the former British Airways West London Air Terminal to provide 398 dwellings, integral covered car parking, retail and office accommodation and a health club.

15.18.2 The original 1960s building comprised reception, waiting and baggage-handling areas for customers of British European Airways (a predecessor of British Airways) as well as several storeys of offices, passengers being taken in special buses from the building direct to their flights from London's Heathrow Airport. Increasing traffic congestion and changing passenger-handling practices caused the building to become obsolete and several years ago its ground-floor accommodation was converted into a large food supermarket with an adjoining car park, although the work posed many constructional and logistical difficulties. Continuity of use of the existing supermarket had to be maintained throughout the course of the work and the multi-storey baggage tower, housing a continuous spiral ramp, which adjoined the office block, could not be adapted for re-use as residential accommodation, so it was demolished and replaced by a purpose-built steel-framed 18-storey residential tower (Figures 15.15 and 15.16). Also, the building is supported on a multitude of steel columns scattered among a main junction of London Underground's railway system, adding a further complication to work to three existing layers of basement to convert them to car parking, plantrooms and the residents' health club. These three basement levels were converted into four new storeys by demolishing two intermediate floors and reinstating three new slabs. New ramps were installed to provide access to the three levels of basement car parking.

15.18.3 Work to the superstructure was no less radical – all external cladding, internal non-structural walls and partitions were removed,

Figure 15.15 Point West, South Kensington, London; a view of the site during the course of rebuilding work. Glazed curtain walling is being stripped from the main concrete-framed structure in preparation for its recladding to suit residential use and scaffolding is being erected around the spiral-ramp baggage tower prior to its demolition and replacement by a new residential tower block. (Photograph: Alan Johnson)

Figure 15.16 Point West, South Kensington, London; conversion work approaches completion; new upper floors have been added to the main quadrangular block which has been reclad in brickwork and refenestrated and, in the foreground, a purpose-designed residential tower has replaced the baggage tower. (Photograph: Alan Johnson)

the structure being stripped down to the basic skeleton of floor slabs, columns and beams. A width of building adequate to accommodate the reception areas of apartments ranged along both sides of a spinal corridor was achieved by adding a continuous strip of floorspace at each level to all four internal faces of the 'racetrack plan' office building and correspondingly reducing the size of the central lightwell to about a third of its original area. This policy also allowed the necessary lifts, firemen's lifts, escape and firefighting lobby-approach staircases to be housed in internal angles between the spinal corridors and the scaled-down central lightwell. Three new storeys were added on top of this 'doughnut' of re-used/extended structure in the form of three-storey dwellings – the majority of the 32 'penthouses'. This new accommodation was created over a new steel-framed structure, the intermediate floors being of joisted timber construction below chipboard decking. The most spectacular penthouse apartment absorbed the three topmost storeys of the new tower block (see Figures 15.17 and 15.18).

15.18.4 Throughout the re-used structure, integration of building services was facilitated by reserving a 650 mm-deep continuous suspended-ceiling void for horizontal service runs, although, because the original structure contained many deep downstand beams, much diamond-drilling of reinforced concrete was necessary to effect continuity of services, all carried out in locations and conditions closely monitored by the local authority Building Control Officers. Partitions between individual rooms and party walls between dwellings all of 'dry wall' construction – party walls comprising two layers of plasterboard to both faces, fixed to galvanized

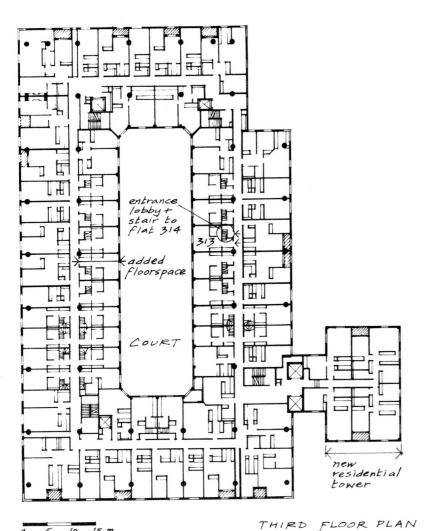

entrance
lobby +
stair to
flat 314

313

added
floorspace

COURT

new
residential
tower

0 5 10 15 m

THIRD FLOOR PLAN

Figure 15.17 The third-floor plan of Point West, South Kensington, London, showing subdivision and extension of the formerly unitary office space to provide more than 60 separate flats (from a total within this scheme of almost 400 dwellings)

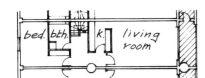

bed. bth. k. living room

FOURTH FLOOR : FLAT 313 LIVING LEVEL

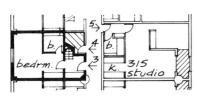

5
bedrm. 4 b. 3 315 studio k.
b.

THIRD FLOOR : FLAT 313 ENTRANCE LEVEL
FLAT 314 ENTRANCE LOBBY

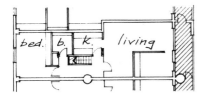

bed. b. k. living

SECOND FLOOR: FLAT 314 LIVING LEVEL

0 1 2 3 4 5m   PLANS OF TYPICAL FLATS

Figure 15.18 Plans of typical flats located on the third floor of Point West, South Kensington, London.

light-gauge steel framing sited either to both sides of a clear central piped services zone or a continuous central membrane of two layers of 15 mm-thick fire-resisting board. A 75 mm-thick layer of mineral wool quilt sound insulation is incorporated immediately to the rear of both plasterboard skins and the overall thickness of any party wall is 230 mm.

15.18.5 The external faces of the remodelled accommodation were reclad with cavity walls consisting of an inner leaf of lightweight blockwork and an outer one of facing brickwork supported on stainless-steel shelf angles bolted to the reinforced concrete structure. The new steel-framed tower block was treated identically. Epoxy-paint finished aluminium-framed double-glazed windows were installed within the masonry skin to provide the necessary natural light and ventilation to habitable rooms, all space heating being by electricity. Only the penthouses are air conditioned, fan-coil units providing this service. The total cost of this large rehabilitation contract was of the order of £75 million.

15.19 Conclusion: the future role of building rehabilitation and re-use

15.19.1 With the increasing pace of technological change, new purpose-built structures will become obsolete more quickly as the life of a building from procurement to obsolescence shortens and increasing attention to environmental concerns may cause it to be thought extravagant to so closely tailor the form of a building to first-user needs that the option of quite simple adaptation to alternative uses is denied. It may be significant that, for several years, the Boeing Aircraft Company in the USA has erected office accommodation which can be readily converted into aircraft-manufacturing space, and in a world of shrinking natural resources, a similar consideration of potential for adaptation to other uses may have to be an ingredient of the thinking of designers of most new buildings in the future. Increasing pressure for the conservation and careful use of resources makes it logical to view existing buildings as a resource – expensive items in terms of the energy consumed in their creation and operation which must therefore be conserved and adapted to new uses, perhaps on several occasions during their lifetimes. In this way, we may witness a return to a philosophy which promotes most sparing use of materials, optimum use of available skills and in-built adaptability – a long-life, loose-fit, low-energy approach of the type that characterized building design in the pre-industrial age.

16
Euro legislation
(Europe – The Legal Dimension)
Richard Dyton

16.1 Introduction

16.1.1 The law affecting those concerned in the maintenance and preservation of buildings has changed significantly in the past few years. This has resulted from a number of decisions of the House of Lords (culminating in that of *Murphy v. Brentwood* (1990)) which have restricted the rights of third parties to sue contractors and consultants in tort.

16.1.2 This major upheaval in the law is set to continue, not by any decision of the House of Lords but by the creeping effect of the Directives and Regulations issued from the European Commission. Indeed, UK law has already and, almost imperceptibly, been changing by these provisions introduced to the law, often without Parliamentary debate by the 'back-door' route of the Standing Order.

16.1.3 This chapter describes the major areas of European legislation which will concern practitioners. Such legislation will, of course, affect all, including those involved in the UK only. However, what follows describes the rules which are (and will be) 'common' throughout the European Community comprising the 12 (soon 16) member states, and thus they are described from the perspective of those who have or will become involved with buildings and projects situated in Continental Europe itself.

16.1.4 As an indication of the opportunities which are present in Europe, Table 16.1 shows the construction output of the major EC countries, with particular reference to renovation and modernization. The approximate statistics are from 1989 and do not include figures for the former Eastern bloc countries, nor do they take into account the economies of Austria, Sweden, Finland and Norway which, at the time of writing,

are about to be formally admitted to the enlarged European Union. They also do not take into account the European Structural Funds which together constitute a massive injection of aid, mainly to the less-developed regions (Greece, Portugal, Eire, Northern Ireland, Spain, Italy, Corsica) to develop, *inter alia*, a national infrastructure. These funds should lead to large-scale construction programmes and further opportunities for UK practitioners. Notwithstanding these omissions, the potential is very large compared to that of the UK alone.

16.2 Public procurement

16.2.1 The principle upon which the EC Public Procurement legislation was conceived was that of 'free movement of persons, free movement of capital, freedom to provide services and the free movement of goods'. It is intended that these freedoms should enable contractors and providers of services to compete without restriction in any member state. The most recent legislation seeks to avoid the practices of the past, whereby central and local government authorities persistently preferred local contractors and consultants. This occurred in a number of ways: some deliberate and some unintentional. For example, an Irish authority was found guilty by the European Court of discriminating in favour of its own nationals because the specification in the tender documents required the products to conform to an Irish standard rather than any other national or international standard which offered equivalent guarantees of safety and suitability.

16.2.2 The law in this area was first adopted by the European Community in 1971 and it has been amended recently to strengthen the freedoms mentioned above.

Table 16.1 Construction output

Country	Renovations and modernization (£ million)	% of total	Total (£ million)
UK	17 004	42.84	36 696
West Germany[a]	34 000	40	85 800
France	18 911	46.72	40 479
Spain[b]	3 400	20	17 000
Italy	20 349	39.8	51 124

Source: *Euroconstruct*.
[a] Figures unavailable for East Germany.
[b] 1988 figure.

16.2.3 Scope of the Directives

The rules on public procurement relate to construction and civil engineering contracts above a threshhold figure of public works exceeding 5 million ECUs (approximately £3.3 million). The Directives also apply to public supply contracts in relation to such as lease, rental and hire purchase.

16.2.4 The legislation affects contracts awarded by 'contracting authorities' which covers local authorities, regional authorities and central government. This has recently been extended further to cover contracts awarded by the 'Utilities' in the energy, water, telecommunications and transport sectors. This even applies where the utility has been privatized as in the UK in the cases of BT and British Gas. The definition is also wide enough to cover private companies procuring with, and on behalf, of the public purse. There are detailed rules in relation to the tender procedures to which the contracting authorities have to conform. This means that all large European construction contracts (except those for private industrial and commercial clients not purchasing on behalf of the public purse) will be open to all EC contractors on the basis of competitive bidding. Most recently Consultants' commissions (i.e. those for engineering, architectural or surveying services) have been added to the list of contracts which must be open to all throughout Europe. For consultants the value of the work must be 200,000 ECU (or £141,431) before the public procurement regime applies.

16.2.5 Advertisement and selection procedure

In the large majority of cases a contracting authority will be obliged to publicize the essential characteristics of the proposed work throughout the Community. This will be done by publication in the Tender Supplement to the *Official Journal of the European Communities* and in the TED (Tenders Electronic Daily) Data Base. The Tender Supplement and the Data Base contains daily updates to all tenders published in the *Official Journal* and is accessible to the public. Practitioners will note that if they wish to obtain more information, they should contact the Office for Official Publications of the EC Sales Department, L-2985 Luxembourg. Subscription to the Tender Supplement can be made by application to HMSO Publications Centre, 51 Nine Elms Lane, London SW8 5DR.

16.2.6 Time limits apply within which the contracting authority must send out contract documents and within which tenderers must reply.

16.2.7 In addition, contracting authorities must refer to European standards rather than national ones unless, as at present, there is an absence of European standards, in which case reference to national technical specifications or equivalents is permissible.

16.2.8 There are two stages in the selection process. First, the contracting authority must consider whether the tenderer is acceptable (possible reasons for unacceptability are such things as bankruptcy or serious misrepresentation in the tender information). Tenderers must give evidence of their economic and financial standing and their technical knowledge or ability. The second stage is that the contract must be awarded on the basis either of the lowest tender or the tenderer making the most 'economically advantageous tender'. The criteria concerned are 'price, time allowed for completion, running costs, profitability and technical merit'.

16.2.9 In some cases, however, the procedures will be 'restricted' or 'negotiated'. The restricted procedures will apply where, for example, costs of complying with totally 'open' procedures will be disproportionate to the total contract value. In such cases the contracting authority can restrict participants who are invited to tender. In rare cases, 'negotiated' procedures will be allowed where the contracting authority can choose its supplier after direct negotiations with a limited number of potential suppliers.

16.2.10 Compliance

There are now in force provisions to allow aggrieved contractors or consultants to enforce their new rights. Previously, tenderers had little opportunity to bring their grievances to court, since the past procedure was that the European Court brought an individual member state to account. However, the new Directive allows disputes to be dealt with at national level in accordance with the review procedures for aggrieved tenderers. Tenderers who feel they have been unfairly discriminated against can bring the full force of the local law to bear on the contracting authority. The remedies available include the right to apply for suspension of the procedures leading to the award of the contract, setting the decision of the authority aside and the right to damages.

16.2.11 Summary

The new rules of public procurement greatly increase the availability of information regarding contract opportunities throughout the community. Contractors, consultants, suppliers and services providers should take advantage of the new compliance legislation in order to enforce their rights if they wish to compete across borders for lucrative contracts and commissions.

16.3 Construction products and European construction standards

16.3.1 The problems associated with public procurement have also been found in the construction products market. The EC discovered that it was fragmented because it consisted of trade barriers in the form of national technical specifications and requirements. Thus in order to achieve acceptance Europe-wide, products previously had potentially to face 12 different national requirements. The new Construction Products Directive seeks to eliminate these technical barriers to cross-border trade by allowing products to conform with '*essential requirements*', and thus to allow them to move between member states.

16.3.2 New approach

Instead of seeking to bring together the important factors of each country and then to draw up the technical details into a single set of European specification, the approach of the Directive is to give the Commission the task of establishing which factors constitute the 'essential requirements' of all the different health and safety regulations drawn up to protect the market. Only after these have been decided will the technical details of those requirements be defined by the CEN (the European standard-making body), who

will effectively delegate this task to the national representatives on the CEN technical committees.

16.3.3 The new approach to construction products has a number of interesting elements:

- The 'essential requirements' to which the product must conform if it is to be granted free movement within the EC are as follows:

  Mechanical resistance and stability
  Safety in case of fire
  Hygiene, health and the environment
  Safety in use
  Protection against noise
  Energy, economy and heat retention

- The 'essential requirements' relate to construction *works* and not to the construction products as such. This means that, as the Directive points out:

  'The essential requirements shall be given concrete form in the Interpretive Documents' which will create 'the necessary links between the essential requirements and the standardisation mandates'.

These Interpretive Documents are currently being drafted by a standing committee and this will lead to European standards implementing the essential requirements in any given product group in more detail. These will then be translated into national standards by the relevant authorities in the EC member states (e.g. the BSI in the UK). There is concern among some consultants that European standards might be unrealistic for certain products because of the differences, for example, in climate between member states. The response to this is that the European standards need not represent fixed specifications, but will be sufficiently flexible to allow for geographical and climatic differences. The final drafts of the Interpretive Documents have now been published and they are breathtakingly wide in the detail specified. Each, however, is specified in performance rather than descriptive terms, thus enabling the necessary flexibility to apply across the entire continent.

16.3.4 Practitioners working abroad and, indeed, within the UK, may find that they are required to conform to European Standards, Structural Eurocodes, or the national equivalent. This means that practitioners must be aware of which products conform to the European Standards. There are a number of ways of achieving conformity with the 'essential requirements' for a particular product:

- The most popular way a product can conform is likely to be self-attestation by the producer. The manufacturer may demonstrate his or her product's compliance by applying the symbol

'CE' to the product or its packaging to indicate the fitness of a product for its intended use. A manufactured product bearing the CE mark will be allowed free movement and use between member states.

- The second way in which it may be demonstrated that a product conforms is by the use of quality management systems by the manufacturer.

- The third method is by third-party certification. This must be done by a body holding an EN 45000 accreditation (the European standard dealing with competence of independent testing authorities). A working group of the CEN is currently looking at the requirements for testing and certifications. The testing body may award either a declaration of conformity or a certificate of conformity. This will note the requirements to which the product conforms and any conditions relevant to the use of the product.

16.3.5 Structural Eurocodes which are design codes (related to requirements for the resistance, serviceability, and durability of structures) have been published by CEN. They give a general basis for the design of a wide range of building and civil engineering works. A number of public bodies in Europe have now adopted the Eurocodes and insist that they must be used for design submissions.

16.3.6 Summary

The scope of this EC legislation is important not only to manufacturers of construction products but also to those who specify and select such products. There is still some confusion as to whether the Directive will also apply to the complete process of installation as well as to the construction products themselves. The UK government, for example, has implemented the Directive in relation to products only. The EC may extend this, however, to cover the process of installation as well. Much depends, of course, on the preparation of the Interpretative Documents and the eventual standardization for an estimated 3000 products. Until then, however, specification and selection must not discriminate on the basis of national standards only – equivalent national or international standards must be specified. Inspectors enforcing local building regulations will not be able to object to products conforming to another member state's national standards if they are equivalent to a European standard or the appropriate standard of that locality.

16.4 Health and safety legislation

16.4.1 The background

New laws affecting health and safety will be implemented throughout Europe during 1994 and onwards. This legislation is extremely important because it signifies a stronger, tougher, approach by the European Commission to health and safety both in the context of those working on site and those occupying buildings after works have been completed.

The two most relevant directives are the Construction Sites Directive and the Workplace Directive. There are also, however, a host of specific measures which are aimed at particular hazards. These include the Carcinogens Directive, the Biological Agents Directive, Exposure to Asbestos Directive, the Directive on Exposure to Harmful Substances, and the Noise Framework Directive.

The original intention behind the European health and safety drive was to eliminate the existence of barriers to trade by harmonizing legislative measures on health and safety at work in the various member states. Now, however, the real agenda appears to be, quite simply, the reduction in the occurrence of accidents and ill-health amongst workers.

16.4.2 Construction Sites Directive

The formal health and safety roles of Planning Supervisor and Principal Contractor will be allocated for all construction activities throughout Europe, including renovations, repair and demolition. It seems likely that the employer will appoint his lead designer as the Planning Supervisor and his main contractor as the Principal Contractor.

16.4.2.1 The Planning Supervisor must be appointed at Feasibility Stage and has a host of duties including the following:

- Preparation of a Health and Safety Plan setting out the overall arrangements for the safe operation of the project.

- Notification of the works to the local equivalent of the Health and Safety Executive.

- Ensuring that all designers, including specialist sub-consultants have considered health and safety issues in their design. For example, designing to eliminate health and safety problems for those who maintain or clean the structure or, in years to come, in relation to the re-pointing of brickwork or the replacement of roof linings.

- The assessment of tenders to determine the adequacy of sums allocated to health and safety measures and the adequacy of time in the content of the project programme.

- The consideration of the impact of variations on the Health and Safety Plan during the contract and the compilation at the end of the project of the Health and Safety File similar to, but in much greater detail than, the Maintenance Manual focusing particularly on health and safety aspects of the materials used together with design implications that will affect those who will be maintaining and cleaning the structure.

16.4.2.2 The Principal Contractor maintains overall responsibility for ensuring safe working methods and practices on site. He also has additional formal duties, such as:

- Dealing with health and safety matters specifically in the tender, i.e. allowing sufficient time in the contract period and sums in the price to comply with health and safety requirements. In addition, the contractors health and safety record may be taken into account when the contract is awarded.

- Developing the Health and Safety Plan in much greater detail with the Planning Supervisor.

- Consideration with the Planning Supervisor of the impact of variations during the contract on the Health and Safety Plan.

16.4.2.3 Failure to comply with the measures will mean serious consequences including criminal sanctions. The Health and Safety Executive in the UK, for example, has threatened to enforce the new provisions strictly and whilst the level of enforcement throughout the different member states is bound to be variable, the threat of criminal and civil action for breach of the rules will be a major consideration for both contractors and consultants working abroad.

16.4.3 The Workplace Directive and other specific measures

The Workplace Directive lays down a list of specific conditions related to matters such as structural stability, fire precautions, ventilation, windows, room dimensions, hygiene facilities and first-aid installation. The effect on both contractors and consultants is clear: the building of all new workplaces will have to comply with the Directive however it is implemented in the particular member state concerned. If a building is not built to comply with the relevant Health and

Safety Regulations then the occupier of a building will seek compensation damages or an indemnity from the designer and/or builder in respect of any sums ordered to be paid to workers or as a result of alterations ordered to rectify the inadequate provision.

Contractors, in particular, must take care to avoid using any construction products which would render them in breach of the specific directives, such as those prohibiting the use of asbestos or other harmful substances. The contractor also has a specific duty to avoid certain practices on site which are considered would be harmful to his workers. Common noise standards, for example, have been established for contract plant and equipment, such as compressors, power generators, concrete breakers, excavators, bulldozers and loaders. This may add additional costs to the tender sum.

16.4.4 Summary

Since the UK's Health and Safety Executive is well ahead of other European states in the implementation of the European health and safety legislation, UK consultants and contractors should have a greater degree of awareness of the requirements and experience in ensuring compliance with the regulations. Consequently this may mean that UK tenders have an adavantage, particularly in relation to prestigious public projects where compliance with health and safety legislation will be most rigorously demanded both by the employer and by the local Health and Safety Executive.

16.5 Mutual recognition of qualifications

16.5.1 Article 48 of the Treaty of Rome secures the freedom of movement of workers within the Community and Article 7 prohibits discrimination on grounds of nationality within the EC.

16.2.2 The initial approach of legislation from the EC on this area was to treat each profession separately, harmonizing, as far as possible, education and training requirements. The approach of the EC is now to adopt recognition of professional qualifications on a more *general* level and this affects those concerned in the maintenance and preservation of buildings in a number of different respects.

16.5.3 Architects Directive

In June 1985 a specific Directive recognized architectural qualifications and experience throughout the EC. Member states were slow to introduce this into their national law (and in Italy and Greece there are still some severe practical impediments). In addition, problems were caused by different regimes in each member state dealing with the registration of architects. For example, ARCUK in the UK strictly controls registration, but in other member states there may only be a voluntary system of registration (Denmark) or even no system whatsoever (Ireland).

16.5.4 The Directive requires member states to recognize any architectural qualifications of a university degree standard and it is implicit in these conditions that the holders of relevant qualifications understand all matters of public concern in architecture and are able to make the appropriate response.

16.5.5 An architect may, therefore, need to demonstrate the following if he or she is to work in another member state within the EC:

- Proof of nationality: normally a passport will suffice.

- Proof of qualification: a diploma granted by one of the competent authorities.

- Evidence of practical training experience: either from ARCUK or the competent body from the country in which the experience was gained. UK architects are urged, in all cases, to consult ARCUK to discuss the exact type of work contemplated abroad, and therefore the type of certificate which ARCUK needs to issue.

16.5.6 General Directive on professional qualifications

This Directive will extend the principles comprised within the Architects Directive to other professional qualifications, such as engineers, quantity surveyors and members of the Chartered Institute of Building. This Directive deals with the general system for recognizing professional education and training in regulated professions for which university courses or an equivalent lasting at least 3 years is required.

16.5.7 Certificates of Experience Directive

Certain member states require evidence of formal training from their nationals in order to allow them to carry out self-employed trades within their territories. Others permit their nationals to set up in business having acquired their skills through apprenticeships and employment experience. The aim of the Certificates of Experience Directive is to allow self-employed tradespeople to work throughout the European Community. There are currently no guidelines available for each state or each trade, but the relevant authority in the UK for those wishing to work abroad is the DTI. The DTI will be required to issue Certificates of Experience and has done so already for plumbers, bricklayers, roofers, heating and ventilating contractors, thatchers and house builders.

16.5.8 Since requirements of each member state do vary considerably, it will be necessary to approach the DTI on a case-by-case basis in order for the relevant Certificate of Experience to be granted.

16.5.9 The European Centre for the Development of Vocational Training is attempting to enable qualifications in member states to be readily compared and the EC Commission has prepared a draft Directive for the mutual recognition of national vocational qualifications. Until this is adopted, however, the best approach is to ensure that Certificates of Experience are obtained.

16.6 The general position in relation to planning and building regulations in Europe

16.6.1 By comparison to the position in the UK, planning control and building regulations in other countries within the EC are based firmly on *regional* planning with a rigid zoning system and a strict plan which is enforced by law. The UK planning system, by contrast, is based upon plans which are guides to the decision-making process rather than blueprints. Thus, the basic difference is that whereas there is far more flexibility in the UK, the position in other EC countries is much more rigid.

16.6.2 There are underlying obvious reasons for this: historical and cultural. Historically, for example, the only countries within the EC whose legal systems are *not* based on the Napoleonic code are the UK and Ireland. In the other EC countries legal decisions are based on statutes and written constitutions rather than 'judge-made' law as in the UK and Ireland.

In addition, in the southern EC countries there is more emphasis on the right of the individual to do what he or she pleases and thus less control. There are also trends such as urbanization and migration, which are particularly associated with the southern European countries, and this has meant that cities are still growing. Particular trends in other countries are noticeable, such as the encroachment of the sea in the Netherlands and in the UK the emphasis to clear slum dwellings.

16.6.3 While this difference in relation to the flexibility of planning is a major and identifiable one, it is becoming less so now that the UK is moving towards stricter simplified planning zones and the other EC countries are allowing their strict plans more adaptability.

16.6.4 The other major general difference is in relation to building control. Whereas in the UK there is a clear difference between planning control and building control, this is not so in Europe. Generally, the end result of a single process is permission to build. In other words, one building permit covers the whole job in Europe, both planning and building regulations. The decision to grant the building permit in Europe is taken from a different perspective than the equivalent in the UK. UK planners make decisions on a case-by-case basis in an attempt to implement a policy. In other EC countries, the planning officers take a much more detailed approach and are more concerned with techniques used by those planning construction than with policy considerations.

16.6.5 A brief outline of the planning process in France, Germany, Italy and Spain

1. *France*

- At the more general level in France there is the Schéma Directeur d'Aménagement et D'Urbanisme (SDAU). This equates broadly to the British structure plan. At the detailed level there is the Plan d'Occupation des Sols (POS), which is similar to but more detailed and on a smaller scale than the British local plan.

- The control of development is comprehensive. Permission is necessary for virtually all types of development, even wooden chalets and garden sheds. The law is applied by way of the 'Permis de Construire' (Building Permit). There is also a lesser form of control by means of a Déclaration Préalable (Preliminary Declaration). In the latter case, an application must still be made but work may proceed after one month if a declaration has not been proposed. Work under this scheme includes the cleaning and maintenance of facades, which under French law is required every 10 years. Also included under this type of application is reconstruction or works on Listed Buildings under the Historic Monuments legislation.

- The decentralization of power in the French system means it is essential to consult the Maire

(Mayor) at a very early stage. Normally this leads to a clear understanding of what is and what is not permissible, since there is little scope for discretion once the POS is agreed. After this, if the application complies with all the local rules and those of the POS, there are very few reasons why permission should be refused. Thus there is more certainty in obtaining planning consent than in the UK. Statistics indicate that 95 per cent of all Permis de Construire applications are granted. There is a time limit for the process but, in fact, most Permis de Construire decisions are reached within 3 months. Construction must commence within two years of the Permis and the whole process must be completed by a Certificate of Conformity.

2. *Germany*

- With the exception of Scandinavian countries, Germany has the most formalized system for ensuring compliance with legal and technical requirements of trade and industry.

- Once again, the Building Permit is the single formal permission required before works can be carried out. The 'Länder' (state governments) must take into account detailed requirements, such as provision for public access (including ambulance and fire services), lighting and ventilation and technical requirements. The local regulations may also contain aesthetic standards.

- The application is normally made by a professionally qualified architect who is registered as an architect with the 'Länd'. There is no specified time within which a decision must be made, but if there is no decision within 'a reasonable time' the applicant can petition the administrative court. Three months is normally considered a reasonable time. The rejection rate is very low, less than 2 per cent, and about 79 per cent of applications are granted with conditions (20 per cent granted unconditionally). The low rejection rate is a consequence of a consultation that takes place before an application is made. The Building Permit is generally valid for 2 years, but it may be extended one year at a time.

3. *Italy*

- There are two types of building approval in Italy. The 'auto rizzazione edilizia' applies to all internal alterations to existing buildings which do not increase the floor area, such as the installation of central-heating plant and lifts, changes to the facades and roofs, external works in connection with an existing building, plantrooms, external storage of materials, demolition, landfills and excavations. This type of approval benefits from the fact that works may proceed 90 days after notification if the authority has not pronounced on the application. The other type of application is the 'concessione edilizia', and this applies to all other types of development. In this case, if a decision has not been reached 60 days from the application it is deemed to have been refused. The approved works must be started within 1 year of the date of approval and must be completed within 3 years. If the work is not completed at this time a new approval must be sought.

- In relation to historic buildings, the regulatory body operates at regional level and its approval must be requested before carrying out any work on a listed building. The architectural and archaeological heritage of Italy is a major factor affecting construction in its towns and cities. If an item of artistic importance is found during construction, this can prevent the development continuing. An example of this is the construction of the Metro line in Rome where completion suffered major delays due to architectural finds during construction.

4. Spain

- In Spain each local authority grants building permits, but approval in relation to technical regulations or standards is the sole responsibility of the architect who also has initial control of the structure plan and town plan via the local College of Architects. The architect is, therefore, extremely powerful in Spain in determining what is built.

- The local authority's building permit may be refused if the project does not accord with planning criteria (which can include aesthetic considerations). Some undeveloped areas will not have a plan, so the planning criteria are not necessarily known in advance and major redevelopments may require changes to them. The local authority will also verify whether specific local regulations (e.g. in relation to fire and health) are satisfied. In large towns specialized planning committees grant the building permits and in small areas it is done by the Mayor, and this may lead to delays in some cases.

16.6.6 Summary

It can be seen that there is no unified approach to planning and building control in the EC at present. It is possible that this may change in the near future, particularly since the signing of the Maastricht Treaty. This may have a particular impact on historic buildings, since included in the new moves to harmonization are areas which relate to 'environmental' and 'cultural' matters. This may possibly include harmonization of the rules relating to listed and historic buildings, but there is no draft directive yet in this area at the time of writing. UK law in this area is well advanced from that of the laws of other member states in any event.

There is, however, draft legislation which may affect this general area in the form of a draft directive on environmental assessment. This is divided into two directorates:

- DG2 Environmental Directorate. This seeks to deal with the urban environment, the future of cities and the assessment of the environment in relation to pollution, etc. This contains many good suggestions but may be of little relevance without the funds to enforce it.

- DG16 Regional Development Fund Directorate is perhaps more important. This may be more powerful, since it has access to the structural funds of the EC. These funds may be allocated to member states to improve infrastructure, and may allow national planning to look to the Commission for aid and assistance. Portugal, for instance, has particularly benefited from these funds.

In the medium term it is clear that there will be a unified planning system for Europe, although the area within its ambit is more likely to be concerned with transport, ecology, tourism, pollution and employment, rather than land use as such.

16.7 The likely future for the construction industry in Europe

16.7.1 Demise of the Construction Specific Directive and the Services Directive

The European Commission has recently put forward draft proposals for a Construction Specific Directive and a Services Directive. The underlying intentions were to harmonize the liabilities and insurance obligations of all those concerned in the construction industries of Europe, and to provide a uniform liability for all those providing services in Europe, including architects, engineers, surveyors and other building professionals.

However, the difficulties experienced by those attempting, bravely, to produce a uniform pattern of liability throughout Europe whilst confronting different insurance and legal systems has meant that the progress for both these major pieces of legislation has been halted. It now seems unlikely that either will be implemented thereby dealing a considerable blow to the concept of the 'level playing field' for contractors and consultants throughout Europe. The principles of freedom of movement and freedom of establishment may seem rather a hollow prize if, despite the public procurement regime and the mutual recognition of qualifications, the UK contractor or consultant is presented with insurmountable practical 'hurdles' to any effective work in those countries. These 'hurdles' could comprise problems of language, the availability of professional indemnity insurance, knowledge of that member's codes of conduct, working knowledge of the procedures of planning authorities, familiarity with local taxation, and an awareness of the relevant legal liability. Thus, these continued practical difficulties (which persist despite the formal rules) may restrict effective movement and the differing implementation, enforcement, and, indeed, respect for the rules which exists from member state to member state may be a significant factor in undermining the former principles.

16.7.2 The Commission's response

The European Commission is, however, far from unaware of these problems. It has sought to encourage cross-border cooperation between private firms and companies to seek to overcome the inherent difficulties. A specific vehicle for this is the European Economic Interest Group (EEIG), which is a type of quasi-partnership established by European regulation. The EEIG is viewed as a separate legal entity distinct from its members. Its intent is to encourage businesses to cooperate across the national frontiers of the European Community by enabling EEIGs to provide services to the cooperating businesses. In addition, the ECs merger control policy specifically allows mergers and associations between smaller entities. This permits the formation of single-project vehicles or one-off joint venture agreements to undertake joint work.

This has resulted in a number of UK practitioners combining with local firms in other member states in order to overcome the types of cultural and other difficulties referred to above. One EEIG, in particular, includes architectural practices from most EC member states. Its aim is primarily to focus on European Commission-based work and its elevated management team looks for suitable projects that match the make-up of the group. Other examples include the partnership formed between a UK architect and a French design and build contractor which won the competition for the rehabilitation of the south-east area of Tarbes, the largest French town in the Midi-Pyrénées region of Toulouse.

16.7.3 Summary

The 'retreat' from Maastricht, the current failure of the ERM and talk of subsidiarity rather than federalism provide indications as to why agreement on the grand scale envisaged by the Construction Specific Directive and Services Directive has not been forthcoming and these provisions are now unlikely to reach fruition. It also provides the indication as to why only the most uncontroversial objectives such as the health and safety measures which do not involve excessive cost have been passed into legislation.

16.7.4 The standardization drive which is now proceeding at full pace after the publication of the final drafts of the Interpretive Documents demonstrates that while the emphasis of European legislation has changed, its vigour has not reduced. Likewise, the impact of the health and safety legislation (judging by the current philosophy of the Health and Safety Executive) is a significant factor to be taken into account by all practitioners working in Europe.

16.7.5 The scope for work within Europe has been given a substantial fillip by the enlargement of the European Community to the European Union of sixteen rather than twelve States. The inclusion of Austria, Sweden, Finland and Norway represents a large potential increase for tendering in Europe by UK contractors and consultants. These countries are comparatively wealthy per capita and generally have large public works projects. The European Union is expected to be enlarged further to around twenty States from 1996.

16.7.6 The opportunities available to UK practitioners are substantial, particularly for medium or larger sized firms. This is because these firms should be able firstly, to analyse effectively the advertisements for work opportunities abroad and, secondly, to establish the necessary local link to enable compliance with local legislation. Those UK practitioners who most readily adapt to the changing legislation of Europe will eventually reap the rewards in terms of a much greater client base and work experience. Such adaptation may be by simply ensuring an up-to-date knowledge of Europe-wide tenders, together with the relevant European standards and health and safety requirements, or it may be merely a greater willingness to take advantage of the more flexible regime for associations and joint ventures across frontiers. Whichever step is taken it must be a positive one otherwise the competition will take the work not only in other European States but in the UK as well.

Acknowledgements

The author is grateful for help with this chapter from Mr Jens Knocke (National Swedish Institute for Building Research), Professor Donald Bishop (Emeritus Professor), Mr Neil Pepperell (RIBA Indemnity Research Limited), Mr Dargan Bullivant, Mr Richard Grover (Estates Management Department of Oxford Brookes University), Ms Stefanie Fischer (Burrell Foley Fischer), Miriam West (BSI Technical Help to Exporters) and Mr Michael Baird (ARCUK Admissions Department).

Further reading

Architectural Practice in Europe, Royal Institute of British Architects

CIRIA Special Publications – *Spain, Portugal, Italy, Germany and France*

The Green Paper on European Standardization, BSI

Knight, C., *European Construction Documents*

Knocke, J., *Post-Construction Liability and Insurance*

Speaight and Stone, *Architect's Legal Handbook*

Uff, J. and Lavers, A., *Legal Obligations in Construction*

17
Statutory inspections and spare parts
E. Geoffrey Lovejoy

17.1 Introduction

17.1.1 Once approval has been obtained under town planning and building control legislation and the building completed and handed over to the owners, lessees or tenants it is incumbent upon the building occupier to ensure that everyone who uses the building is kept safe. This duty may be shared by a number of tenants, be the responsibility of a person or corporate body or be administered by a managing agent. However, it may be arranged in detail on a day-to-day basis the building occupier still has the legal duty of care.

17.1.2 Quite apart from the matters discussed in Chapter 12 regarding responsibility for cleaning and access it will be necessary for certain statutory duties to be carried out to ensure that the building or process can continue to operate lawfully.

17.2 Hoists and lifts

Hoists and lifts need to be thoroughly examined by a competent person at least once every 6 months and the report of the results of the examination entered in the general register. Passenger lifts need to have two ropes, fitted with a gate which needs to be closed for the lift to be operated. Over-running must be prevented by efficient automatic devices.

17.2.2 Teagle openings shall be securely fenced and properly maintained. Secure handholds shall be provided on either side and the fencing to the door or opening kept in position except when hoisting or lowering.

17.2.3 Suspended access equipment should be similarly examined as for hoists and lifts.

17.2.4 Cranes, etc. shall be of good construction, sound material, adequate strength, free from patent defect and properly maintained. They shall be thoroughly examined by a competent person at least once in every period of 14 months. The outcome of the examination must be recorded. All lifting equipment must be plainly marked with its safe working load.

17.3 Confined spaces

Confined spaces are addressed in the provisions of Section 30 of the Factories Act 1961. Access should be certified by a responsible person as safe for entry. Provisions of adequate supplies of air, breathing apparatus and ropes are required and effective steps must be taken to prevent any ingress of dangerous fumes.

17.4 Steam boilers and receivers

Steam boilers need to be thoroughly examined and the examination report recorded at intervals of 26 months or 14 months for specified classes of boilers. Similar provisions are specified for steam receivers and containers and air receivers, normally at 26 monthly intervals.

17.5 Window cleaning

17.5.1 The provision of belts and harnesses for the use of window cleaners will normally be the responsibility of the cleaning contractor. The inspection and examination of such articles will be the responsibility of the building occupier only if they are specific to the building. In that case the examination should be carried out as recommended every 3 months.

17.5.2 Where provided, access eyes or hooks should be installed in accordance with BS 5845 (Specification for permanent anchors for industrial safety belts and harnesses) and examination carried out as recommended in that standard.

17.6 Fire warnings

Every means of giving warning in case of fire shall be tested or examined at least once in every period of 3 months or when required by an inspector.

17.7 Spare parts

17.7.1 A senior maintenance surveyor was asked how he set about seeking replacement components to curtain walls more than 10 years old. He replied that experience had shown that enquiries of manufacturers rarely proved successful and the safe assumption was that such components were unobtainable.

17.7.2 It is, indeed, a moot point whether the provision of spare parts for replacement purposes is, or ought to be, a significant consideration at the time of the design and construction of a building, or is better left to chance, which may enable shrewd or lucky purchase at the time of need, or may necessitate piecemeal or expedient work. In practice, scarcely a mention of the issue is to be found in the guidance provided by the standard references. Thus, the *Job Manual* of the RIBA[1] refers to warranties from nominated sub-contractors and suppliers, but makes no specific reference to replacement parts. The *Registered House-Builders Handbook*[2], the *National Building Specification*[3] and *Specification*[4] make no allusion to spare parts as such – though the last makes mention of second-hand bricks in a context which identifies these as retrieved from demolished work.

17.7.3 To his credit, Lee in *Building Maintenance Management*[5] has a section on stock control, but this relates to a concept of holding some quantity of standard items in an organized maintenance programme, rather than individual items. A DES performance specification includes the naive statement that 'items subject to regular wear must be easily replaceable and some guarantee must be given on the availability of spare items for the 60 year life.'[6]

17.7.4 By contrast, the *Terotechnology Handbook* (DOI, 1978) *does* include as a consideration in preparing a specification the question 'Will spares and/or service be available for the intended life of the unit, and at what estimated cost?'

17.7.5 The advent of metrication of the building industry in 1970 gave weight to aspects of the problem in that obsolescence of all Imperial components was an inevitable and early

consequence. In its train, provision for adaption of many of these components was seen to be necessary.

17.7.6 There was commercial justification for the production of such adaptors: but the result was in conflict with opportunity for rationalization which the more passionate advocates of metrication foresaw. Indeed, the transitional stage threatened a situation where the supplier and merchant might stock all three – Imperial, metric and adaption components.

17.7.7 Some philosophical justification may be seen for *not* attempting to provide the replacement in parts in all cases. Thus, it may be argued that many old buildings gain in character and interest by representing a palimpsest of new work overlaying the older with a disregard for the niceties of precise matching of material or detail.

17.7.8 Arguments afford cold comfort for the maintenance manager faced either with the need for replacement *during* a prescribed limited life, or having to accept a need to prolong that life beyond its time on behalf of a client (whether owner or occupier) who is not persuaded of the validity of the philosophical argument. The motor car or gas cooker manufacturer may be expected to honour an undertaking to keep a stock of spares for (say) a decade but how often is the question even asked of a system building manufacturer? Even so, how meaningful would be the answer, given the rate of technological change, fluctuation of demand and, indeed, the propensity of firms in the building industry to go out of business?

17.7.9 There are other reasons that makes straight replacement impracticable in some cases. Regulations and standards may change. The shortcoming which gives rise to the need to replace may have occurred so often as to necessitate redesign. Many elements may, in any case be purpose-made. Records may not exist or may be misleading. The need for replacement is often seen as an opportunity to refine or adapt the detail – sometimes to overcome an initial shortcoming, sometimes to meet a new need resulting, perhaps, from a change of use, always aggravating the replacement difficulty. Yet again, there might be seen risk in putting the initial supplier in a monopoly position and commercial merit may be seen in seeking the nearest available compatible component or material. Practical problems may well arise from the order of disassembly of the original and insertion of the replacement differing in sequence from the first construction – for few designers attempt to foresee and fewer succeed in foreseeing such issues.

17.8 Classification of materials

We may attempt a classification of materials, components and fitments in ascending order of difficulty in finding and adequate replacement.

17.8.1 Decorative and protective coatings: asphaltic and bituminous roofings

These are inherently simple to replace. The main difficulty posed is in diagnosing when local patching or making good ceases to be economical and overall renewal is justified. With local repairs a problem of matching may arise, for there must be a probability that the new tone and surface will contrast with the weathered adjacent work. The abutment line must be evident unless good fortune or judgement enables relief or some other feature to conceal the junction.

17.8.2 Concrete, brick, stone, mortar, plaster, plain glass and sheet metals

In each case a repetition or reconstruction of the original feature is likely to be relatively easy, but a 'match' of texture or tone is less so. If the adjacent work is being cleaned at the same time the problem is lessened. For instance, if plaster is in any case to be decorated the problem scarcely arises.

17.8.3 Timber

A repetition of the original detail is likely to be possible, although perhaps expensive in labour. Particular problems include satisfactory functional joints of new to old work both for joinery and carpentry. In the former case, the special difficulty may be weathering, sightliness or fire resistance; in the latter case issues of joint strength and fixings. A match of appearance is extremely difficult due to the non-availability of most of the grades of timbers formerly used.

17.8.4 Metal and plastic components and fitments, patterned glass

Where these are standard, problems are unlikely although a precise match may be impossible and a compatible approximation may have to be accepted. But obsolescence does occur, as the above discussion on metrication shows.

17.8.5 Specially fabricated components and fitments are more likely to pose problems, for the economics of production are likely to operate against a competitive price for single or small run replacement units. Patterned and coloured glass is at risk of being unobtainable.

17.8.6 Compound fitments and equipment

Standard products are likely to pose few problems so long as the manufacturers or fabricator remains in business, the model remains in production or the components are themselves standard (see paragraph 17.8.4). The more complex is the piece of equipment or the more sophisticated the design, the more at risk is the replacement. The designer should give special consideration to the latter of these criteria, and would be well advised to insist on the use of British Standard parts where possible.

17.8.7 Purpose-made products give rise to the greatest difficulties. The manufacturer or fabricator cannot or may not be relied upon to store spare parts. If the user is to do so, problems arise in judging what parts to store, in what quantity, and where. Where functional inadequacy gives rise to the need for replacement, an early and difficult decision to replace all the units may have to be made. In this case, of course, any stored parts are wasted.

References

1. *Job Manual*, RIBA (revised 1977)
2. *Registered House-Builders Handbook*. National House Building Council (revised 1984),
3. *National Building Specification*, National Building Specification Ltd (1973)
4. *Specification*, Architectural Press (published annually)
5. Lee, R., *Building Maintenance Management*, Crosby, Lockwood Staples
6. Sliwa, J., Need for a new approach (Performance Specification 3), *Architect's Journal* (3 March, 1978)

18
Maintenance manuals and their use
Jacob Blacker

18.1 A brief history

In February 1966 the author of this section was commissioned by the Building Centre Trust to design and collate and edit a Building Owners' Maintenance Manual. The production of this much-discussed project was the result of a Building Centre Forum at which Edward D. Mills formulated the need for such a manual based on RIBA recommendations. He produced a mock-up manual which was to form the basis for the manual that was finally published. The author's own involvement in this project was the result of his response and sustained interest in this forum and his own need to compile a manual for an ILEA school, while in partnership with Erno Goldfinger.

Since the first edition there have been three updated versions. The Manual has been widely used for methodical information recording for many buildings. This has been a successful enterprise over the years and has helped to create an awareness of better planned maintenance of buildings.

During recent years there have been signs that with the coming of age of facilities management and extensive use of computer software, a new approach should be investigated for the collecting and editing of maintenance information.

A greater variation in building construction and the advent of hi-tech buildings plus the growth of corporate ownership demands a greater choice in the planning of a building's life. To this end a Guide Book is being prepared by Jacob Blacker based on a step-by-step basis showing not only how to choose the appropriate manual but what are the cost and management variables. The previous manual is not invalidated nor out of date and the comprehensive checklists reprinted here will be incorporated into the new Guide book.

The information outlined in the following chapter represents a reduced summary of the seven chapters of the proposed new Guide Book.

18.2 The process

18.2.1 The building full cycle

Preparation of any guide as part of the process of building procurement must involve an examination of the entire cycle of events from inception to completion and then onwards to the daily maintenance. The diagram included in the guide is based on the chronological process of events, indicating the involvement of the various organizations and individuals in the process at each stage of development.

There are five basic stages:

1. Design time
2. Construction time
3. Defects liability period
4. Maintenance contracts and management
5. Feedback

18.2.2 Design time

In the setting up of the professional team traditionally the architects are the first appointed consultants and they in turn appoint the other consultants (e.g. Structural Engineer, Mechanical Engineer, Quantity Surveyors, Landscape Architects, etc.). Feedback information from maintenance and facilities managers has beneficially involved the facilities managers in the early design team's decision-making process. This should result in design decisions that eliminate maintenance problems and unnecessary expenditure in the life cycles of the building.

Decisions made at this early stage of the project will affect the budget, the nature of the building type, the degree of flexibility in the planning of possible future users of the building and the attitude towards maintenance and maintenance budgets.

The architects coordinate the requirements of the various consultants and authorities and specialist suppliers to complete the design process.

These proposals, once authorized by the building owner and the authorities, are translated into the drawings and schedules until all the relevant information is ready for construction.

18.2.3 Construction time

Tenders are obtained and a contractor appointed or a contract management company who will coordinate a variety of specialist contractors. The construction will proceed and the various consultants and the architect will monitor and inspect the programme of the works. Most contracts have an agreed *start date* and *practical completion date* and a *final completion date*.

18.2.4 Defects liability period

This is often the critical time when the building owner or occupant takes possession of the building. It is often a period when the occupant is still under the trauma of moving and is anxious for the business of the company to proceed undisturbed. This is often when the greatest damage to the building's systems is detected. The reason for this is the lack of instruction in how to operate and maintain some of the vital services and controls.

18.2.5 Maintenance contracts and management

Although the names change, the process that the maintenance manager has to go through is exactly the same as that of the Architect, i.e.

Priority decision process
Specification and instruction
Tender
Contract and implementation.

18.2.6 Feedback

This is often the most difficult and least achieved of the processes of building procurement. There should be a simple process involving the maintenance management informing the building's architect of problems and successes of the building's maintenance needs. There is a tendency for the maintenance team to consider that their operations are post-construction and therefore feedback will only cost clients, money and complications. The creation of maintenance instructions and specifications before the building user's practical completion becomes the only way of obtaining an element of *feedback* continuity.

18.3 The new guide

Achievement of a successful full-cycle operation in the realm of building procurement involves the entire design team and specialist consultants irrespective of who leads the team. This will ensure that the proper procedures and decisions are taken to establish an appropriate form of Maintenance Manual and software program and budget.

The new Maintenance Manual guide will incorporate a detailed selection process to ensure that the decisions affecting maintenance budgets are made at an early stage in the design programme (see Figure 18.1).

18.3.1 Selecting a manual

The selection of an appropriate manual is a task that should be undertaken seriously as it will not only reflect on the maitenance budget and life of the building but will also influence the architect in the design decision-making process. Often the owner will not have the expertise to initiate the process and will rely on the professional team to advise and call in a facilities manager.

The selection of a manual can only be made when the design policy has been established and the intentions of the building owners have been clarified. The needs for manuals can be divided into three basic types of new or existing buildings:

Group or corporate
Individual
Parts of a building and leaseholds

18.3.2 Group or corporate building

Many companies such as hotels which are designed on a more predictable specification will have corporate policies on the type of services and materials to be used and this will enable the editor of the Manual to adapt and synchronize it with the company's own manual.

18.3.3 Individual buildings

These are probably the most common of all building types and require the greatest amount of information on the building and its installations. By its nature this will be the most expensive manual to produce. Particular care should therefore be taken to produce accurate 'as-built' drawings and information and to instruct all contractors to provide information for maintenance in the prescribed manner making it easier to collect and edit the data.

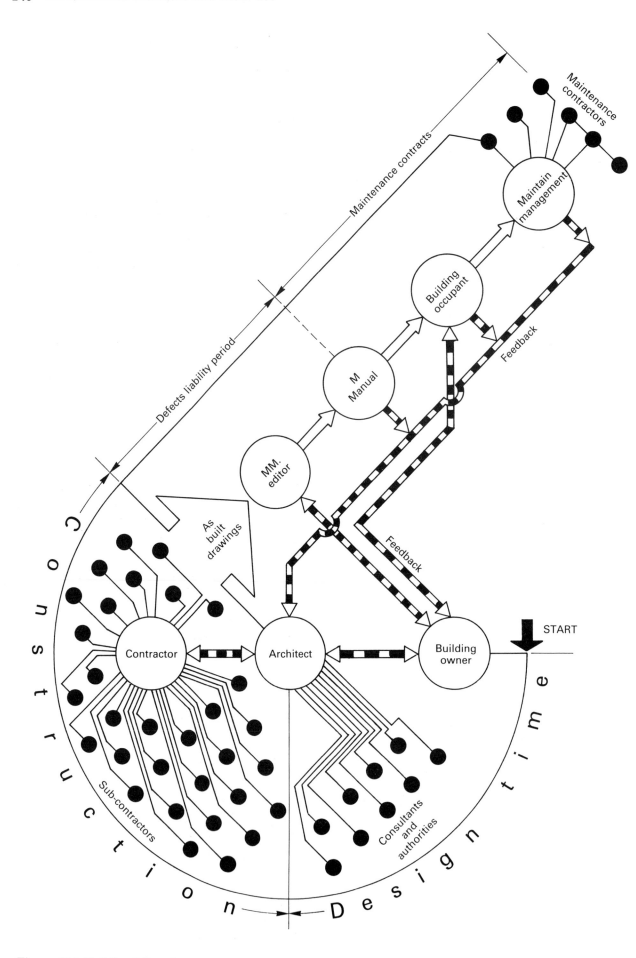

Figure 18.1 Building full cycle

18.3.4 Parts of a building and leaseholds

This manual concerns the limited information on maintenance that will relate to the tenant's responsibilities for maintenance of part of the building. Obviously, an 'Internal Decoration' lease agreement will only involve information about decorations and fittings and not about the structure. It would, however, be prudent to include a small manual on responsibility in each lease agreement for clarification.

18.4 Commission and select

After deciding on the manual type the next step is for a survey to be carried out on information and drawings that are available. Accurate information is crucial to economic planned maintenance. The preparation of building surveys can often be an expensive part of the Manual.

18.4.1 The remaining items in this section are:

1. *Energy Audits* For the economic function of the mechanical and electrical installation.
2. *Technical Audits* For the performance of the building's fabric related to repairs and costs.
3. *Maintenance Planning* Determines the type of organization needed together with the most suitable contracts – term contracts, unit rates, lump sums, etc.
4. *Facilities or Estate Managers* Is a full-time manager needed or will existing staff cope?
5. *Cost Control and Budgets* The need to establish planned maintenance budgets. Who controls the purse strings? Software spreadsheets and control.

18.4.2 Technical specifications

The maintenance instruction sheet is the controlling mechanism and, if properly written and considered, should be capable of being used as part of a tender specification. If a manual is not commissioned at the beginning of a project then each time a tender is asked for a specification will need to be prepared. It obviously pays to do the job properly, and professional help is essential.

18.5 Computer software

There are several products on the market and a list will be published with the new manual. Cost and compatibility are the two important factors so that computer programs can be easily adapted to

suit individual needs. Most programs include complete survey data collection, planning and data entry and plant inventory.

18.6 Data collection

The guide will show step by step how to prepare information for the manual at each stage of a project. The entire guide is based on the chronological order of events in the building process. It also stipulates who should perform each task, starting with the site purchase to design, approvals, construction, the defects liability period and through to maintenance management and user comfort.

There is no short cut to the collection of survey data. Collection and collation require that, when gathered, information has to be entered in a consistent format, suitable for the software program. The checklists form the basis for data collection and retrieval for specifications and instructions.

18.7 Checklist

This will consist of an abbreviated form of the checklists that were prepared for the *Building Centre's Maintenance Manual* by Jacob Blacker. These checklists cover the following:

1. Description of the building.
2. Description of the building's services.
3. General information.
4. The building fabric.
5. Mechanical and electrical installation.
6. Guides, regulations and laws.
7. Occupant's guide notes and cleaning guides.

18.8. Selection and storage

This will contain advice about information for use and for guidance and storage:

1. 'As-built', architects' and structural engineers' drawings.
2. Diagrams of the mechanical and electrical systems of the building and mechanical and electrical engineers' drawings.
3. Photographic records.
4. CAD program selection and storage.

18.9 Summary

The key to planned maintenance lies in the accuracy of the information provided. A systematic approach to all technical information is an essential ingredient for achieving this end. The checklists included in this chapter have been designed to communicate various levels of information to the various individuals that form the building team.

18.9.1 There is information for the company executive who may have little knowledge of buildings as well as the operative who has to carry out the work.

18.9.2 New methods of construction have enabled a whole series of high-tech buildings whose maintenance requirements are often specific to those particular buildings, and there is no doubt that the architects who designed them should be consulted prior to any maintenance work commencing.

18.9.3 Computerized information can be a relatively simple matter if the decision on a new project is taken early enough. The cost of data collection on-site is often an expensive process that cannot be avoided in the preparation of manuals for existing buildings.

The new manual will contain a detailed checklist on a stage-by-stage process for the compilation of each manual type. All the data sheets and checklists contained in the *Building Centre's Maintenance Manual* will be re-used in the new manual.

Extracts from *The Building Centre: Maintenance Manual and Job Diary* (1981) reproduced by permission of the Building Centre Trust.

		Preface
		Introduction
1. Information	1.1	Instructions for the compilation and completion of the manual
	1.2	Description of the building
	1.3	Emergency contacts
	1.4	Ownership and Contract consultants
	1.5	Authorities, consents & approvals
	1.6	Subcontractors & suppliers
	1.7	Maintenance contracts
2. General Building Fabric	2.1	Floor loadings and restrictions
	2.2	General maintenance instructions
	2.3	General maintenance log sheets
	2.4	Annual guide chart
	2.5	Inspection reports
	2.6	Manufacturers' instruction leaflets index
	2.7	Fittings and component schedule index
3. Building Services	3.1	Services capacity, location and restrictions
	3.2	Services instructions
	3.3	Services log sheets
	3.4	Manufacturers' instruction leaflets index
	3.5	Fittings and component schedule index
	3.6	Annual summary chart
	3.7	Commissioning records and test certificates list
4. Drawings	4.1	Consultants check list
	4.2	Drawing list
5. Cleaning and User Guide	5.1	User guide notes
	5.2	Cleaning instructions
	5.3	Cleaning guide chart

1. Information

.1   Instructions for the compilation and completion of the Manual

A step by step recommended process for the editor/architect in accordance with the stages of the RIBA plan of work and according to the sequence of the work. At the head of each instruction sheet, a reference is made to the book *"Building Maintenance and Preservation"*. This book should be used in conjunction with the various instructions.

1.1.1 Sketch design

Discuss type and size of manual with consultants.

Recommend manual type to client, obtain approval and agree fee for the work.

Purchase copies of the "Maintenance Manual" from The Building Centre.

Incorporate relevant sheets into architect's job book or diary.

Inform all consultants to include this service in their contract for services.

Obtain additional copies of the Maintenance Manual and issue to consultants.

Inform quantity surveyor to include for this service in the budgets and cost reports.

1.1.2 Working drawings

During preparation of working drawings inform all consultants and the quantity surveyor to include a clause in their specifications for the preparation of the manual. The clause should include such items as time limits for data sheets and information, penalties for non-completion, commissioning records, record drawings, test certificates, schedules of components, photocopies of legal documents, "as built" record drawings.

Record and photocopy all consents and approvals.

Instruct clerk of works/site agent to keep records of all site works for as-built records as the work proceeds.

Collect all names and addresses from job book or diary and keep record of changes.

Invite all subcontractors to confirm their willingness to tender for maintenance contracts.

1.1.3 Contract period

At the first site meeting instruct main contractor to provide a list of all subcontractors and suppliers names and addresses, collect records on site of all as-built works particularly below ground, commissioning records and test certificates. Make it clear that all information must be available at handover date and that a retention may be held against completion until this information has been provided.

Issue all consultants with copies of check lists from the manual with a note asking them to keep the records up to date as the work proceeds.

Regularly remind the contractor of his obligations to keep records and to chase all specialists, especially towards the end of their contracts.

After commissioning of the plant and equipment such as heating and ventilation, air conditioning, pumps, etc. the building owner should be instructed by the specialist to ensure that the new operatives are fully conversant with the method of operating the equipment, its controls, and expected teething troubles.

1.1.4 Post practical completion period

Remind subcontractors that the information must be presented in the prescribed manner.

Issue reminders for the collection of all as-built drawings and check that all check list items are included, with particular reference to access ducts and removable panels.

Edit all information from consultants and main contractors and sub-contractors. Add co-ordinated reference numbers to leaflets and instructions.

Send final draft to consultants for final checking.

Inform the building owner of his liabilities for maintenance, security of the building and commencement date of maintenance contracts.

Ensure that the building owner informs his insurance company of the increased liability on taking possession of the building.

Prepare and check the description of all design objectives and systems and prepare explanatory diagrams of each system.

Collect photocopies of all consents, commissioning records, test and fire certificates and other relevant documents.

Prepare set for microfilming, with clients consent, and deposit a copy in clients bank or safe.

Send complete sets to client according to requirements and order replacement copies of the Maintenance Manual from The Building Centre, 26 Store Street, London WC1E 7BT.

1.2 **Information**
Description of the building and site

1.2.1 References	RIBA plan of work stage 10 *"Building Maintenance & Preservation"*, Chapter 1
1.2.2 Explanation	It is essential for the architect and consultants to make their building design objectives and limitations clear. Appropriate maximum and minimum design criteria and loadings for all useable areas and systems of the building should be specified to avoid misuse of the building and misunderstanding of the building's capability. Buildings change ownership and new owners will not have had the personal experience and contact in the design process.
1.2.3 Instructions	Select and collect appropriate design drawings and photographs of general arrangements of the building and its systems and design features. According to the complexity of the building, additional diagramatic drawings could be prepared either as plans and sections or as three dimensional diagrams, similar to the drawings now required by the Water Authority for water meter applications. These diagrams are isometric or axonometric drawings showing the building form, the no. of floors and the location of major elements or equipment and a single line diagramatic drawing explaining the systems and control mechanisms. Location of the various system controls should be clearly marked in the diagrams.

1.2.4 Check lists

Building description
- Full name & address
- Ordnance survey references
- Ownership boundaries
- Boundary wall responsibilities
- Traffic & circulation
- Internal circulation
- Fire routes & escapes
- Leasehold or freehold
- Water table depth
- Unusual ground conditions
- No. of car parking spaces

Diagrams
- Water supply storage & distribution
- Heating & hot water system
- Mechanical systems & ventilation
- Electrical systems
- Sound relay & telephone
- Drainage & sewers
- Structural systems
- Room numbering
- Colour codes
- Computer
- Video & TV

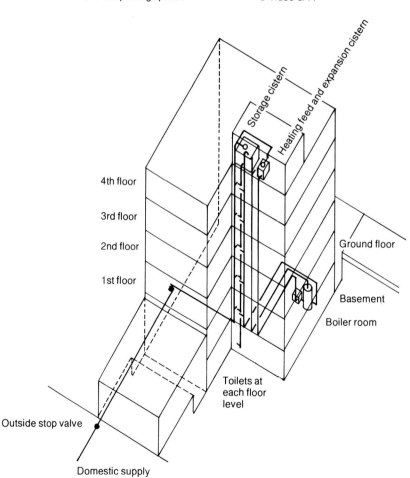

1.3 Information
Emergency contacts

1.3.1 References	*"Building Maintenance & Preservation"*, Chapters 7.4.18, 11.2.21, 11.4.2, 11.4.3
1.3.2 Explanation	The building occupant should be responsible for the compilation and updating of this list. In the event of an emergency the building owner/occupant should always be notified and with the consent of the building owner or authorised agent. Always notify the relevant specialist or consultant as emergency actions are often carried out with great speed. The results of these actions often result in serious consequences. A register of keyholders for the building should be kept updated.
1.3.3 Instructions	Check with consultants or specialist the name and number of the individual to be contacted in case of emergency. Prepare a formal list of keyholders & deposit with the local police and security company (if applicable).

1.3.4 Check list

● Fire	● Handyman	● Boiler
● Theft	● Lift service	● Fuel supply
● Local police station	● Heating & hot water	● Roof
● Local hospital	● Plumbing	● Telephone
● Security control	● Drainage	● Water board
● Insurance co.	● Glazier	● Gas board
● Adjoining owner	● Mech. handling	● Elec. board
		● Keyholders

1.3.5 Specimen

Item	Name, Address & Telephone no. and times available
Plumbing	J.R. Bender, 12 Left Lane, Barkworth, SW7. 01-690 4646 9am – 9pm 01-460 8686 9pm – 9am

1.4 Information
Ownership & Contract consultants

1.4.1 Reference	*"Building Maintenance & Preservation"*, Chapters 12.21.2, 12.21.3, 12.21.4
1.4.2 Explanation	Apart from the advantage of having the names and addresses of the various consultants involved with the building for the purposes of alteration to the building, the secondary function of these lists is to allow the user to inform the designer of materials or concepts that are either failing or succeeding. This is an instrument for feedback to the designer.
1.4.3 Instruction	The files of the architect, the contractor and building owner will usually contain the required names, addresses and telephones. These should be checked so as to be as up to date as possible.

1.4.4 Check list

- Building owner
- Surveyor
- Solicitor
- Insurance co.
- Agent
- Landlord
- Leaseholder
- General contractor
- Project management
- Contract supervisor

- Architect
- Quantity surveyor
- Structural engineer
- H & V consultant
- Electrical consultant
- Drainage consultant
- A.V. consultant
- Mech. engineer

- Energy consultant
- Graphic designer
- Interior designer
- Designer/transport
- Catering designer
- Landscape designer
- Fire protection

1.4.5 Specimen

Occupation	Name, Address & Telephone no.
Architect	L. Darvinici, 4 Lovenham Terrace, Wiltfull on Thames, Surrey CZ7 XYZ. 01-490 4904

1.5 Information
Authorities consents and approvals

1.5.1 Reference	*"Building Maintenance & Preservation"*, Chapter 15

1.5.2 Explanation

During the life of a building, its use and function will alter, as well as the Regulations and Building Acts that govern it. There exists a statutory obligation on the part of the building owners-occupier to comply with the Acts of Parliament. Non-compliance can affect insurance claims and policies, particularly when disaster strikes. Easy access to documents, reference numbers and dates saves a great deal of time in locating old information at the offices of an authority. The check list serves also as an aide memoire to the building owner whenever a change of use or physical alteration takes place. Many consents have time limits or special conditions and these should be checked at regular intervals.

1.5.3 Instructions

Obtain and collate copies of all consents and relevant associated documents. These can usually be found in the architect's job file. Extract information from these – check for validity and enter the information onto the proforma sheet.

1.5.4 Check lists

Ownership
● Freeholder
● Landlord
● Insurance company
● Funding agency

Legal:
● Rights of way
● Easements
● Rights of light
● Dedication of land
● Road closures
● Party wall awards

Authorities Consents
● Outline planning consent
● Full planning consent
● Building regulations
● District Surveyor
● Advertisement town planning
● Licencing Authority
● Scheduled Ancient Monuments
● Royal Fine Arts Commission
● Listed building consent
● Conservation area consent
● Local amenity groups

Statutory Undertakers
● Gas Board
● Electricity Board
● Water Board
● The Post Office

Acts of Parliament
● Public Health Acts 1936, 1961, 1973
● Housing Acts 1963
● Factory Act 1961
● Offices, Shops, Railways & Premises Act 1963
● Fire Precautions Act 1971
● Explosives Act 1923
● Petroleum Cons. Act 1928
● Town & Country Planning Acts 1966, 1971, 1972, 1977
● Health & Safety at Work Act 1974
● Coal Mining (subsidence) Act 1957
● Chronically Sick & Disabled Persons Act 1970
● Local Government Act 1972
● Control of Pollution Act 1974
● Clean Air Acts 1956, 1968
● Electricity Act 1947, 1957
● Gas Act 1972
● Water Act 1973
● Highways Act 1959, 1971
● London Building Acts 1939
● Building Regulations
● Public Health Acts
● Thermal Insulation Act 1957
● Civic Amenities Act 1967, 1978
● Licencing Act 1969, 1967

1.5.5 Specimen

Item of Consent	Authority's Name & Address & Ref. No.	Date	Tel. No.
Full planning consent	Probox District Council, Bury Buildings, Probox, Surrey RN684.Ref. RDC/103/4/P/57		0496 39801

1.6   **Information**
Subcontractors and suppliers

1.6.1 Reference	*"Building Maintenance & Preservation"* Chapters 17, 12.12, 12.13
1.6.2 Explanation	Subcontractors should be listed from the architects and contractors file. Building companies will sometimes cease to function for a variety of reasons, and often in these cases make arrangements for other companies to take over their servicing and maintenance commitments. It is therefore important for management to keep a watchful lookout for items and installations under guarantee. Subcontractors often undertake maintenance contracts after completion of the main contract. Assuming that they peformed satisfactorily there would be obvious advantages to this practice. These items should be marked accordingly with an * and referred to Part 1, Section 6 of this manual.
1.6.3 Instructions	Collate names and addresses and collect, where available, performance spccifications for installation and attach these to Part 1, Section 6. Obtain confirmation from those firms willing to carry out maintenance contracts and mark accordingly.

1.6.4 Check lists

Subsoil
- soil investigation
- piling
- tanking
- dewatering
- foundations
- geotechnics

Drains
- stoneware
- pitch fibre
- concrete
- headings
- cast iron
- plastic
- sewers
- manholes

External
- boundary walls
- fences and gates
- landscape
- water pumps and watering
- lighting and illuminations
- security and alarms
- roads and kerbs
- paths and paving
- trees and planting
- sewer pumps
- road signs and graphics
- play equipment

Services dry
- electrical installation
- sound relay
- alarms and security
- intercom
- telephones
- lift, passenger
- lifts, goods
- electronics
- lifting gear
- thermostats
- light fittings
- clocks
- lightning conductors
- audio visual
- film
- document conveyors
- mechanical handling equipment
- computer
- specialist equipment
- pumps
- solar heating

Services wet
- plumbing and drainage
- refuse disposal
- heating and hot water
- air conditioning
- sprinklers
- fire fighting equipment
- tanks fuel and water
- pipe systems
- sewage disposal
- cold water services
- ventilation
- gas installation
- hosereels
- catering equipment
- kitchen fittings

Primary and secondary elements
- bricks and blocks
- concrete, institu
- concrete, precast
- doors, internal
- doors, external
- doors, special
- window cleaning equip.
- window cradles
- manhole covers
- staircases
- flooring, internal
- flooring, external
- roofing
- structural concrete
- structural steel
- structural timber
- pavement lights
- roof glazing
- wall coverings
- wall surfaces
- partitions
- windows
- glazing
- decorations, internal
- decorations, external
- ducts
- duct covers
- joinery fittings
- cupboards, special
- cupboards, standard
- paints and polyurethanes
- polish stains
- sanitary fittings
- ironmongery
- w.c. cubicles
- roller shutters
- furniture
- metal work

1.6.5 Specimen

Item	Name & address of supplier/subcontractor	Telephone no.
Piling	A. B. Borehole, Drill Estate Romford, Essex.	039 64321

The Building Centre: Maintenance Manual and Job Diary

1.7 Information

Contract and maintenance contracts information

1.7.1 Reference	*Building Maintenance & Preservation* Chapters 12.9, 12.10, 12.11, 12.12, 12.13, 12.14
1.7.2 Explanation	There are two levels of information to be recorded. First the static information regarding the original contract for the building and, second, the changing list of names and addresses for maintenance contracts. The static information is useful for retrieval of information of past events and the current list of contractors should be kept in a usable order related to the project and check list below, and updated according to change.
1.7.3 Instructions	Static information: refer to architects or contractors file for retrieval. Maintenance Contracts: The building owner together with the architect and maintenance surveyors should recommend a list of regular maintenance contracts. Names, addresses and telephone numbers of all contractors and subcontractors to be entered and updated. As new contractors are introduced, large organisations with their own inhouse labour should list the individuals and departments' names and extension numbers. Tender Documents: should include information extracted from the manual i.e. drawings etc. See check list "Documents for Tender". Include in the specification copies of all catalogues and trade literature to be provided when a contractor instals new products or systems in the building.
1.7.4 Check lists	Documents for Tender

Documents for Tender
- performance specification
- data sheets & manufacturers' commissioning records
- test certificates
- instructions sheets & nos.
- systems diagrams
- consultants design objectives

- floor loadings & restrictions
- items under guarantee

- components list or schedule
- consultants drawings
- extent of the work
- maintenance documentation
- time schedules
- services location & restriction chart

Recommended check list for regular maintenance contracts
- cleaning
- chimneys & flues
- redecoration
- partitions
- planting
- floor & door springs
- fire fighting equipment
- window cleaning gear
- alarms & systems
- fuel supplies

- plumbing & drains
- mechanical handling equipment
- lifts & escalators
- heating & ventilation
- pumps and sumps
- rainwater goods
- light fittings
- equipment & machines
- landscape, internal & external

1.7.5 Specimen

Item	Name & address of maintenance contractor	Telephone no.
Window gear	A. J. Climbers of 29 Slipway Drive, S.E.	0268 94328

2.1 General Building Fabric
Floor loading and restriction chart

2.1.1 Reference	*"Building Maintenance & Preservation",* Chapters 8.1.3, 8.4, 10.2, 10.3, 10.4. 13.1, 13.5,
2.1.2 Explanation	Whatever the original function or age of a building there exist limitations on its use from structural, spatial and mechanical specifications, as well as the compliance with various building regulations. Many regulations exist governing the building, its use and potential. Infringement of these can affect the safety and efficiency of the building. Should accidental damage occur whilst these infringements are taking place, insurance claims and premiums can be affected.
	As a general guide, the structural engineer should always be consulted before taking action affecting any structural changes or the placing of heavy objects such as safes, machines, etc.
2.1.3 Instructions	For new buildings the information can be obtained from the drawings of the consultants and from the fire and planning consents included in this document.
	When floor areas are quoted these should have their limitations defined, e.g. when they are used for calculation of rentals per sq.M., i.e. "skirting to skirting" or "plaster to plaster" dimensions. Where older or historic buildings are involved, a surveyor or engineer should research and establish safe loadings and restrictions as a guide for the building owner.

2.1.4 Check list

- prepare floor plans
- colour or tones for floor zones
- mark fire lobbies & doors
  fire equipment
- floor fire rating

- check occupancy
- specify measurement definitions
- floor levels
- area totals
- occupancy totals

2.1.5 Specimen

Location	Floor level	Floor loading	Area in M² Gross	Nett	User restriction	Maximum occupancy	Fire rating
Block A	3rd floor	100P.S.F.	40,000	35,000	Offices	400	2HR.

2.2 General Building Fabric
General maintenance instructions

2.2.1 Reference	*"Building Maintenance & Preservation"* Chapters 5, 8.41, 8.42.
2.2.2 Explanation	The purpose of the instruction sheet is to make clear to the maintenance operative the items of work to be done, the criteria and acceptable tolerances, and the frequency of each operation. The instruction number is used as a cross reference between log sheets and annual guide charts.
	Instructions for each part of the building can involve several operations and these should be separately numbered so that the log sheet will show which criteria have and have not been met.
2.2.3 Instructions	Check all instructions with appropriate specialist and consultants.
	Determine frequency and targets with the maintenance manager and check all locations.
	Ensure that numbering system is in accordance with the latest practice.
	Collate all manufacturers' leaflets and add number system.
	Add leaflet code no. in the instruction column where appropriate, to avoid duplication of instructions.
	The abbreviation for types of maintenance required are as follows: **I** = Inspect **M** = Maintain **E** =Expected Renewal "Inspect" refers to a visual inspection and checking of installation and meter readings. "Maintain" involves operation of work and adjustments. "Expected Renewal" is an estimate of the life of a product or item of the building and does **not** refer to a renewal date. Many items vary enormously from their expected life.

2.2.4 Check list

- construction joints
- expansion joints
- sealants and gaskets
- chimneys
- staircases
- structural frames
- walls, internal
- walls ,external
- windows, timber
- windows, metal
- windows, glazing
- flooring, internal
- flooring, external
- sanitary fittings

- roofs, pitched
- roofs, flat
- doors, internal
- doors, external
- doors, fire
- doors, special
- ironmongery
- refuse disposal
- plumbing & drainage
- storage system
- roads & paviors
- gates & fences
- decorations, internal
- decorations, external
- signs & numbering

I = Inspect   **M** = Maintain   **E** = Expected renewal

2.2.5 Specimen

Item & Location	Instr. No.	Instructions with operational criteria & targets defined and I, E or M frequency
Gutters to Main Roof	81.1 2 3	I = 3 monthly M = clear debris, flush downpipes 6 monthly E = 7 years check joints gaskets & brackets for renewal

2.3 General Building Fabric
General maintenance log sheet

2.3.1 Reference	*"Building Maintenance & Preservation"* Chapters 5, 8.
2.3.2 Explanation	Log sheets for the maintenance operative to complete. Information can be entered onto the Annual Guide Charts by management. The most critical column on this sheet is the recording of the planned targets achieved, e.g. if there are three instructions and only two of the targets are met then the column will show only those and the remedial action column will reflect the targets that have not been met.
2.3.3 Instructions	Compare instruction numbers and check the items with consultants.
	Prepare copies of log sheet proforma for distribution to maintenance managers. These should be issued together with appropriate instruction sheets related to the work.
	Issue copies of the Annual Guide Chart to the management.

2.3.4 Check list

- construction joints
- expansion joints
- sealants & gaskets
- chimneys
- staircases
- structural frames
- walls, internal
- walls, external
- windows, timber
- windows, metal
- windows, glazing
- flooring, internal & external
- roofs, flat & pitched

- doors, internal, external
- doors, fire & special
- ironmongery
- refuse disposal
- plumbing & drainage
- storage systems
- roads & paviors
- gates & fences
- decorations, internal
- decorations, external
- sanitary fittings
- signs & numbering

2.3.5 Specimen

Instr. no.	Item or material	Date of inspection	Planned targets achieved	Remedial action taken	Initial
801	Guttering	4.4.44	801: 1 & 2	801.3 Replaced one section of gutter.	

2.4 General Building Fabric
Annual building maintenance guide chart

2.4.1 Reference	*"Building Maintenance & Preservation"* Chapters 12.14, 12.19, 8.4 and pages 84-85.
2.4.2 Explanation	The purpose of this chart is to assist management in the assessment of financial and labour projections for the financial year. The use of symbols for projected and actual work functions will enable managers to have an instant visual check on progress of the work. Information from the chart can be easily programmed for computer work schedules.
2.4.3 Instructions	Instruct managers in the use and function of the chart. Add reference numbers to all leaflets. Prepare information for computer schedule. Liaise with cost analysis and budgetary controls to assess frequencies of maintenance. The following symbols are recommended for use in the chart. **Key:** □ Planned maintenance ■ Completed maintenance ◇ planned remedial work ◆ completed remedial work Inspection report sheets are provided for distribution to the surveyors. Information collected in this format will assist in the easy retrieval of work items for the planned maintenance chart. The Report sheet is similar to the example on page 84-85 of the book *"Building Maintenance & Preservation"*

2.4.5 Specimen

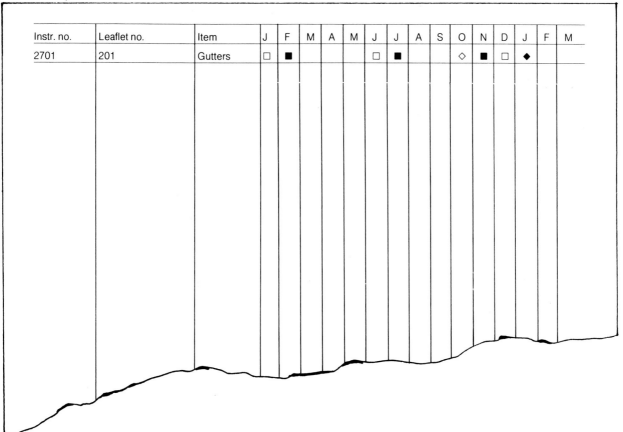

Instr. no.	Leaflet no.	Item	J	F	M	A	M	J	J	A	S	O	N	D	J	F	M
2701	201	Gutters	□	■				□	■			◇	■	□	◆		

2.5 General Building Fabric
Inspection Reports

2.5.1 Reference	*"Building Maintenance & Preservation"*. Chapter 8.4 and pages 84 & 85.
2.5.2 Explanation	The inspection report should be used for regular annual inspections. The results of these inspections should be transferred to the cyclical maintenance instructions. After 3 or 4 years, the inspections can be linked to the timing of redecorations and should occur before each redecoration period commences. The inspection report can be used as an addition to the log sheets.
2.5.3 Instructions	Distribute inspection report sheets to the maintenance surveyor prior to the agreed inspection dates. Transfer information onto instruction sheets and maintenance Guide Charts (see 'I' for inspect). This will enable managers to keep a control on the timing and programming of work. According to the different management systems adopted by individual companies, the log sheets can be used for the purpose of inspection and so eliminate the need for a separate report. It should however be made clear that the report sheets are primarily for general inspection by a surveyor.

2.5.4 Check lists

Exterior decoration
- timber surfaces, paint/stain
- wall coatings
- other surfaces

Internal decoration
- ceilings
- walls
- other surfaces

Roofing flat/pitched
- finish
- insulation
- structure
- dampproofing
- ceiling
- underfloor voids

Upper floors
- finish
- screed
- structure
- ceiling
- suspended ceiling

Staircases
- structure
- treads
- finishes
- balustrades
- soffits

Windows & doors (External)
- glazing
- construction
- ironmongery
- floor springs/risers
- finish
- jointing

- partitions etc.
- structure
- finish
- doors (internal)
- glazing
- construction
- ironmongery
- floor springs/risers
- fixed furniture
  construction
  finish

Sanitary & drains
- drainage
- plumbing
- fittings sinks
  basins
  baths
  urinals
- sewage disposal
- manholes
- gutters & downpipes

External works
- boundary walls
- fencing
- paths & paving
- landscape
- roads & curbs
- drainage
- manholes
- gullies

Mechanical installation
- heating and ventilation
- heat source
- controls
- distribution/pipes
- ducts
- radiators
- insulation

Electrical installation
- switchgear
- distribution
- fittings & equipment
- burglar alarms
- lightning conductors
- lifts

Gas installation
- carcassing
- fittings
- equipment

2.5.5 Specimen

Item & Location	Condition			Item & Location	Condition		
	sound	suspect	defective		sound	suspect	defective
	+	0	−		+	0	−

The Building Centre: Maintenance Manual and Job Diary

2 General Building Fabric
.6 Manufacturers instruction leaflet index

2.6.1 Reference	*"Building Maintenance & Preservation"*, Chapters 14.2 & 8.7.
2.6.2 Explanation	This sheet is an index for the available manufacturers' trade literature. These data sheets may or may not contain maintenance instructions. Their value is just as important as they will usually contain a full description of the material and its components.
	Information on data sheets is the responsibility of the supplier.
2.6.3 Instructions	Collect and collate data sheets & leaflets and cross-reference with new leaflet numbers.
	Obtain written guarantees from manufacturers and record dates of issue.
	Do not alter information on the data sheets without the written consent of the manufacturer or supplier.

2.6.4 Check list

- construction joints
- expansion joints
- sealants and gaskets
- chimneys
- staircases
- structural frames
- walls, internal
- walls, external
- windows, timber
- windows, metal
- windows, glazing
- joinery
- metalwork

- flooring, internal & external
- roofs, flat & pitched
- doors, internal, external
- doors, fire & special
- ironmongery
- refuse disposal
- plumbing & drainage
- storage systems
- roads & paviors
- gates & fences
- decorations, internal, external
- sanitary fittings
- signs & numbering

2.6.5 Specimen

Leaflet no.	Subject	Manufacturer/Supplier	Cat. no. & guarantee	Date of issue
1101	Gutters	A.V. Drainer Co.	A3706	4.4.44

2.7 General Building Fabric
Fittings and components schedules and index

2.7.1 Reference	*"Building & Maintenance & Preservation"*, Chapter 8
2.7.2 Explanation	This sheet is intended as a quick reference and index for architect's components and finishes schedules, such as doors & ironmongery, windows, sanitary fittings, light fittings, furniture and general finishes. Copies of the schedules should be attached to this section.
2.7.3 Instructions	Obtain copies of architects schedules and manufacturers component schedules.
	Determine the guarantee periods of each component.
	Cross-reference with leaflet numbers.
	Where possible have reductions of schedules made, but ensure that they are still legible.

2.7.4 Check list

- doors
- ironmongery
- windows
- panels
- partitions
- sanitary fittings
- internal light fittings
- internal finishes
- furniture
- office machines
- external finishes
- planting lists
- external light fittings
- lettering

2.7.5 Specimen

No.	Item & location	Arch. Schedule no.	Guarantee period	Leaflet no.
1201	Ironmongery, internal	A30507.01	5 years	A1106

3. Building Services

.1 Services capacity and loading restrictions

3.1.1 Reference	*"Building Maintenance & Preservation"*, Chapters 6.1.2, 6.1.3, 6.1.4, 6.2, 6.3.

3.1.2 Explanation

Most building services have been designed to cope with specific maximum loads and capacities, together with limitations on user performance. These limitations have been built into the brief for the building by the architect and consultants as a result of factors and instructions from the Planning Authorities, Building Regulations, Acts of Parliament and cost limitations and specialised requirements of the building owner.

Some of the building's services will have reached their maximum capacities at the completion date of the building, and some will reach maximum as a result of subsequent changes and additions. Items that are particularly affected in this way are boilers, electricity supply, drainage, water supply and ventilation requirements. It is therefore important to update these charts whenever changes in the use of the buildings are effected.

3.1.3 Instructions

Obtain written confirmation from the various consultants for the information required. For this purpose a copy of the chart should be circulated and returned to the architect.

The first chart should state the maximum capacities allowed for in the building systems. This is followed by the floor by floor loadings and restrictions. The permanent or fixed maximums will be a useful reference at the top of each page, enabling the maintenance engineer to check the totals as changes to buildings occur.

3.1.4 Specimens

Services Maximum Capacity

Lifts Loading Max.		Water Capacity		Sewer Diameter peak flow	Heating Boiler size	Gas Main	Elec.	Air Handling
Goods	Passenger	Mains	Tanks					
3 ton	16 persons	40mmØ	2000 litres	600mmØ 5000 1 pm	200,000 BTU	70mmØ	400 KVA	40,000 cfm

Services Loading & Restrictions Chart

Location/Zone	Max. heat temp at 1° Ext.	Elec. capacity in KVA	Ventilation air changes	Other services availability					
				Gas	Water	Drains	Tel	Telex	Comp

3.2 Building Services
Building services instruction

3.2.1 Reference	"*Building Maintenance & Preservation*", Chapter 6.
3.2.2 Explanation	Instructions for operations of work to be carried out should be condensed into precise statements for each operation. Criteria and targets for acceptability levels should be established so that the inspector or operative need only mark or tick the appropriate column in the log sheet. Instructions for each item of maintenance can involve several operations. They should be numbered so that the log sheet will show which of the targets have and have not been met. The main objective is to reduce the amount of paperwork and written statements to a minimum. Log sheets and annual guide charts are cross-referenced by the Instruction no. The abbreviation of **I** = Inspect, **M** = Maintain, **E** = Expected Renewal Time should assist control of manpower and alert the building owner or manager when action is required.
3.2.3 Instructions	Operational instructions may be covered by standard manufacturers literature for individual components, or by subcontractors or suppliers specially prepared maintenance manuals. In some cases the instructions will be straightforward and simple, and will be prepared by consultants. Obtain and collate the various instruction sheets and manuals, and add the appropriate index no. The instructions for these sheets should include the reference to the appropriate data sheet, manual, test certificate or commissioning record.

3.2.4 Check list

- window cleaning equipment
- window control equipment
- hot & cold water
- gas installation
- electrical installaton
- air conditioning
- boilers & instrumentation
- lighting systems
- clockwork
- telephones
- telex
- intercom
- video
- computer
- T.V.
- security
- mechanical handling
- lifts, goods
- lifts, passenger
- hosereels
- sprinklers
- fire fighting
- pumps
- solar heating

3.2.5 Specimen

Item & location	Instr. no.	Instructions with operational criteria & defined targets, commissioning records & frequency
Boilers, Basement	1512	Check pressure 50 npsf – 60 pm Oil gauge 25° – 27° Water composition salt 20-22%, calcium 13%-13.5%

3.3 Building Services

General Services maintenance log sheet

3.3.1 Reference	*"Building Maintenance & Preservation"*, Chapter 6.
3.3.2 Explanation	The Service Log Sheet has the same function as the General Building Log Sheet and the explanation and instruction notes should be referred to.
	For services maintenance, however, there are likely to be more detailed instructions and consequently more entries. The accurate keeping of records is consequently more important.
	Log sheets are for the maintenance operative to complete. Information can be entered into the Annual Guide Charts by management. The most important column on this sheet is the recording of the "planned targets" achieved, e.g. if there are 3 instructions and only 2 of them have been met, then the "remedial action" column will reflect the targets that have **not** been met. Accurate records can by this method influence the wording of instructions an also help to determine if the original instructions are accurate.
	The quantity of log sheets issued to the operatives will depend on the no. of contractors maintaining the building. As a general guide it would prudent to issue each trade or operation with their own sheets, properly titled an coded, e.g. electrical, boilers, pumps, etc. For this purpose three copies of the proforma are included.
3.3.3 Instructions	Prepare copies of all leaflets and instruction manuals and renumber to suit.
	Obtain information from consultants that the various instructions are accurate and complete.
	Issue standard proformas to maintenance operatives and obtain copies for retention in the Manual.

Abbreviations to be used:
◇ Inspect   **I** fill in solid when achieved
○ Maintain   **M** fill in solid when achieved
○ Expected life **E** fill in solid when achieved

3.3.4 Check list

- window cleaning equipment
- window control equipment
- hot & cold water
- gas installation
- electrical installation
- air conditioning
- mechanical handling equipment
- lifts
- hosereels
- sprinklers
- fire-fighting equipment
- lighting systems
- clockwork
- telephones
- telex
- intercom
- video
- computer
- T.V.
- security
- pumps
- solar heating
- waste disposal

3.3.5 Specimen

Instr. no.	Item	Equipment	Inspection date	No. of planned targets	Remedial action taken
1609	Heating	Boiler Pumps	4.4.79	7 out of 10	◇ Renewed Gaskets ◇ Changed oil ◇ Rewired stats

The Building Centre: Maintenance Manual and Job Diary

3.4 Building Services
Manufacturers instruction leaflets index

3.4.1 Reference	*"Building Maintenance & Preservation"*, Chapters 6, 14 & 18.
3.4 2 Explanation	Service manufacturers' leaflets are usually related to individual items of equipment and not installed systems. Service consultants will need to collate these and decide on the appropriate order and numbering. For the large and complex building it may be advisable to include a copy of the services system diagrams at the head of each section with identification of the various items of equipment. The primary function of this sheet is as an index to manufacturers' instructions.
3.4.3 Instructions	Collect and collate sets of leaflets as required and retain one copy for the architect's file.
	Instruct consultants to co-ordinate leaflets and add any of their own instruction sheets or diagrams if necessary.
	Leaflets should be numbered to allow for cross-reference with instruction sheets.

3.4.4 Check list

- window cleaning equipment
- window control equipment
- hot & cold water
- gas installation
- electrical installation
- air conditioning
- boilers & instrumentation
- mechanical handling
- lifts – goods, passenger
- hosereels
- fire fighting
- solar heating
- lighting systems
- clockwork
- telephones
- telex
- intercom
- video
- computer
- T.V.
- security
- sprinklers
- pumps
- waste disposal

3.4.5 Specimen

Leaflet no.	Subject	Cat.no.	Date of Issue
1506	Boilers & Instruments	B7093	14.4.80

3.5 Building Services
Fittings components and schedules index
(Electrical and Mechanical)

3.5.1 Reference	*"Building Maintenance & Preservation"*, Chapter 14.
3.5.2 Explanation	Ordering of spare parts for mechanical services can be a time-consuming process. The keeping of schedules and component lists up to date should provide easy access to this information. For items for which spares no longer exist, i.e. existing buildings, a list of existing manufacturers equipment should be prepared, with manufacturers references and catalogue numbers included when available.
3.5.3 Instructions	Co-ordinate schedules from the architect, mechanical and electrical consultants and renumber accordingly. Obtain from suppliers and specialists all component schedules and spare parts lists and cross-reference with index numbers.

3.5.4 Check list

Schedules
● architect
● mechanical consultant
● electrical consultant
● drainage consultant
● lighting consultant

Subcontractor
● components
● mechanical
● heating & ventilation
● electrical
● machines
● equipment

3.5.5 Specimen

Item	Contractor & Consultants Schedule No.	Leaflet no.	Guarantee Period	Comments

3.6   Building Services

Annual summary guide chart

3.6.1 Reference	*"Building Maintenance & Preservation"* Chapter 6.
3.6.2 Explanation	The purpose of this chart is to assist management in the assessment of financial and labour projection for the financial year and cost budget.
	For the management of more than one building, information from these charts can be transferred onto master charts. This will also serve as a useful guide for the work progress.
	The use of the simple outline shapes for 'Inspect', 'Maintain' and 'Expected Life' will add another dimension to its function

3.6.3 Instructions

Obtain information from the Maintenance Instruction Sheets for frequency: **I, M** or **E** and fill in the appropriate symbol in the relevant month for planned maintenance dates. As the work proceeds, the information can be obtained from the log sheets and the relevant shapes marked in the time scale of actual work done, and the shape filled in colour for visual reference. The following symbols are recommended:

☐ Planned Inspection   ■ Inspection Completed
○ Planned Maintenance   ● Maintenance Completed
◇ Planned Renewal   ◆ Renewal Completed

3.6.4 Check list

- window cleaning gear
- window gear
- hot & cold water
- heating
- air conditioning
- ventilation
- boilers & flues
- instrumentation
- pumps

- lighting systems
- electrical
- clockwork
- telephones
- telex
- intercom
- video
- computer
- T.V.

- fire alarms
- hosereels
- sprinklers
- fire fighting
- security
- solar heating
- waste disposal

3.6.5 Specimen

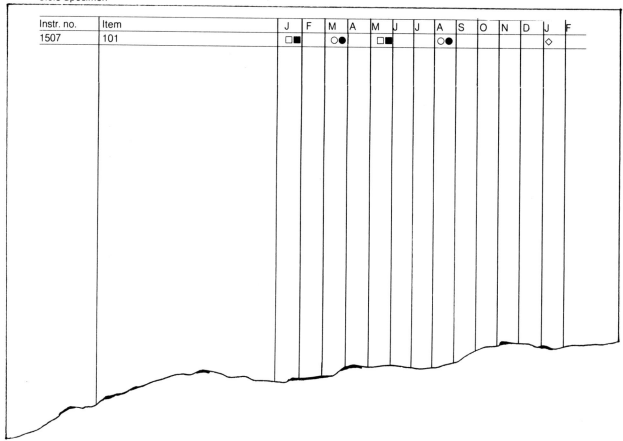

Instr. no.	Item	J	F	M	A	M	J	J	A	S	O	N	D	J	F
1507	101	☐■		○●		☐■			○●					◇	

3 Building Services

.7   Commissioning records and test certificate index

3.7.1 Reference	*"Building Maintenance & Preservation"*, Chapters 6.34, 6.36 & 14.
3.7.2 Explanation	All large installations should be issued with commissioning records and test certificates. These are vital records for the maintenance contractor as they will determine the performance specifications and quality of performance at the time of completion of the building and therefore enable a proper performance monitoring record to be kept. Commissioning records usually refer to systems and test certificates to components.
3.7.3 Instructions	Instruct all subcontractors and suppliers to provide both certificates as part of their contract and deposit these with the consultants for checking at the time of completion of the installation. Copies of all records and certificates should be attached, code numbered and cross-referenced.

3.7.4 Check list

- window cleaning gear
- window gear
- hot & cold water
- heating
- air conditioning
- ventilation
- boilers & flues
- instrumentation
- pumps

- lighting systems
- electrical
- clockwork
- telephones
- telex
- intercom
- video
- computer
- T.V.

- fire alarms
- hosereels
- sprinklers
- fire fighting
- security
- solar heating
- waste disposal

3.7.5 Specimen

Commissioning Record Sheet no.	Systems	Components	
	Item & location	Test Certificate no.	Item & location
C1601	Boiler Firing Heating	T1601	Tubes to boiler

4 .1 Drawings
Consultants drawing check lists

4.1.1 Reference	*"Building Maintenance & Preservation"* Chapters 6, 9 & 10.
4.1.2 Explanation	The purpose of these check lists is to ensure that the various consultants incorporate all necessary information for maintenance on their drawings. Complete sets of drawings of appropriate scales and detail will allow for easy information retrieval. As changes to the building occur, alterations can be marked on the drawings and redistributed. For this purpose a set of negatives or microfilm should be deposited with the client's bankers or solicitors for safe-keeping.
4.1.3 Instructions	Decisions about the degree of maintenance information to be put on the drawings should be made at an early stage of design details by the architect and design team. The responsibilities of each party should be made clear at the outset, so that work can be planned economically to produce necessary and sufficient information on time and in a co-ordinated manner. This same instruction should be included in the invitation to tender for subcontractors work requiring drawings.
	Cross references between architects, consultants and contractors drawings are important when tracing information after completion of the building. CI/SfB or other systems of numbering can be used, but simpler systems based on the architect's layout drawing numbers, with the use of different prefixes for the various consultant's drawings have been found to work well.
	Buildings that have relatively simple systems and construction can be covered by a simple manual, a set of architect's and consultant's drawings and manufacturer's leaflets.
	Larger buildings or developments generally have more complex systems that will usually need the preparation of special drawings and schematic axonometrics of the special systems and services installation.
4.1.4 Check list	Check list for as-built record drawings to be provided for the maintenance of the buildings by all consultants.

Drawings	Scale
● Site location plan & external services	1:500 0r 1:200
● All floor plans & sections and underground services plans	1:100 or 1:50
● Typical cross-sections through the building, special sections showing windows, toilets, access panels, ducts, stairs, drainage & waterproofing	1:50 or 1:20
● Typical construction showing ducts, access, construction & joinery	1:10 or 1:50
● Door, ironmongery, window, sanitary fittings, light fittings, lettering, component lists. Finishing schedules, commissioning records, test certificates, graphics, fittings, machines, furniture.	schedules
● The Clerk of the Works should record on site all obstructions, water levels found on site, and the exact location of all underground services, abnormal foundations and special underground conditions encountered.	site record drawing

The various consultants that generally provide drawings will be:
- ● Architect
- ● Structural Engineer
- ● Services Engineer
- ● Public Health Engineer
- ● Acoustic Consultant
- ● Electrical Consultant
- ● Energy Consultant
- ● Landscape Consultant

Architect's drawings. Structural Engineer's, drawings.

Architect		Engineer
● date	● meter cupboards	● floor loading
● building lines	● access panels	● direction of spans
● north point	● vehicular access	● direction of main steel
● road names	● removable structural	● points of heavy loads
● sheet no.	panels	● piles & pile caps
● ancient lights windows	● non-structural walls	● non load bearing areas
● plot radio	● moveable partitions	● notes on maximum
● density	● solid floors	loading
● water table	● hollow floors	● guide notes on
● sewer levels	● ducts	future alterations
● sewer flood levels	● inspection covers	● schematic three-
● river flood zones	● overflows	dimensional drawings
● interceptor levels	● rising mains, water	explaining systems.
● bench marks	● rising mains, electricity	● notes on cutting of
● air flight paths	● extracts & flues	holes in structure
● fire escape routes	● room numbers	
● fire doors	● cladding materials	
● fire compartments	● door types	
● fireman's lift	● lightning conductors	
● fireman's lobby	● security alarms	
● flooring types	● accessibility for maintenance (see Chapter 11)	

4.1 Drawings
Consultants drawing check lists

Electrical Consultant
- main incoming cables
- main transformers & switchgear
- power factor correction equipment
- earthing
- emergency cut-out switches
- isolators, master switches & fuses
- low voltage distribution system
- power outlets
- supplies to lifts & other mechanical equipment
- small power systems to sockets & trunking
- emergency power systems

- lighting supplies & switches
- emergency lighting
- fire & security alarms
- sound relay, public address & telephones
- closed circuit television
- special computer equipment
- extra low voltage control wiring
- lightning conductors & suppression
- CO_2 extinguishers for electrical equipment

Acoustic consultant
- special panels
- absorbtion levels
- decibel rating
- machine noise levels
- acceptable noise – internal
- acceptable noise – external

- areas of special treatment
- sound relay systems
- acoustic controls & panels
- reverberation times
- schematic drawings

Public Health/Drainage Consultant
- water mains & stopcock meters
- pumps & storage tanks
- hot & cold water systems
- pipe layouts & calorifiers
- valves, stopcocks , drain points & overflows
- water treatment
- drinking water supplies
- sanitary fittigs

- foul & surface water drains
- manholes interceptor access points
- emergency pumping equipment
- sewage storage & pumps
- fire fighting equipment, hydrants, pumps
- sprinklers, mains, wet risers, dry risers
- hosereels, pumps, control valves

Service Engineer
- ducts & filters
- fire dampers & controls
- air diffusers
- gas pipework & meters
- valves & pressure controls
- electrical power supplies to equipment & switch controls
- areas of low & high air pressure
- fuel tanks & valves
- meters, submeters
- stopcocks for water, gas & fuel
- air conditioning units
- boilers & calorifiers
- automatic & manual controls
- max. min, temperatures

- design temperature range
- humidity levels
- machine noise ratings
- min. air change
- chillers & condensers
- heat exchangers & recovery equipment
- solar heating
- cooling towers
- air handling units
- extract fans
- control panels
- thermostats & sensors
- pipe runs & valves
- drain points & heat emitters

Energy Consultant
- thermal resistivity of:
- external surfaces
- floors
- walls
- roofs

- energy consumption levels, see chapter 3 according to fuel being used
- solar systems
- heat gain-loss & ventilation
- diagramatic drawings
- explanation of systems

Landscape
- depth of statutory services
- street furniture, seats
- lamps, playground equipment
- soft landscape
- trees

- grass
- watering instructions
- storage of equipment
- planned schedule for maintenance
- plant list

5.1 Cleaning and User Guide
User guide notes

5.1.1. Explanation

These notes are prepared from a series of simple instructions used by the GLC and other local authorities for the general use of the occupants, tenants and cleaners.

They are not meant for the guidance of the professional team.

Many of the items mentioned are common sense and will be, in any case, adhered to by most responsible occupants.

However, it is as well to have these notes as a check list for landlords to provide to tenants and for general use for managers of maintenance organisations.

Particular importance is attached to the fire, safety and care notes as the safety of the occupants is their highest risk.

Occupants are in constant daily contact with the building and its parts and should be encouraged to report faulty and tell-tale signs to their managers, and so avoid the often costly and disastrous results of neglecting these signs.

Building owners and occupants should ensure that a copy of these notes is deposited with the individual who is in charge of each floor of a building.

5.1.2 Instructions

First Aid maintenance

Keep a small supply of essential tools and maintenance materials on site for minor or urgent repairs.

Windows, glass & doors

Remove broken glass in windows and doors and cover openings temporarily with hardboard or stout card.

Ironmongery

Loose screws should always be tightened. Replace or refix simple items such as coat pegs, broken handles, locks and bolts as necessary, clean and oil.

Water supply

Clearly label all stopcocks and draw-off cocks correctly and where possible all pipe runs, either with labels or colour codes to B.S. standards. A supply of washers of the correct size and type should be kept on a string or wire attached to stop cocks and ball valves. Keep the area around valves and sprinkler stop cocks clear at all times. Duct panels should be hinged and not screwed for immediate access.

Gas & electricity supply

Keep access to intake and meters free at all times. Keep a spare supply of fuse wire and cartridge fuses adjacent to fuse board.
When gas apparatus is moved check that all apparatus is fully capped off.

Dry rot and woodworm

Immediately report tell-tale timber dust droppings and small holes of woodworm or any fungal growth in the building. Fungus can live in the brickwork without flowering for long periods and then attack timber at a later date.
Sudden changes in material colours often indicate moisture penetration. Do not cover up damp patches but report them immediately.

Flood and burst pipes

In the event of burst pipes or flood, turn off all stop cocks and all electric water heaters. Flooding can easily penetrate the electrical conduit system – turn off all affected area electrical switches and have electricians check out the system.

Rain water flooding and drainage

Check out and clear all gulleys and gutters, and lift manhole covers to see if the outlets are blocked, on a regular basis to avoid flooding.

Paths and roads

Watch out for unevenness of surface and subsidence.

Roofing, pitched

Check pitched roofs for missing tiles or slates and subsidence in surface. Check & clean gutters, gullies and outlets.

Roofing, flat

Watch out for discolouration of ceilings and walls due to leaks, clean outlets regularly. Check roof surface for "ponding"

Walls and structure

Record and monitor all cracks, both large and small and notify structural engineer of all large cracks.

Services

Make sure that there is a regular service and maintenance contract for **all** items of equipment and services i.e. heating, pumps, mechanical, machines, electrical installation. Do not add loading to any system without checking that the existing system can handle the additional load.

Energy saving

Close doors and windows to keep warmth in.
If rooms are too hot adjust thermostats or turn off radiators.
Turn off heating in unoccupied rooms.
Report excessive draughts
Report any dripping water taps, hot or cold.
Switch off unnecessary lights.
Switch off office equipment when not in use.
Report any malfunction of thermostats.
Add insulation when necessary.

5.1 Cleaning and User Guide
User guide notes

Fire Safety and Care	– Be careful – the safety of human life is at stake.
Means of escape	Staircases, lobbies and corridors must be kept clear of obstructions at all times. Self closing doors must **NOT** be permanently fixed open or propped open. Chains, locks or padlocks should **NOT** be fitted (during occupation times) to any escape doors, panic bolts. Corridors or escape passages should **NOT** have chemicals or inflammable materials stored on their route. Cupboards on these routes should not exceed 3m length without a 1m gap. Always check new fire arrangements when alterations to layouts are made. All doors designated fire doors should not have lockable fastenings that require keys to open in the direction of escape. Cabin hooks should not be fitted except to doors where large crowds are passing through and then the fastenings should be removed afterwards. All decorative finishes or display pictures in fire corridors should be of non-combustible materials. There should be no display materials within 600mm of fire risk equipment such as radiators, cookers, meters etc. Check all fire extinguishing equipment regularly and report defects immediately. First Aid equipment should be placed for easy access to all and clearly labelled as such.

5.2 Cleaning and User Guide
Regular cleaning instructions

Window cleaning	
Glass	Wash down with clean water and leather with damp chamois leather.
Anodised aluminium	Clean with warm soapy water applied with a cloth or leather. Dry and polish with a soft dry cloth.
Bronze	Apply thin coating of paraffin oil, allow to dry thoroughly and polish with a soft cloth.
Stainless steel	Clean with warm soapy water, dry off with a leather and polish with a soft cloth.
Lacquered metal	Clean with warm soapy water applied with a cloth or leather. Dry and polish with a soft dry cloth.
Ironmongery	
Anodised aluminium	Clean with warm soapy water applied with a cloth or leather. Dry and polish with a soft dry cloth.
Chromium plate	Clean with warm soapy water applied with a cloth or leather. Dry and polish with a soft dry cloth.
Bronze	Apply thin coating of paraffin oil, allow to dry thoroughly and polish with a soft cloth.
Brass	Apply thin coating of metal polish. Allow to dry and polish with a soft cloth.
Stainless steel	Clean with warm soapy water, dry off with a leather and polish with a soft cloth.
Lacquered metal	Clean with warm soapy water applied with a cloth or leather. Dry and polish with a soft dry cloth.
Walls	
Fairface brickwork	Brush down with a stiff dry brush.
Cement glaze	Wash down with warm water and weak detergent. Rinse with clean water and leather down with chamois leather.
Glazed wall tiles	Wash down with warm water and weak detergent. Rinse with clean water and polish with a soft cloth.
Plastics laminates	Remove dirty marks with cellulose polish. Wipe over with a damp cloth and dry with a soft cloth.
Polished hardwood	Apply reviving polish with a damp leather, rub well in and leave no surplus.
Glazed brickwork	Wash down with warm water and weak detergent. Rinse with clean water and polish with a soft cloth.
Granite	Wash down with warm water and weak detergent. Rinse with clean water and polish with a soft cloth.
Marble	Wash down with warm soapy water. Rinse with clean water and dry with a soft cloth. Apply a thin coating of wax furniture cream and polish with a soft cloth.
Vitreous enamelled panels	Wash down with warm water and weak detergent. Rinse with clean water and polish with a soft cloth.
P.V.C.-coated fabric	Wash down with warm water and weak detergent. Remove dirty marks by brushing with a soft brush and rinse with clean water.
Floors	
Coloured asphalte	Apply water/wax emulsion polish with a damp cloth or mop and, when dry, lightly buff with a polishing machine. Remove accumulations of polish with detergent in hot water, rinse with clean water and, when dry, re-apply polish.
Linoleum	Remove dirty marks with a damp cloth or mop. Scuff marks which cannot be removed by this method may be removed by light rubbing with a fine steel wool pad dipped in liquid polish. Apply paste or liquid wax polish, rub into surface with a dry cloth and, when dry, buff with a polishing machine. A water/wax emulsion polish gives greater non-slip properties. Apply with a damp cloth or mop and, when dry, buff with a polishing machine. Remove accumulations of polish with detergent in hot water, rinse with clean water and, when dry, re-apply polish or dry rub with a fine steel wool pad and re-apply polish.
Rubber	Apply water/wax emulsion polish with a damp cloth or mop and, when dry, lightly buff with a polishing machine. Remove accumulations of polish with paste cleaner using warm water, rinse with clean water and, when dry, re-apply polish.
Terrazzo	Lightly scrub with warm water using a gritty cleansing powder, rinse with clean water and squeegee to remove surplus water.
Thermoplastic tiles	Remove dirty marks with a damp cloth or mop. Scuff marks which cannot be removed by this method may be removed by light rubbing with a fine steel wool pad dipped in liquid polish. Apply water/wax emulsion polish with a damp cloth or mop and, when dry, lightly buff with a polishing machine. Remove accumulations of polish with detergent in hot water, rinse with clean water and, when dry, re-apply polish.
Hardwood block or strip	Apply paste or liquid wax polish, rub into surface with a dry cloth and, when dry, buff with a polishing machine. Remove accumulations of polish by dry rubbing with a medium steel wool pad or wet rub with a medium steel wool pad, dipped in liquid polish and, when dry, buff with a polishing machine.
Quarry tiles	Wash with warm water and detergent and rinse with clean water. Squeegee to remove surplus water.

5.2 Cleaning and User Guide
Regular cleaning instructions (continued)

Concrete tiles	Wash with warm water and detergent and rinse with clean water. Squeegee to remove surplus water.
Concrete granolithic	Scrub with warm water and detergent. Rinse with clean water. Squeegee to remove surplus water.

Paintwork internal
Washable distemper	Brush lightly with a dry fine hair broom.
Gloss paint, semi-gloss paint, flat oil paint	Wash down with warm water and weak detergent. Rinse with clean water and leather down with chamois leather.
Emulsion paint	Wash down with warm water and weak detergent. Rinse with clean water and leather down with chamois leather.

Furnishings
P.V.C. upholstery	Clean with warm soapy water, applied with a soft brush. Rinse with clean water and dry with a soft cloth.
Fabric upholstery	Vacuum clean and brush with a stiff upholstery brush.
Polished wood	Apply wax furniture cream and polish with a soft dry cloth.
Painted wood	Wash down with warm water and weak detergent. Rinse with clean water and polish with a soft cloth.
Painted metal	Wash down with warm water and weak detergent. Rinse with clean water and polish with a soft cloth.

Toilets
Sanitary fittings	Clean with porcelain cleaning powder and warm water. Rinse with clean water and polish with a soft cloth.
Copper	Apply thin coating of metal polish. Allow to dry and polish with a soft cloth.
Stainless steel	Clean with warm soapy water and dry off with a leather and polish with a soft cloth.
Mirrors	Wash down with clean water and leather with damp chamois leather.
Chromium plate	Clean with warm soapy water applied with a cloth or leather. Dry and polish with a soft dry cloth.

5.3 Cleaning and User Guide
Regular cleaning instructions (recommended frequency)

Item	Location	Operation	Recommended frequency
General			
Ash trays	All rooms	Empty and dust and replace	Daily
Waste paper bins	All rooms	Empty and dust and replace	Daily
Litter bins	All rooms	Empty and dust and replace	Daily
Furniture			
Polished wood and metal	All rooms	Dust	Daily
		Polish	Fortnightly
Painted wood and metal	All rooms	Dust	Daily
		Wash	Annually
P.V.C. covered upholstery	All rooms	Dust	Daily
		Wash	Monthly
Fabric upholstery	All rooms	Vacuum clean	Monthly
Plastic	All rooms	Dust	Daily
		Wipe over	Monthly
Mirrors and glass in bookcases, etc.		Dust	Daily
		Wash	Monthly
Floors			
Asphalte			
Linoleum		Sweep and lightly buff	Daily
Rubber	All rooms		
Thermoplastic tiles		Polish	Fortnightly
Hardwood			
Terrazzo			
Quarry tiles	Lavatories	Sweep and wash	Daily
Concrete tiles	All other rooms	Sweep	Daily
Concrete		Wash	Weekly
Granolithic			
Carpets and rugs	All rooms	Vacuum clean	Daily
Walls			
Glazed tile	Lavatories	Dust	Weekly
		Wash	Monthly
Glazed brickwork	All other rooms	Dust	Weekly
		Wash	Six-monthly
Fairfaced brickwork	All rooms	Brush down	Annually
Cement glaze			
Gloss paint			
Emulsion Paint			
Semi-gloss paint	All rooms	Dust	Weekly
Flat oil paint			
Plastic (Wareite, Formica)		Wash	Annually
Granite			
Marble			
P.V.C.-coated fabric			
Polished hardwood	All rooms	Dust	Weekly
		Revive polish	Annually
Ceilings			
Washable distemper	All rooms	Dust	Monthly
Gloss paint			
Semi-gloss paint	All rooms	Dust	Monthly
Flat paint		Wash	Annually
Emulsion paint			
Door furniture			
Metal finishes			
Anodised aluminium			
Chromium			
Bronze	All rooms	Dust	Daily
Brass		Clean	Weekly
Lacquered metal			
Sanitary fittings	All rooms	Clean	Daily
Glass	All rooms	Clean	Monthly

19
Directory of Organizations
Pauline Borland

The following list has been based on information supplied by organizations in the UK and compiled by The Building Centre Trust. This is believed to be up to date at the time of going to press but the editor and publishers cannot accept responsibility for any errors or omissions.

Almshouse Association
Billingbear Lodge, Wokingham, Berkshire RG11 5RU
Tel: 0344 52922   Fax: 0344 862062

Information and advice to members by telephone, letter, visit by enquirer or site visit.
Journal; publications; technical advice; training courses.

The aims of the Association are to assist and advise the Trustees of Almshouses on all matters concerning the Trusts they administer with special reference to the conversion and restoration of existing almshouses, many of which are listed as of architectural or historic interest, and the building of new almshouses.

Ancient Monuments Society
St Annes Vestry Hall, 2 Church Entry, London EC4V 5HB
Tel: 071-236 3934

Information and advice to the general public and members by telephone or letter.
Three newsletters and a journal annually.

Its purpose is the study and conservation of historic buildings of all ages and types. Each year transactions containing papers on many aspects of conservation and architectural history are published and the Society receives 7000 consultations on listed buildings. It can offer advice on sources of money, new uses for redundant buildings, the dating and value of historic buildings and contest public enquiries.

Architectural Salvage
c/o Hutton and Rostron, Netley House, Gomshall, Nr Guildford, Surrey GU5 9QA
Tel: 0483 203221   Fax 0483 202911

Information and advice to the general public and subscribers by telephone, letter.
Technical advice; registers.

Architectural salvage is a scheme which aims to encourage the sale and exchange of materials which would otherwise be lost. For a small fee customers may register their requirements and vendors the items they wish to sell. Buyers and sellers are then put in touch with each other.

Association for Studies in the Conservation of Historic Buildings
c/o Margaret and Richard Davies and Associates, 20a Hartington Rd, Chiswick, London W4 3VA
Tel/fax: 081–994 2803

Information to the general public and members by letter. Publications; technical advice; lectures; seminars, training courses

The Association was founded in 1986.
Membership is by invitation and is open to people or bodies professionally engaged in work relating to the conservation of historic buildings. The Association aims to keep its members informed of all aspects of the conservation of historic buildings by providing a forum for meetings, lectures and discussions and arranging visits to buildings and places of interest. It retains close connections with other bodies in this field and is actively concerned with the setting up of training courses in the conservation of historic buildings.

BRE Advisory Service
Building Research Establishment, Bucknalls Lane, Garston, Watford WD2 7JR
Tel: 0923 664664   Fax: 0923 664098

Information and advice to the general public, Trade, Professions, and Associate Members by

telephone, fax, letter, visit by enquirer, site visit.
Publications; technical advice; library or reference
facilities.

The BRE Advisory Service provides information
about its research and the interpretation of results.
It gives information and advice on most aspects of
construction and fire-related topics. Design
problems can also be discussed by consultation or
an adviser can visit a site or offices to study a
problem at first hand.

BRE Fire Research Station

Melrose Avenue, Borehamwood, Herts WD6 2BL
Tel: 081-953 6177   Fax: 081-207 5299
Moving to BRE Garston in September 1994.
Address as above

Information and advice by telephone, letter, visit
by enquirer or site visit.
Publications; technical advice; library or reference
facilities; films, lectures/seminars.

The Fire Research Station, part of the Building
Research Establishment, has a staff of about 120,
including graduates in physics, chemistry,
engineering, mathematics, statistics, economics and
architecture and undertakes research into all
aspects of fire. The aim of the Fire Research
Station is to minimize losses.

Brick Development Association

Woodside House, Winkfield, Windsor, Berkshire
SL4 2DX
Tel: 0344 885651   Fax: 0344 890129

The Brick Development Association is dedicated
to the task of promoting the use of brick in our
environment in the most effective and attractive
way. It offers a range of services including free
technical advice on brickwork, a range of technical
publications, technical seminars and award
schemes. Its magazine, *Brick Bulletin*, is
recognized worldwide as the foremost journal of
contemporary brickwork and is available free to
specifiers and other UK professionals. The BDA
also offers lectures and other educational services
to students as well as technical input to events
which provide continuing professional
development for practising architects, engineers,
builders, clerks of works and others.

British Board of Agrément

PO Box 195, Bucknalls Lane, Garston, Watford
WD2 7NG
Tel: 0923 670844   Fax: 0923 662133

Information and advice to the general public,
subscribers and members by telephone, letter, visit
by engineer, site visit.
Publications and technical advice; lectures;
seminars

The Board is principally concerned with the
testing, assessment and approval of products used
both in new building and for maintenance. The
subjects for assessment are normally new or
innovatory products but some traditional products
are also covered. Agrément Certificates are issued
for each product found to be satisfactory for use
and these are available from either the
manufacturer concerned or purchased from the
Board direct. Agrément Certificates give a totally
independent opinion of the performance in use of
the product and whether it complies with relevant
Building Regulations, Standards and Codes of
Practice. The BBA is a non-departmental public
body whose council members are appointed by
The Secretary of State for the Environment.

British Cement Association

Century House, Telford Avenue, Crowthorne,
Berkshire RG11 6YS
Tel: 0344 762676   Fax: 0344 761214

Information and advice to all enquirers based in
the UK by telephone, letter, visit by enquirer or
site visit.
Journal; publications; technical advice; library and
reference facilities; films; lectures/seminars;
training courses.

British Cement Association offers to users of
cement and concrete a service of technical
information and advice based on the work of its
research station combined with wide practical
experience and the collection of information on a
worldwide basis. It is an independent trade
association financed in part by the manufacturers
of Portland cement in the UK.

British Ceramic Confederation

Federation House, Station Road, Stoke-on-Trent
ST4 2SA
Tel: 0782 744631   Fax: 0782 744102

Information and advice to members only.

The Confederation is the representative body for
manufacturers of ceramic products in the United
Kingdom. It comprises the National Federation of
Clay Industries (*q.v.*), The British Ceramic
Manufacturers' Federation, and direct members.

British Constructional Steelwork Association Ltd

4 Whitehall Court, London SW1A 2ES
Tel: 071-839 8566   Fax: 071-976 1634

Information and technical advice to the general
public and members by telephone or fax.
Publications, library.

The Association promotes the use of steel in
building and offers technical advice on steelwork

through a range of publications, lectures and seminars.

British Flat Roofing Council
38 Bridlesmith Gate, Nottingham NG1 2GQ
Tel: 0602 507733   Fax: 0602 504122

Information: to architects, specifiers, contractors, building owners and educationalists.
Advice: Technical Telephone Advisory Service
Education: Conference and Seminar Service
Publications: A definitive technical handbook *Flat Roofing: Design and Good Practice* (price on request) and Technical Information Sheets

BFRC is an independent non-profit-making organization which aims to demonstrate the significant, practical, economic and aesthetic advantages of flat and low-pitched roofs. It promotes research including the testing and use of new waterproofing materials and wishes to raise standards of design, construction, performance and maintenance of the roof in service.

British Plastics Federation
6 Bath Place, Rivington Street, London EC2A 3JE
Tel: 0839 506070   Fax: 071-457 5038

Information and advice to the general public and members by telephone, letter, visit by engineer.
Publications; registers.

The aims of the British Plastics Federation are:

1. To continue and develop a profitable industry with benefits to companies, employees, customers and suppliers to the UK economy
2. To develop the UK production of plastics
3. To develop the UK consumption of plastics
4. To increase the UK exports of plastics
5. To increase the UK produced share of UK domestic demand for plastics
6. To overcome the constraints on the industry's growth.

British Standards Institution
2 Park Street, London W1A 2BS
Tel: 071-629 9000   Fax: 071-629 0506
Moving in September 1994 to 389 High Road, Chiswick, London W4 4AJ
Linford Wood, Milton Keynes MK14 6LE
Tel: 0908 221166   Fax: 0908 320856

Information to the general public, subscribers and members by telephone, letter, visit by engineer.
Journal; publications; technical advice; library or reference facilities; lectures; seminars; training courses.

Responsible for the preparation and promulgation of voluntary national standards for all sectors of industry and trade. Some 700 new and revised standards are published each year comprising dimensional, performance and safety specifications, methods of test and analysis, codes of practice and glossaries of terms. BSI also operates a variety of self-financing and information services for British and foreign customers relating to standards, regulations and certification systems.

British Wood Preserving and Damp-Proofing Association
6 Office Village, 4 Romford Road, Stratford, London E15 4EA
Tel: 081-519 2588   Fax: 081-519 3444

Information and advice to the general public; subscribers and members by telephone, letter, visit by enquirer.
Publications; technical advice; library or reference facilities; registers; lectures.

The British Wood Preserving and Damp-Proofing Association is a scientific and advisory association. It is a body which collects information on the preservation and fire-proofing of timber, plus the damp-proofing of buildings, and on the methods of applying preservatives and fire retardants. It also sponsors scientific research into the use of preservatives, fire retardants and damp-proofing materials. It makes available the results of its researches to all enquirers by the publication of leaflets, a technical advice service and specialist lectures. It is completely impartial in its outlook and in the advice it gives.

British Woodworking Federation
82 New Cavendish Street, London W1M 8AD
Tel: 071-580 5588   Fax: 071-436 5398

Information and advice to members and specifiers of woodworking products by telephone and letter.
Publications; technical literature.

The British Woodworking Federation is a trade organization representing the interests of manufacturers of timber components, principally for use in building. Many members are involved in the production of components for use in the maintenance and refurbishing of existing buildings.

Building Centre
26 Store Street, London WC1E 7BT
Tel: 071-637 1022   Fax: 071-580 9641

Permanent exhibition centre. Information to specifiers and the general public on building products and other related topics.
Newsletter.

The Building Centre is an independent organization receiving no government subsidy and is controlled by The Building Centre Trust, a registered charity. Through its displays of building products and data services, The Building Centre provides professional industry specifiers and the general public with information from building product manufacturers, trade associations, and other statutory bodies.

Building Centre Trust
26 Store Street, London WC1E 7BT
Tel: 071 637 1022   Fax: 071 580 9641

The Building Centre Trust was established in 1963 as an independent charitable organization to provide support for educational, research and cultural activities in architecture and building. It occupies a central neutral position providing an opportunity for all relevant parties to work together for the benefit of the industry and those it serves. The Trust considers applications for funding from both organizations and individuals within the industry. It has also initiated a number of major projects. In each case the common practice is to provide initial seed finance for the underwriting of risk for projects which are concerned with education and the selection and specification of building components. Projects are expected to prove their effectiveness and value within an agreed time-scale. The Trust also prefers to work in partnership with other bodies.

Building Cost Information Service
85–87 Clarence Street,
Kingston-upon-Thames, Surrey KT1 1RB
Tel: 081-546 7554   Fax: 081-547 1238

Information and advice to subscribers only in form of loose-leaf data sheets or on-line via computer.
Information sheets on: Tender level and cost indices; Labour hours and wages; Statistics and economic indicators; Publications Digest; Cost Studies; Detailed cost analyses; Concise cost analyses.
Publications; reference facilities.

The Royal Institution of Chartered Surveyors' Building Cost Information Service is a collaborative venture for the exchange of building cost information, so that subscribers involved in design and construction can have ready access to the best available data for building economy.

Building Employers Confederation
82 New Cavendish Street, London W1M 8AD
Tel: 071-580 5588   Fax: 071-631 3872

Information and advice to the general public and members by telephone, fax and letter.
Bulletins, publications; technical advice; reference facilities; films; lectures/seminars; training courses.

Membership of the BEC is open to all reputable firms engaged in the building process. Under the Confederation's 'umbrella' are affiliated organizations covering the industry's main specialist operations. The main function of the BEC is to assist and advise its members, and to represent them in negotiations and representations, particularly with national and local government and the trade unions.

Building Maintenance Information
85–87 Clarence Street, Kingston-upon-Thames, Surrey KT1 1RB
Tel: 081-546 7555   Fax: 081-547 1238

Information and advice to subscribers only by means of loose-leaf data sheets and publications. Information on: General and background information; Cost indices; Publication Digests; Case studies; Occupancy cost analyses; Design/performance data sheets; R&D papers; Maintenance estimating data.
Publications; library and reference facilities; seminars.

BMI is not an enquiry service. It collects and disseminates information from many relevant sources, but to a large extent its unique data are derived from records, operating circumstances and experiences of its own subscribers. BMI encourages better communications between upper management and maintenance management and also between property occupiers and the design team.

Building Research Establishment
Bucknalls Lane, Garston, Watford WD2 7JR
Tel: 0923 664664   Fax: 0923 664098

Journal; publications; library or reference facilities; films; lectures/seminars; training courses.

The Building Research Establishment is the main government organization concerned with research and development for the construction industry, including the prevention and control of fire. It comprises the Building Research Establishment at Watford, the Fire Research Station at Borehamwood, the Fire Laboratory at Cardington, and the Scottish Laboratory at Glasgow. The work of the Establishment is broad in scope, ranging from fundamental research to the study of practical problems which the construction industry faces as a result of current change and innovation. The Overseas Division holds information on building research centres and similar organizations throughout the world.

Building Research Establishment Scottish Laboratory
Kelvin Road, East Kilbride, Glasgow G75 0RZ
Tel: 03552 33001   Fax: 03552 41895

Information and advice to the professions, general public and members by telephone, letter, visit by enquirer, or site visit.
Publications (Digests, Information Papers, etc.); technical advice; reference facilities; films/videos; lectures/seminars.

The Laboratory is responsible for a programme to assess the performance of buildings in more severe climatic conditions and also for the Scottish BRE Advisory Service. Research topics include the performance and durability of walls and roofs, the effects of improving insulation standards, behaviour of thin renderings, methods of insulating, condensation and moisture problems, waterproofing of walls, building on contaminated land and energy-efficient housing.

Building Services Research and Information Association
Old Bracknell Lane West, Bracknell, Berks RG12 7AH
Tel: 0344 426511   Fax: 0344 487575

Information: (free to members) by telephone, letter or fax; IBSEDEX on-line database.
Publications; technical advice; library

BSRIA provides a research testing and consulting service for the mechanical and electrical services of a building covering heating, air conditioning and ventilation, plumbing and sanitation, power and lighting and communications.

CADW: Welsh Historic Monuments
Brunel House, 2 Fitzalan Road, Cardiff CF2 1UY
Tel: 0222 465511   Fax: 0222 450859

Information and advice to the general public by telephone, letter, visit by enquirer or site visit.
Journal (Annual report); publications; technical advice or reference facilities.

The council advises on grants towards the repair and restoration of buildings which are of outstanding architectural or historical interest.

Chartered Institution of Building
Englemere, King's Ride, Ascot, Berkshire SL5 8BJ
Tel: 0344 23355   Fax: 0344 23467

Information and advice available to members by telephone, letter, visit by enquirer and site visit.
Journal; publications; library and reference facilities; lectures and seminars.

The object of the Institute is first, to promote the science and practice of building and second, to establish and maintain standards of competence and conduct in persons engaged or about to be engaged in building science and practice or in education and research connected with them.

Chartered Institution of Building Services Engineers
Delta House, 222 Balham High Road, London SW12 9BS
Tel: 081-675 5211   Fax: 081-675 5449

Information and advice to the general public and members by telephone or letter.
Journals; publications; technical advice

The promotion for the benefit of the general public of the art, science and practice of such engineering services as are associated with the built environment and with industrial processes. The advancement of education and research in the above, and publication of the results of such research.

Civic Trust
17 Carlton House Terrace, London SW1Y 5AW
Tel: 071-930 0914   Fax: 071-321 0180

Enquiry and advisory service: advice and information available.
Library: available to the public by appointment.
Publications: list available.

The Civic Trust aims to stimulate action for the improvement of the built environment throughout the UK and encourage quality in town planning and architecture. The Trust is a registered charity founded in 1957 by Lord Duncan-Sandys. The Trust fosters high standards of planning, design, restoration, and new building. It makes annual awards for good development of all kinds. The Regeneration Unit has been set up to implement practical schemes of environmental improvement, economic growth and community involvement in several towns and inner-city areas following pioneering projects in Wirksworth and Calderdale. Thorne, Ripon, Loftus, Brigg and Ulverston are among 30 other projects currently in progress. There are Associate Trusts in the North East of England, Scotland and Wales (q.v.). The Trust is also a founder member of the government's UK 2000 initiative and has project offices in Liverpool and Bristol.

Civic Trust for Wales – Treftadaeth Cymru
4th Floor, Empire House, Mount Stuart Square, Cardiff CF1 6DN
Tel: 0222 484606   Fax: 0222 482086

Information and advice to Civic Societies, local government, the general public and subscribers by telephone, letter, visit by enquirer or site visit.
Bi-monthly newsletter; publications; technical advice (by liaison); library or reference facilities; registers of Civic Societies; films; lectures/seminars.

An environmental charity associated with the Civic Trust, the Civic Trust for Wales seeks to stimulate environmental awareness and improvement of the built environment encouraging community action to this end.

Civic Trust: North East

Floor 4, MEA House, Ellison Place, Newcastle upon Tyne NE1 8XS
Tel: 091-232 9279   Fax: 091-230 1474

Information and advice to the general public and subscribers.
Publications; technical advice.

A charitable trust devoted to the physical improvement of the towns, cities and villages of the North-East region. The Trust operates an advice service in matters of conservation and preservation, improvement of the overall environment and produces books and leaflets.

Cleaning and Support Services Association

73–74 The Hop Exchange, 24 Southwark Street, London SE1 1TY
Tel: 071-403 2747   Fax: 071-403 1984

Information and advice to the general public, subscribers, and members by telephone, letter, visit by enquirer or site visit.
Publications; technical advice; lectures/seminars; training courses.

The CSSA is a trade association aiming to improve the image of cleaning and support services for both client and constituent members. It aims also to improve service through a strict code of ethics and standards so that its logo acts as a client guarantee.

College of Masons

42 Magdalen Road, Wandsworth Common, London SW18 3NP
Tel: 081-874 8363   Fax: 081-871 1342

Information and advice to the general public, subscribers and members by telephone, letter, or visit by enquirer.
Technical advice; lectures.

To promote in every way possible the widest interest in the craft of masonry and to encourage and assist members in the study of all aspects appertaining to their craft. From time to time the college organizes lectures, discussions, and visits to places of interest. Membership covers all aspects of stone, marble and granite, from quarrying, working, cleaning, restoration, lettering, building, to ecclesiastical constructions.

Concrete Repair Association

PO Box 111, Aldershot, Hants GU11 1YW
Tel: 0252 21302   Fax: 0252 333901

Information: to trades and professions by telephone, letter or fax.
Publications; specifications; technical documentation; seminars.

CRA aims to promote the development of the practice of concrete repair; the advancement of education and technical training in concrete repair and the representation of the Association's members in matters of common interest and improvements in liaison with professional bodies, local authorities, and specifiers.

Conservation Unit

Museums and Galleries Commission, 16 Queen Anne's Gate, London SW1H 9AA
Tel: 071-233 4200   Fax 071-233 3686

Conservation Information Network (on-line) Register, publications, training grants.
Library open to conservators and students by prior arrangement.

Based within the Museums and Galleries Commission, The Conservation Unit was established in 1987 as a forum for conservation advice and support throughout the public and private sectors, intended for providers and users of conservation services. It has charitable status and is funded by The Office of Arts and Libraries.

Copper Development Association

Orchard House, Mutton Lane, Potters Bar, Herts EN6 3AP
Tel: 0707 650711   Fax: 0707 642769

Information and advice to the general public and members by telephone, fax or letter.
Publications; technical advice.

The Association encourages the use of copper and copper alloys and promotes their correct and efficient application. Its services, which include the provision of technical advice and information, are available to those interested in the utilization of copper in all its aspects. The Association also provides a link between research and the user industries and maintains close contact with other copper development organizations throughout the world.

Council for the Care of Churches

83 London Wall, London EC2M 5NA
Tel: 071-638 0971   Fax: 071-638 0184

Information to the general public and members by telephone, letter or visit by enquirer.
Journal; publications; technical advice; library or

reference facilities by appointment; registers; lectures/seminars.

The Council advises on the construction, care and adaptation of churches and their furnishings and provides information on their architectural and historic qualities. Specialist committees of the Conservation Committee, include Monuments and Furnishings, Paintings and Stained Glass. The Cathedrals Fabric Commission for England advises on proposals affecting the fabric and furnishings of cathedrals, and on applications to planning authorities relating to cathedral precincts.

Crafts Council
44a Pentonville Road, London N1 9BY
Tel: 071-278 7700   Fax: 071-837 6891

Information and advice to the general public, professional organizations by telephone, letter or visit by enquirer.
Publications; registers; lectures/seminars.

The Crafts Council exists to promote the work of Britain's artist craftsmen and to encourage an interest in their products.

DoCoMoMo (International Working Party for Documentation and Conservation of Buildings, Sites and Neighbourhoods of the Modern Movement)
Eindhoven University of Technology, BRB Postvak 8, PO Box 513, 5600 MB Eindhoven, The Netherlands
Tel: 31 (40) 472433 Fax: 31 (40) 434248

DoCoMoMo-UK
The Building Centre, 26 Store Street, London WC1 7BT
Tel: 071-637 0276   Fax 071-637 0276

Information and advice to members and subscribers. Newsletters; conferences; lectures; publications; exhibitions.

DoCoMoMo exists to press for the documentation and conservation of the best examples of the architectural heritage of the Modern Movement and to promote a greater understanding of the ideas behind it. Architects, engineers, art and architecture historians and administrators are especially welcome to join.

Dry Stone Walling Association of Great Britain
YFC Centre, National Agricultural Centre, Stoneleigh Park, Kenilworth CV8 2LG
Tel: 021 378 0493

The DSWA provides an information and advice service to the public and to its members. This service includes: a regularly updated *Register of*

Professional Wallers covering the whole of the UK; technical specifications for traditional dry stone walls; Cornish hedges and simple retaining walls; a series off DSWA-published books (available by mail order) including *Building & Repairing Dry Stone Walls, Building Special Features in Dry Stone* and *Better Dry Stone Walling* along with technical books from other publishers; practical craft skills certification scheme, details of courses, etc.

The Association was formed in 1968 and is an expanding charitable organization which seeks to ensure the best craftsmanship of the past is preserved and that the craft has a thriving future.

When writing for information, a stamped, addressed envelope is much appreciated.

Ecclesiastical Architects' and Surveyors' Association
c/o David Clark, Scan House,
29 Radnor Cliff, Folkestone CT20 2JJ
Tel: 0303 254008, 0227 459401   Fax: 0227 450964

Information and advice to members by telephone, letter or fax.
Technical publications; lectures/seminars.

The Ecclesiastical Surveyors' Association was founded in 1872 as the association of 'surveyors of ecclesiastical dilapidations'. Membership includes diocesan surveyors, architects and surveyors appointed to the cathedrals and cathedral official residences; those holding appointments as church architects under the Inspection of Churches Measure 1955 and who have as part of their practices the care and repair of churches; also those known to have knowledge and experience of church architecture and the conservation of historic buildings, or the design and construction of new churches, clergy houses, and other church buildings. It is the aim of the Association to serve as a forum for architects and surveyors professionally engaged in work relating to church property in its widest aspects.

Electrical Contractors' Association
ESCA House, 34 Palace Court, London W2 4HY
Tel: 071-229 1266   Fax: 071-221 7344

Information and advice to the general public (on a limited basis) and members by telephone, fax or letter.
Journal; publications; technical advice; library or reference facilities; registers; films; lectures/seminars; training courses.

Select membership of firms with proved commercial standing and technical competence. All installation work complies with IEE Wiring

Regulations. All members covered by Guarantee of Work and Contract Completion Guarantee Schemes. Membership includes firms in the UK and Eire (except Scotland which has its own association).

English Heritage
23 Saville Row, London W1X 1AB
Tel: 071-973 3000   Fax 071-973 3001

Information and advice to the general public by letter.

English Heritage (The Historic Buildings and Monuments Commission for England) is the largest independent organization in the country. It is the government's official adviser on conservation legislation concerning the historic environment and provides the major source of public funds for rescue archaeology, town schemes, and repairs to historic buildings and ancient monuments. English Heritage is also responsible for the preservation and presentation of some 400 historic properties in the nation's care. Until 1984, these functions were carried out by the Department of the Environment.

Eurisol-UK Mineral Wool Association
39 High Street, Redbourn, Herts AL3 7LW
Tel: 0582 794624   Fax: 0582 794300

Information and advice to the general public by telephone, letter, visit by enquirer.
Publications; technical advice; lectures; seminars.

The aim of the Association is to further the cause of energy conservation in the UK by promoting the significant improvement of the nation's thermal insulation standards for domestic, industrial and commercial buildings and industrial plant. Eurisol-UK believes that insulation has a major contribution to make to any National Energy Conservation Plan, and that the first step in any such plan should be to eliminate waste in fuel/energy usage, reduce carbon dioxide pollution, thereby saving the nation millions of pounds.

Europa Nostra United with the International Castles Institute (IBI)
Lange Voorhout 35, 2514EC The Hague, The Netherlands
Tel: 31(70) 356 0333   Fax: 31(70) 361 7865

Europa Nostra/IBI is a key European non-governmental organization in the field of conservation whose aims are:

1. The protection and enhancement of the European architectural and natural heritage.
2. High standards of architecture, and of town and country planning.

3. The improvement of the European environment.

In order to achieve those aims Europa Nostra/IBI seeks to influence public opinion and authorities on various levels (local, national and international). It does this through resolutions, publications, conferences, open days, workshops, special targeted campaigns, scientific studies, as well as by the Annual Award Scheme. The organization also awards the IBI Medals of Honour and has a Europa Nostra Restoration Fund.

Federation of Master Builders
Gordon Fisher House, 14–15 Great James Street, London WC1N 3DP
Tel: 071-242 7583   Fax: 071-404 0296

Information and advice to members of the Federation by telephone, letter or visit by enquirer.
Journal; publications; technical advice; library or reference facilities; register of members.

The Federation is an employers' association comprising mainly small and medium-sized building firms, many of whom specialize in repairs and maintenance work and some in conservation. A register of its members is kept in each of its ten regions. Among its aims the Federation strives to encourage excellence in building construction and good practice in the conduct of business.

Federation of Plastering and Dry Wall Contractors
82 New Cavendish Street, London W1M 8AD
Tel: 071-580 5588   Fax: 071-436 5398

Information and advice to the general public and members by telephone, fax, letter or visit by enquirer.
Journal; publications; technical advice.

Fire Protection Association
140 Aldersgate Street, London EC1A 4HX
Tel: 071-606 3757   Fax 071-600 1487

Information free to members by telephone, letter or visit.
Publications; journal; training; lectures/seminars.

The Fire Protection Association is a national fire-safety organization of the United Kingdom, one of twenty similar bodies existing worldwide for the promotion of greater fire safety, and supported by The Association of British Insurers and Lloyd's.

Identifying and drawing attention to fire dangers, and the means by which their potential for loss can be minimized, FPA services are designed to assist fire, security and safety professionals in industry, commerce and the public sector, achieve and maintain the highest standards of fire safety management within the premises for which they are responsible.

The FPA is one of the four operating units of The Loss Prevention Council, with the LPC Technical Centre, The Loss Prevention Certification Board and The National Supervisory Council for Intruder Alarms.

Galvanizers Association
Wrens Court, 56 Victoria Road, Sutton Coldfield, West Midlands B72 1SY
Tel: 021 355 8838   Fax: 021-355 8727

Information and advice to the general public and members by telephone, fax, letter, visit by enquirers or (where essential) site visit.
Publications; technical advice; library or reference facilities; films; lectures/seminars.

The Galvanizers Association provides technical and marketing services to the UK galvanizing industry.

Georgian Group
37 Spital Square, London E1 6DY
Tel: 071-377 1722   Fax: 071-247 3441

Information and advice to the general public and members by telephone, letter, visit by enquirer or site visit.
Annual report; newsletters; lectures/seminars.

The Georgian Group exists:

1. To save Georgian buildings, monuments, parks and gardens from destruction or disfigurement, and, where necessary, to encourage their appropriate repair or restoration and the protection and improvement of their setting.
2. To stimulate public knowledge of Georgian architecture and town planning, and of Georgian taste as displayed in the decorative arts, design and craftsmanship.
3. To promote the appreciation and enjoyment of all products of the classical tradition in England, from the time of Inigo Jones to the present day.

Institute of Advanced Architectural Studies
University of York, The King's Manor, York YO1 2EP
Tel: 0904 433987   Fax: 0904 433949

Information and advice to specialists.
Publications; technical advice; library or reference facilities; lectures/seminars; training courses.

The Institute of Advanced Architectural Studies is a post-graduate, mid-career and research organization. It offers short courses each session

on aspects of building conservation and maintenance and also full-time post-professional courses in conservation studies.
Publications are available on conservation aspects including *Conservation: A Critical Bibliography*.

Institute of Maintenance and Building Management
Keets House, 30 East Street, Farnham, Surrey GU9 7SW
Tel: 0252 710994/724491   Fax: 0252 737741

Information to building professionals by telephone, letter or fax.
Publications; bi-monthly journal and yearbook (chargeable to non-members); technical advice; seminars/lectures.

IMBM aims to improve the effectiveness of building maintenance in the public and private sectors.

Institute of Plumbing
64 Station Lane, Hornchurch, Essex RM12 6NB
Tel: 0708 472791   Fax: 0708 448987

Information and advice to members by telephone or letter.
Journal; publications; technical advice; registers.

The Institute of Plumbing is a professional body registered as an educational charity, committed to the advancement of the science and practice of plumbing in the public interest. The Institute has over 14 000 members, the majority of whom are enrolled on the Register of Plumbers established by the Worshipful Company of Plumbers in 1886. The Register is maintained by BSI through a BS 5750 Quality Assurance Scheme. The Institute is a nominating member of the Engineering Council and can register Fellows (FIOP) and Members (MIP) as Incorporated Engineers (IEng) and Engineering Technicians (EngTech), respectively.

International Union of Architects
51 rue Raynouard, Paris 75016, France
Tel: 33 (1) 45 243688   Fax: 33 (1) 45 240278

Landscape Institute
6–7 Barnard Mews, London SW11 1QU
Tel: 071-738 9166   Fax: 071-738 9134
Registrar: Peter Broadbent, OBE

The Landscape Institute is the professional body for landscape architects, landscape managers and landscape scientists. Its object is to promote the highest standard of professional service in the application of the arts and sciences of landscape architecture and management.

The Institute has a reference library which is open to the public by appointment. Its journal

Landscape Design is published ten times per year. The Institute will nominate practices to undertake commissions from clients and will respond to requests for information and advice.

Lead Sheet Association

St John's Road, Tunbridge Wells TN4 9XA
Tel: 0892 513351   Fax: 0892 535028

Objectives: To promote and encourage the use of lead sheet in building applications throughout the United Kingdom.
Services: The Technical Information Bureau of the Association (originally part of The Lead Development Association) provides technical advice to the construction industry on all aspects of the use of lead sheet in building, produces technical literature and is active in promoting training in high-quality leadwork, particularly through its Training Centre at West Kent College of Technology.

Mastic Asphalt Technical Advisory Centre Mastic Asphalt Council and Employers Federation

Lesley House, 6–8 Broadway, Bexleyheath, Kent DA6 7LE
Tel: 081-298 0411   Fax: 081-298 0381

Information and advice to specifiers, designers and contractors by telephone, letter, visit by enquirer or site visit (limited).
Publications; technical advice; register of member companies; films; lectures/seminars; training courses.

Trade association and certified employers organization. Promotion of mastic asphalt including provision of free technical advice and limited site inspection service to users and specifiers, maintaining the interests of their members.

Men of the Stones

c/o Mrs Dianne MacKenzie-Ross, Secretary, 25 Cromarty Road, Stamford PE9 2TQ
Tel: 0780 53527

Information and advice to subscribers and members by telephone or letter.
Technical advice; lectures; training courses.
Information to the general public by letter only.

Aims to advocate a greater use of natural stone for building and to encourage study, practice, apprenticeship to and appreciation of the constructional arts and crafts of architecture including stonemasonry, sculpture and carving to provide information and assist with training in this field. Training courses are by arrangement with the Orton Trust Masonry Training Centre, Northamptonshire.

Mortar Producers Association Ltd

74 Upper Holly Walk, Leamington Spa, Warwicks CV32 4JD
Tel: 0926 338611   Fax: 0926 315413

Information and advice to the general public, subscribers and members by telephone, letter, visit by enquirer or site visit.
Publications; technical advice; lectures/seminars; research; sponsorship and involvement with British and European standards for mortar and related materials.

The Association was formed in 1970 to conduct research into lime-based and factory-made mortars. It publicizes their superior qualities by way of data sheets and publications relating to research that has been carried out. It provides delegates for all appropriate British and European standards committees and is a founder member of the European Mortar Industry Organization.

National Association of Scaffolding Contractors

82 New Cavendish Street, London W1M 8AD
Tel: 071-580 5588   Fax: 071-436 5398

Information and advice to members by telephone, fax, letter or visit by enquirer.
Publications; technical advice; training courses.

National Cavity Insulation Association Ltd

PO Box 12, Haslemere, Surrey GU27 3AH
Tel: 0428 654011 Fax 0428 651401

Information and advice to the general public and members by telephone, fax or letter.
Consumer publications; technical advice; data sheets; members' list.

The National Cavity Insulation Association advises and protects those who commission a cavity wall insulation service from its members, who represent around 70 per cent of the industry. It is dedicated to raising industry standards: all members are Agrément Board approved or British Standard registered and the NCIA has its own Customer Protection Plan. Services include an advisory bureau and information literature.

National Corrosion Service

National Physical Laboratory, Teddington, Middlesex TW11 0LW
Tel: 081-977 3222   Fax: 081-943 2989
Telex: 262344 NPLG

Information and advice to the general public and industrial clients by telephone, telex, fax, letter, visit by enquirer or site visit.
Publications; technical advice and consultancy; library and reference facilities.

The National Corrosion Service is an advisory and consultancy service operated by the Department of Trade and Industry to give advice to UK industry on problems of corrosion and to provide assistance in their solution. It has a broad remit, and can provide a wide range of services in many sectors of industry. It is independent of commercial interests and provides a rapid and confidential service.

National Council of Building Material Producers

26 Store Street, London WC1E 7BT
Tel: 071-323 3770   Fax: 071-323 0307

Information and advice to the general public, subscribers and members by telephone, letter or visit by enquirer.
Publications.

BMP is a confederation of trade associations, federations and companies and is the collective representational body for producers and manufacturers of building materials, components and fittings. The council is constituted to represent the common interests of its members to government, the EEC Commission, the BSI, the CBI and other bodies including trade and professional organizations.

National Federation of Clay Industries Ltd

Federation House, Station Road, Stoke-on-Trent ST4 2SA
Tel: 0782 744631   Fax: 0782 744102

The National Federation of Clay Industries is the representative body and employers' organization for manufacturers of clay products, roofing and floor tiles, land drains and refractory goods. It is part of the British Ceramic Federation (*q.v.*). In addition to its representational functions, it provides information and advice to its members on most matters concerning the production of heavy clay products.

National Society of Master Thatchers

High View, Little Street, Yardley Hastings, Northants
Tel: 060 129 293

Information and advice to the general public and members by telephone and letter.
Technical advice.

The Society was formed to represent all member thatchers on matters appertaining to their general good and to negotiate on their behalf. It upholds the highest standards of craftsmanship and regulates the intake of apprentices and trainees.

Paint Research Association

8 Waldegrave Road, Teddington, Middlesex TW11 8LD
Tel: 081-977 4427   Fax: 081-943 4705
Telex: 928720

Information and advice to subscribers and members by telephone, telex, fax, letter, visit by enquirer or site visit.
Journal (*World Surface Coatings Abstracts*); newsletter; special publications; technical advice; library or reference facilities; lectures/seminars; training courses.

The PRA is the leading centre for coatings technology. It caters for coatings in the widest sense, bringing together not only the manufacturers of paint, its raw material suppliers and users but also makers of printing inks, adhesives, sealants, wood preservatives and related products. Its research activities, funded by the UK government, the EC and industrial sponsors, are dedicated to expanding the technical capability and earning capacity of its members and sponsors. It is also an intelligence centre, equipped to appreciate market demands, legal requirements and the significance of discoveries and advances made elsewhere. Information, analytical and consultancy services are available to members and non-members on a fee basis. Further details, additional information and quotations can be obtained from the Managing Director.

Royal Commission on Ancient and Historical Monuments in Wales

Comisiwn Brenhinol Henebion yng Nghymru, Crown Building Plas Crug, Aberystwyth, Dyfed SY23 2HP
Tel: 0970 624381

Information and advice to the general public by telephone, letter or visit by enquirer (letter preferred). Opening hours 10 am–4 pm weekdays.
Publications; technical advice; library or reference facilities.

The Commission has a responsibility for recording monuments, and is willing to give an opinion on age, building type and function. It does not offer technical advice on *methods* of conservation.

Royal Incorporation of Architects in Scotland

15 Rutland Square, Edinburgh EH1 2BE
Tel: 031 229 7545   Fax: 031 228 2188

Information and advice available to the general public and members, by telephone, letter and visit by enquirer.
Publications, technical advice.

The Royal Incorporation is a learned society existing in Scotland to promote good architecture. Membership is open to architects practising in Scotland and the Incorporation regulates the affairs of both its own members and those of the RIBA in Scotland. The Incorporation will also give advice on the selection of an architect, or any matter relating to the practice of architecture in Scotland.

Royal Institute of British Architects
66 Portland Place, London W1N 4AD
Tel: 071-580 5533   Fax: 071-255 1541

Information and advice available to the general public and members by telephone, fax, letter or visit by enquirer.
Journal, publications, technical advice, library and reference facilities, registers, films, lectures and seminars.

The Clients Advisory Service is a free information service designed to assist clients on the selection of architects for specific jobs both in the UK and overseas. It maintains comprehensive records and by modern retrieval methods can provide lists of suitably qualified architects for any design, planning or consultancy situation.

Royal Institution of Chartered Surveyors
12 Great George Street, London SW1P 3AD
Tel: 071-222 7000   Fax 071-334 3800
Telex: 915443 RICS G

Information and advice available to the general public, members and subscribers to the library and cost information services by telephone, telex, letter, fax or visit by enquirer.
Journal, publications, technical advice, library and reference facilities, registers, lectures and seminars.

The Royal Institution of Chartered Surveyors (founded in 1868) is a professional body comprising 65 000 corporate members qualified by examination and experience in the various branches of surveying. Many members, particularly those in the Building Surveying Division, are concerned with the maintenance, conservation and rehabilitation of various types of buildings, and chartered surveyors with the cost and economics of such work.

Royal Town Planning Institute
26 Portland Place, London W1N 4BE
Tel: 071-636 9107   Fax: 071-323 1582

Information and advice available to members by telephone, letter and visit by enquirer.
Journal, publications, technical advice, library and reference facilities, registers, lectures, seminars and training courses.

The RTPI's objectives are to advance the science and art of Town Planning in all its aspects (including local regional and national planning for the benefit of the Public). It is also a watchdog on standards of professional practice and conduct and is the examining body for the profession.

Rural Development Commission
141 Castle Street, Salisbury, Wiltshire SP1 3TP
Tel: 0722 336255   Fax: 0722 332769

The Rural Development Commission advises the government on economic and social matters affecting rural areas and takes measures to further their development. Its prime aim is to stimulate job creation and the provision of essential services in the countryside. Through its network of local offices, the Commission provides a wide range of help including expert advice on finance, management, productivity, marketing, premises, training, rural transport, rural tourism, village shops, or obtaining planning permission. To be eligible for business advice and help, a company should normally have no more than 20 skilled employees and be located in a village or a small town of less than 10 000 population.

Save Britain's Heritage
68 Battersea High Street, London SW11 3HX
Tel: 071-228 3336   Fax: 071-223 2714

Information and advice to the general public by telephone, letter or visit by enquirer.
Publications.

Save Britain's Heritage campaigns for the retention and rehabilitation of worthwhile old buildings and areas. Set up in 1975 by a group of journalists, architects and planners, SAVE issues press releases on threatened buildings, carries out research and produces reports on conservation problems. SAVE does not have members but welcomes 'correspondents' from every part of Britain able to send details of threats to our architectural heritage, and photographs wherever possible.

Scottish Civic Trust
24 George Square, Glasgow G2 1EF
Tel: 041 221 1466   Fax: 041 248 6952

Information and advice to the general public; local amenity societies and local authorities by telephone, letter, visit by enquirer, site visit.
Publications (yearbook and newsletter); *Bulletin of Buildings at Risk* and *Sources of Financial Help for Scotland's Historic Buidlings*; technical advice; videos; lectures/seminars.

The Scottish Civic Trust aims to encourage public interest in the good appearance of town and country and to inspire generally a sense of civic pride; high quality in architecture and planning; the conservation of buildings of artistic distinction or historic interest; the elimination and prevention of ugliness, whether from bad design or neglect; informed and constructive participation in planning matters.

Society for the Protection of Ancient Buildings
37 Spital Square, London E1 6DY
Tel: 071-377 1644   Fax: 071-247 5296

Information and advice to the general public and members on repair problems and methods of repair.
Publications; technical advice for owners and professionals; lectures/seminars; training courses for owners.

SPAB is the oldest national conservation group concerned with historic architecture and has always taught the principles and techniques of 'conservative repair': the use of traditional materials in their correct way but not counterfeiting period, style or detail. William Morris founded the 'anti-scrape' society in 1877 and SPAB still adheres to his manifesto.

Steel Window Association
The Building Centre, 26 Store Street, London WC1E 7BT
Tel: 071-637 3571   Fax: 071-637 3572

Information and advice to architects, specifiers, contractors and the general public, by telephone, letter or visit.
Publications and technical advice; seminars, library and reference facilities

The Steel Window Association's aims are to promote the use of steel windows and allied products, to represent the industry to public and private sector bodies and individuals, to monitor and arrange the provision of technical services so as to maintain standards of performance and to ensure the quality of standards are agreed between the industry and the specifier.

Stone Federation
82 New Cavendish Street, London W1M 8AD
Tel: 071-580 5588   Fax: 071-631 3872

Information and advice to the construction industry and members by telephone or letter.
Technical advice; reference facilities; films

The Stone Federation comprises firms in either Full or Associate membership engaged in the quarrying, working, fixing, cleaning or restoration of natural stone for building. It represents the masonry trade in all aspects of the use and conservation of stone to improve and maintain the nation's architectural heritage.

Suspended Access Equipment Manufacturers Association
82 New Cavendish Street, London W1M 8AD
Tel: 071-580 5588   Fax: 071-436 5398

Information to the construction industry by telephone, letter or fax.
Publication; technical sheets; register.

The Association provides information and advice to its members and users of access equipment.

TRADA; Timber Research and Development Association
Stocking Lane, Hughenden Valley, High Wycombe HP14 4ND
Tel: 0494 563091   Fax: 0494 565487
Telex: 83292 TRADA
Subsidiaries: TRADA Technology Limited; TRADA Quality Assurance Services Limited

Information and advice to the general public, subscribers and members by telephone, telex, fax, letter, visit by enquirer or site visit.
Journal; publications; technical advice; library or reference facilities; films; lectures/seminars; training courses.

TRADA is an independent research association whose work is to further the correct use of timber in building, construction and packaging. It has fire and load test facilities and undertakes consultancy in the disciplines of architecture and structural engineering.

Twentieth Century Society
70 Cowcross Street, London EC1M 6BP
Tel: 0171–250 3857   Fax: 0171–250 3022
Information to members and non-members by letter.
Publications; newsletter; lectures; reports; journal.

The Society aims to promote interest in and preserve the best buildings not only of the 1930s but from 1914 onwards, including post-war.

UPKEEP
Apartment 39, Hampton Court Palace, East Molesey, Surrey KT8 9BS
Tel: 081-943 2277   Fax: 081–943 9552

Permanent exhibition 'Care of Buildings'.
Information and advice to the general public.
Courses, leaflets and press statements.

UPKEEP is an independent educational charity whose aim is to promote good standards of repair, maintenance and improvement of all types of buildings, particularly ordinary houses and flats.

Victorian Society
1 Priory Gardens, London W4 1TT
Tel: 081-994 1019   Fax: 081-995 4895

Information and advice to members by telephone
or letter.
Journal; publications; technical advice;
lectures/seminars; training courses.

Founded in 1958, the Society's aim is to preserve
the best of Victorian and Edwardian architecture,
and also to study the art and history of the period.
It is particularly concerned to protect important
nineteenth- and early twentieth-century building,
both public and private, industrial monuments and
historic areas.

Weald & Downland Open Air Museum
Singleton, Chichester, Sussex PO18 0EU
Tel: 0243 63 348   Fax: 0243 63 475

Information to the general public and construction
professionals.
Publications; lectures.

The Museum houses rescued historic buildings
from south-east England, including houses, barns,
workshops and agricultural buildings. It provides
specialized training seminars and workshops in
association with the Joint Centre for Heritage
Conservation and Management. The Museum, a
registered charity, depends on the support of
individuals and trusts to continue its work.

Worshipful Company of Carpenters
Building Crafts Training School, 153 Great
Titchfield Street, London W1P 7FR
Tel: 071-636 0480

Founded in 1893 by the Worshipful Company of
Carpenters, the School is administered by
members of the Court of the Worshipful
Company. It has close links with other City
Livery Companies. The School organizes full-time
courses for specialized and advanced training in
woodwork and stonemasonry. Such courses are
recognized by the Construction Industry Training
Board and thereby qualify for grant-aid. Specialist
craft courses are continually under development.
Where possible, such courses are of short duration
and tailored to suit the construction industry's
requirements for new production, maintenance
and restoration work for the public and private
sectors of the industry. One of the objects of the
School is to develop courses of training which will
comply with the requirements of the Industrial
Training Act and provide the construction
industry with a facility for specialist craft training
which is supplementary to training given in
technical colleges.

**Worshipful Company of Glaziers and
Painters of Glass**
Glaziers Hall, 9 Montague Close, London Bridge,
London SE1 9DD
Tel: 071-403 3300   Fax: 071-407 6036

A service of information and advice to the general
public is provided by The Glass Information
Officer – also a reference library.

The Glaziers Trust sponsors education in the craft
and helps in conservation of medieval stained glass
and glass of historic interest. The London Stained
Glass Repository, situated in Glaziers Hall,
rescues glass from redundant churches and other
buildings and finds new homes for it. For any
service, first contact should be made with the
Clerk.

Zinc Development Association
42 Weymouth Street, London W1N 3LQ
Tel: 071-499 6636   Fax: 071-493 1555
Telex: 261286 ZILECA G

Information and advice to the general public and
members by telephone, telex, fax, letter, visit by
enquirer or site visit.
Publications; technical advice; library or reference
facilities; films; lectures/seminars.

The Zinc Development Association is a
non-trading body supported by the main
producers of the metal. It has a technical service
which is available to users and potential users of
the metal.

Directory of international organizations

Algeria
Union of Algerian Architects
BP100, 16050, Kouba, Algiers

Argentina
Federación Argentina de Entidades de Arquitectos
Achaval Rodriguez 50, Nueva Cordoba
500-Cordoba
Tel: 54 (51) 39494

Australia
Royal Australian Institute of Architects
2a Mugga Way, Red Hill, ACT 2603
Tel: 062 73 1548

SBIC Australia Pty Limited
PO Box 33
Strawberry Hill Post Office
NSW 2012 Sydney
Tel: 61(2) 318 2988
Fax: 61(2) 319 0565

Austria
Bundesingenieurkammer
Bundesfachgruppe-Architekten Karlsgass 9,
A-1040 Vienna

Bahamas
Institute of Bahamian Architects
PO Box 1937, Nassau

Bangladesh
Institute of Architects
Bangladesh House No. 51, Road No. 5,
GPO Box 3281, Dacca 2

Barbados
Barbados Institute of Architects
The Professional Centre, Noranda,
Collymore Road, St. Michael

Belgium
Fédération Royale des Sociétiés d'Architectes de
Belgique
21 rue Ernest Allard, Brussels 1000

Conseil National de l'Ordre des Architects
160 rue de Livourne, Boîte 2, 1050 Brussels

CSTC – Centre Scientifique et Technique de la
Construction
53 Rue d'Arlon, 1040 Brussels
Tel: 32(2) 230 6282
Fax: 32(2) 230 9125

Bermuda
Institute of Bermuda Architects
PO Box HM 2230, Hamilton 5

Bolivia
Colegio de Arquitectos de Bolivia
Casilla de Correo 8083, av. 16 de Julio 1490, 5
Piso, La Paz

Botswana
Botswana Institute of Development Professions
PO Box 827, Gaborone
Tel: Gaborone 267 53502

Brazil
Instituto de Arquitectos do Brasil
rua Bento Freitas 306, 4 CEP 01220, Sao Paulo

The Building Centre of Brazil
Av. Nilo Peçanha, 50/1904 Centro, CEP 20.020,
Rio de Janeiro
Tel: 55 (21) 2622704
Fax: 55 (21) 2622704

Brunei
Pertubuhan Ukur Jurutera Arkitek Brunei
PO Box 577, Bandar Seri Begawan, Brunei
Darussalam

Bulgaria
Union des Architects Bulgares
3 rue Evlogui Guerguiev, Sofia 1504

Burma
Burma Society of Architects
186 Phayre Street, Rangoon

Cameroon
Ordre des Architectes du Cameroun
BP 926, Yaounde

Canada
Royal Architectural Institute of Canada
55 Murray Street, Suite 330,
Ottawa K1N 5M3
Tel: 1 (613) 241 3600
Fax: 1 (613) 241 5750

Designers Walk Inc.
168 Bedford Road, Toronto, Ontario M5R 2K9
Tel: 1 (416) 961 1211
Fax: 1 (416) 928 9683

McGraw-Hill Information Service Ltd
270 Yorkland Blvd North York, Ontario M2J 1R8
Tel: 1 (514) 496 3118
Fax: 1 (514) 496 3123

Chile
Colegio de Arquitectos de Chile,
Casilla 13377 – Alameda O'Higgins 115,
Santiago de Chile
Tel: 56 (2) 39 1269

China
The Architectural Society of the People's Republic
of China
Pai Wang Chuang, West District, Beijing

CBTDC
19 Che Gong Zhuang Street, 100044 Beijing
Tel: 86 (1) 899 2613
Fax: 86 (1) 802 2832

CIS
Russian Federation Research Insitute for
Information in Building Materials' Industry
(VNIIESM)
16 Leningradskoe Shosse
Moscow 125171
Tel: 7(095) 150 8517
Fax: 7(095) 292 6511

Russian Federation Research Institute for New
Technologies and Information in Construction
(VNIINTPI)
38 Gorky Street
Moscow 125047
Tel: 7(095) 251 1795
Fax: 7(095) 250 2558

Union of Architects of the CIS
3 Ulitsa Schusyeva
Moscow 103889
Tel: 7(095) 203 8060

Union of Architects of the Russian Federation
22 Ulitsa Schusyeva
Moscow 103001
Tel: 7(095) 291 5578

Colombia
Sociedad Colombiana de Arquitectos
Carrera 6a. #26-85.2 Piso, Apdo Aereo 27,765,
Bogota
Tel: 57 (1) 241 97 14
Fax: 57 (1) 283 1989

Costa Rica
Colegio de Ingenieros y Architectos
PO Box 120, San José

Cuba
Union Nacional de Arquitectos e Ingenieros
Humboldt N 104, Esquina Infanta, Vendada,
Havana
Tel: 53 (7) 79 7531

Cyprus
The Cyprus Civil Engineers and Architects
Association
Anis Kominis 12, PO Box 1825, Nicosia

Kibris Turk Muhendis Ve Mimar
Odalari Birligi, 1 Sehit Ibrahim Ali Sok, Lefkosa,
Mersin 10

Czech Republic
The Society of Czech Architects
Letenska 5, 11845 Prague 1
Tel: 42 (2) 53 97 44

ABF Architecture and Building Foundation
Vaclavske Namesti 31, 11121 Prague 1
Tel: 42 (2) 2 366 355
Fax: 42 (2) 2 350 959

USI
Nârodni Trida 10, 11687 Prague 1
Tel: 42 (2) 204807
Fax: 42 (2) 206201

Denmark
Danska Arkitekters Landsforbund
Bregade 66, 1260 Copenhagen K
Tel: 45 (33) 13 12 90
Fax: 45 (33) 93 12 03

Byggecentrum
Dr Neergaardsvej 15, DK 2970 Hørsholm
Tel: 45 76 7373
Fax: 45 76 7669

Dominican Republic
Colegic Dominicano de Ingenieros y Arquitectos
Calle Padre Billini 58, Zona Colonial, Aptdo 1514,
Santo Domingo

Ecuador
Colegio Nacional de Arquitectos de Ecuador
Casillo 7261, Nunez de Vela, no 500 Sector
Inaquito, Quito

Egypt
Society of Egyptian Architects
30.26 July Street, PO Box 817, Cairo
Tel: 20 (2) 815264

El Salvador
Asociacion Salvadorena de Ingenieros y
Arquitectos
75a Avenida Norte 632, San Salvador

Grupo Salvadoreno de Arquitectos
Edificio Duenas 509, San Salvador

Ethiopia
Ethiopian Association of Engineers and Architects
PO Box 5308, Addis Ababa

Fiji
Fiji Association of Architects
Fiji Professional Centre, 21 Desuoeux Road
PO Box 1015, Suva

Finland
Finnish Association of Architects
Eteläesplanadi 22A, SF-001 30 Helsinki
Tel: Helsinki 640 801

RTS
Rakennustietosäätiö
The Building Information Institute
PO Box 1004, SF 00101 Helsinki
Tel: 358 (0) 694 4911
Tel: 358 (0) 694 1897

France
Conseil National de L'Ordre des Architectes
7 rue de Chaillot, 75016 Paris
Tel: 33 (1) 47 23 81 84

Section Française de l'UIA
7 rue de Chaillot, 75016 Paris
Tel: 33 (1) 45 24 36 88
Fax: 33 (1) 45 24 02 78

Batimat
22 rue de Président Wilson, 92532 Levallois Perret
Cedex
Tel: 33 (1) 47 56 5000
Fax: 33 (1) 47 56 08 18

CATED
Centre d'Assistance Technique et de
Documentation du Bâtiment et des Travaux
Publics
Domaine de St Paul, 78470 St Rémy les
Chevreuse
Tel: 33 (1) 30 85 24 59
Fax: 33 (1) 30 85 24 66

CEBTP
Centre Experimental de Recherches et d'Etudes
du BTP
Domaine de St Paul, 78470 St Rémy les
Chevreuse
Tel: 33 (1) 30 85 23 20
Fax: 33 (1) 30 85 23 24

Centre Infobatir
Quai Achille Lignon, 69459 Lyon Cedex 06
Tel: 33 78 93 17 89
Fax: 33 72 44 07 88

Gabon
Ordre des Architectes du Gabon
PO Box 872, Libreville
Gabon

Germany
Bund Architekten der DDR, Breitestrasse 36, 102
Berlin

Bund Deutscher Architekten (BDA)
Ippendorfer Allee 14b, Bonn 53127
Tel: 49 (228) 28 50 11

Bauzentrum München
Radlkoferstrasse 16, D-8000 München 70
Tel: 49 (89) 5107 441
Fax: 49 (89) 5107 166

Heinze GmbH
Bremerweg 184, D-3100 Celle
Tel: 49 (5141) 500
Fax: 49 (5141) 50104

IRB
Informationszentrum RAUM und BAU der
Fraunhofer-Gesellschaft
Nobelstrasse 12, 7000 Stuttgart 80
Tel: 49 (711) 970 2510
Fax: 49 (711) 970 2507

Ghana
Ghana Institute of Architects
PO Box M272, Accra

Gibraltar
Gibraltar Society of Architects
23 Sunnyside House, Naval Hospital Road

Greece
Association des Architects Grecs
3 Ipitou Street, Athens

Chambre Technique de Grece
4 Karageorgi Servias, Athens 125
Tel: 30 (1) 3254 591

Guatemala
Colegio de Arquitectos
7a Calle 'A' #7-11 Zona 9

Federacion Centroamericana de Arquitectos
O Calle 15-70, Coloniz El Maestro

Guernsey
Guernsey Society of Architects
'Roseneath', Grange, St Peter Port

Guyana
Guyana Society of Architects
PO Box 10606, Georgetown

Hong Kong
Hong Kong Institute of Architects
15th Floor, Success Commercial Building,
245-251 Hennessy Road, Wanchai
Tel: 852 (5) 8336323

Hungary
Magyar Epiteszek Kamarajaes
Dienes Laszlo u 2, 1088 Budapest VIII
Tel: Budapest 118 24 44

ETK
Építésügyi Tájékoztatási Központ
Information Centre of Building
PO Box 83, 1400 Budapest 7
Tel: 36 (1) 1422183
Fax: 36 (1) 1427337

Iceland
Arkitektafelag Islands
Freyjugotu 41, 101, Reykjavik
Tel: 354 (1) 11465

Byggingapjonustan
Hallveigarstigur 1, Box 1191-121, 101 Reykjavik
Tel: 354 (1) 29266
Fax: 354 (1) 25380

India
Indian Institute of Architects
Prospect Chambers Annex, 3rd Floor,
Dr D. N. Road, Fort, Bombay 400 001
Tel: 91 (22) 204 6972

Indonesia
Indonesia Institute of Architects
Jl. Raya Pasar, Minggu KM16, Pancoran, Jakarta
12780

Iran
Société des Architectes Iraniens
584 Pahlavi Avenue, Tehran

Iraq
Union of Iraqi Engineers
Committee of Architects, Mansoor, Baghdad

Ireland
Royal Institute of the Architects of Ireland
8 Merrion Square, Dublin 2
Tel: 353 (1) 76 17 03

Israel
Association of Engineers & Architects in Israel
200 Dizengoff Road, PO Box 3082, Tel-Aviv
63462
Tel: 972 (3) 240 274

Israel Association of Architects
22 Gottlieb Street, Tel Aviv 64392

BCI
Building Centre of Israel
PO Box 39027, Ramat Aviv 61390, Tel Aviv
69975
Tel: 972 (3) 425221
Fax: 972 (3) 416930

Italy
Associazione Nazionale Ingegneri ed Architetti
Italiani
24 Piazza Sallustio, 00187 Rome

Consiglio Nazionale degli Architetti
via Sta Maria dell'Amima 10, 00186 Rome
Tel: 39 (6) 689 6009

Centro Edile spa
via Rivoltana 8, I-20090 Segrate, Milan
Tel: 39 (2) 7530951
Fax: 39 (2) 7530057

Ivory Coast
Conseil National de l'Ordre des Architectes de
Côte d'Ivoire
BP 278, Abidjan 17

Jamaica
Jamaican Institute of Architects
PO Box 251, Kingston 10
Tel: (809) 92 680 60

Japan
Japan Institute of Architects
Kenchiku-ka Kaikan, 2-3-16 Jingumae,
Shibuya-ku, Tokyo 150
Tel: 81 (3) 408 71 25

The Building Center of Japan
30 Mori Building, 3-2-2 Toranomon, Minato-ku,
105 Tokyo
Tel: 81 (3) 3434 7155
Fax: 81 (3) 3431 3301

Center for Better Living
19 Akasaka, 1-Chome, Minato-ku, Tokyo 107
Tel: 81 (3) 3568 4901
Fax: 81 (3) 3582 2013

The Japan Building Center
2-6-4 Ginza, Chuo-ku, Tokyo
Tel: 81 (3) 3562 2691
Fax: 81 (3) 3567 5889

Jersey
Association of Jersey Architects
21 Parade Road, St Helier Tel: 0534 31391

Korea (North)
Union des Architects de la RPDC
Rue Jongro Junguëk, Pyong-Yang

Korea (South)
Korean Institute of Architects
1-117 Tonhsung-dong Chongno-gu, CPO Box
1545, Seoul
Tel: 82 (2) 744 8050

Lebanon
Ordre des Ingenieurs et Architectes
De Beyrouth et Tripoli, PO Box 113118, Beirut

Lesotho
Lesotho Architects, Engineers and Surveyors
Association
PO Box MS 1560, Maseru 100

Luxembourg
Ordre de Architectes et des Ingenieurs-Conseils
8 rue Jean Engling, L1466 Luxembourg
Tel: 352 42 2406

Malawi
Malawi Institute of Architects
PO Box 941, Lilongwe

Malaysia
Malaysia Institute of Architects
Perturbalian Akitek Malaysia
4/6 Jalan Taligsi, PO Box 10855, Kuala Lumpur
50726
Tel: 60 (3) 298 4136
Fax: 60 (3) 718 5705

Malta
Chamber of Architects and Civil Engineers
Federation of Professional Bodies
Alamein Road, Medisle Village,
St Andrews STJ 14

Mauritius
Mauritius Society of Architects
c/o Ministry of Works, Phoenix

Mexico
Federacion de Colegios de Arquitectos
de la Republicana Mexicana
Av. Veracruz 24, Mexico City DF 06700
Tel: 52 (36) 570 0007

Mongolia
Union of Architects of the MPR
PO Box 4128, Ulan Bator

Morocco
Ordre des Architectes du Maroc
Washington Square, Résidence Moulay Ismail –
C5, Rabat
Tel: 212 (7) 263 05

Netherlands
Bond van Nederlandse Architekten
P.Bus 19606, Postrekening 71518, 1006
Amsterdam
Tel: 31 (20) 228111

Bouwcentrum
PO Box 299, 3000 AG Rotterdam
Tel: 31 (10) 4309 219
Fax: 31 (10) 412 1115

NBD
Nederlandse Bouw Documentatie
PO Box 23, 7400 GA Deventer
Tel: 31 (5700) 10844
Fax: 31 (5700) 42761

UICB
PO Box 299, 3000 AG Rotterdam
Tel: 31 (10) 430 9208
Fax: 31 (10) 412 1115

New Zealand
New Zealand Institute of Architects
13th Floor, Greenock House, 102-112 Lambton
Quay,
PO Box 438, Wellington
Tel: 64 (4) 735 346

Nigeria
Nigerian Institute of Architects Professional
Centre
(Plot PC 38)
20 Odowu Taylor Street, Victoria Island, PO Box
178, Lagos

Norway
Norske Arkitekters Landsforbund
Josefines Gate 34, 0351 Oslo 3
Tel: 47 (2) 602290

Norsk Byggtjeneste A.S.
PO Box 1575, Vika, N-0118-Oslo 1
Tel: 47 (2) 833 690
Fax: 47 (2) 834 233

Pakistan
Institute of Architects of Pakistan
36-T, Gulberg 2, Lahore

Panama
Sociedad Panamena de Ingenieros y Arquitectos
Avenida Manuel Espinosa B,
Apartado Postal 7084, Panama 5

Papua New Guinea
Papua New Guinea Institute of Architects
PO Box 1278, Port Moresby

Paraguay
Asociacion Paraguaya de Arquitectos
Alberdi 454, 2° Piso, Of.16 Casilla Postal 1526,
Asuncion

Peru
Colegio de Arquitectos del Peru
Apartado 5972, Avenida San Felipe 999, Lima 11
Tel: 51 (14) 71 37 78

Philippines
United Architects of the Philippines
Cultural Centre of the Philippines, Roxas
Boulevard,
Metro Manila
Fax: 63 (2) 832 3711

Poland
SARP
UL Foksal 2.00.959 Warsaw
Tel: 48 (22) 27 87 08

Centralny Osrodek Informacji Budownictwa
U1 Senatorska 27, Warsaw 00950
Tel: 48 (22) 272 449
Fax: 48 (22) 404 228

Portugal
Associacao dos Arquitectos Portugueses
Rue Barata Salguerio 36, Lisbon 2
Tel: 351 (1) 395 1401

Puerto Rico
College of Architects of Puerto Rico
Apartado Postal 73, San Juan 00902

Romania
Union des Architectes de la Republique de
Roumanie
18-20 rue Academiei, 70109 Bucharest

Saudi Arabia
ICMC
Information Center for Material and Construction
PO Box 8934, Riyadh 11492
Tel: 966 (1) 46 24674
Fax: 966 (1) 46 30538

Senegal
Ordre des Architectes du Senegal
Ministère de L'Urbanisme et de l'Habitat, BP
253, Dakar
Tel: 221 22 79 12

Sierra Leone
Sierra Leone Institute of Architects
PO Box 1189, Freetown

Singapore
Singapore Institute of Architects
20 Orchard Road, SMA House 02-00, Singapore
0923
Tel: 65 3388977

South Africa
Institute of South African Architects
PO Box 2093, Houghton, Johannesburg 2041
The Pines, 9 Gordon Hill, Parktown 2193
Tel: 27 (11) 486 1683

CSIR Information Services Construction Industry
Information Centre
PO Box 395, Pretoria 0001
Tel: 27 (12) 841 4807
Fax: 27 (12) 841 4755

Spain
Consejo Superior do los Colegios de Arquitectos
de España
Paseo de la Castellana 12.4°, Madrid 28046
Tel: 34 (1) 435 22 00
Fax: 34 (1) 575 3839

Centro Informativo de la Construcción (CIC)
Roger de Lluria 117, 08037 Barcelona 37
Tel: 34 (3) 215 7738
Fax: 34 (3) 215 8415

ITEC
Carrer Wellington 19, 08018 Barcelona
Tel: 34 (3) 309 3404
Fax: 34 (3) 300 4852

Sri Lanka
Sri Lanka Institute of Architects
120/10 Wijorama Mawatha, Colombo 7

Sudan
Sudanese Institute of Architects
Dar el Muhandis, PO Box 6147, Khartoum

Surinam
Unie van Architeken in Surinam
Postbus 1857, Paramaribo

Swaziland
Swaziland Association of Architects
PO Box A387, Swarzi Plaza, Mba bna

Sweden
Svenska Akitekters Riksforbund
Norrlandsgatan 18.111 43 Stockholm
Tel: 46 (8) 679 7230
Fax: 46 (8) 611 4930

AB Svensk Byggtjanst
S-17188 Solna
Tel: 46 (8) 734 5000
Fax: 46 (8) 734 5099

Byggcentrum Göteborg AB
Göteborgsvägen 97, S-43137 Mölndal
Tel: 46 (31) 27 2400
Fax: 46 (31) 27 0007

Norrlands Byggtjanst
Kungsgatan 73, S-90245 Umea
Tel: 46 (90) 125910
Fax: 46 (90) 134223

Switzerland
Schweizerischer Ingenieur und Architekten Verein
Postfach, 8039 Zurich

Bund Schweizer Architekten
Frobelstrasse 33.8032, Zurich

DOCU
Schweizer Baudokumentation
Ch-4249 Blauen
Tel: 41 (61) 761 4141
Fax: 41 (61) 761 2233

Schweizer Baumaster-Centrale
Talstrasse 9/Borsenblock, CH-8001 Zürich
Tel: 41 (1) 211 7688

Syria
Ordres des Ingenieurs et Architectes Syriens
Azme Square, Dar al Mouhandiseen Building,
PO Box 2336, Damascus

Taiwan
National Union of Architects Association (ROC)
9F1 396 Kee-Lung Road, Section 1, Taipei 10548

Tanzania
The Architectural Association of Tanzania
PO Box 567, Dar es Salaam

Thailand
Association of Siamese Architects
1155 Phaholyothin Road, Bangkok 10400
Tel: 66 (2) 278 1666

Trinidad & Tobago
Trinidad & Tobago Institute of Architects
PO Box 585, Port of Spain, Trinidad

Tunisia
Ordre des Architectes de Tunisie
1 Rue de l'Arabie Saoudite, Tunis 1002

Turkey
Chamber of Architects of Turkey
Konur Sokak 4, Yenisehir-Ankara

YEM
Yapi-Endüstri Merkezi
The Building & Industry Centre
Cumhuriyet Caddesi 329, 80230 Harbiye-Istanbul
Tel: 90 (1) 247 4185
Fax: 90 (1) 241 1101

Uganda
Uganda Society of Architects
PO Box 9514, Kampala
Tel: 256 (41) 33853

United States of America
American Institute of Architects
1735 New York Avenue NW, Washington, DC 20006
Tel: 1 (202) 626 7300
Fax: 1 (202) 783 8247

National Housing Center
15th and M Street, NW Washington, DC 20005
Tel: 1 (202) 822 0520
Fax: 1 (202) 822 0316

Uruguay
Sociedad de Arquitectos de Uruguay
Casilla de Correo, 176, Montevideo
Tel: 598 (2) 900 259

Venezuela
Colegio de Arquitectos de Venezuela (CAV)
Apartado 78140, La Urbina, Sector North 1070 A, Caracas

Vietnam
Union des Architectes de Vietnam
23 bid Dinh Tien Hoang, Hanoi

Yugoslavia
Savez Arhitekata Jugoslavije Knesa Milosa 9/1, 11000 Belgrade

Zaire
L'Ordre des Architectes du Zaire
BP 3657 Kinshasa-Gombe

Zambia
Zambia Institute of Architects
PO Box 34730, Lusaka
Tel: 260 (1) 75751

Zimbabwe
Institute of Architects of Zimbabwe
Riembarta, PO Box 3592, 256 Samora Machel Av. East, Harare

20
Bibliography

1. Design and maintenance of buildings

Building economics: appraisal and control of building design costs and efficiency (I. H. Seeley), Macmillan, London (1983).

Building maintenance (I. H. Seeley), Macmillan, London (1987).

Designing for Health and Safety in Construction, H & S Executive.

Developments in building maintenance (E. J. Gibson), Applied Science Publishers Ltd, London (1979).

Planning (10th) Edition. ed. E. D. Mills. Butterworths.

The relationship between design and maintenance (R & D Paper), DoE.

The Property Services Agency Library Publication 'Current information on maintenance' has 647 entries on this subject which include all Journal articles and reports etc., together with BRE and other publications.

2. The economics of maintenance

Building design of maintenability (B. Feldman), McGraw-Hill Book Co., New York (1975).

Building economics: appraisal and control of building design, cost and efficiency (I. H. Seeley), Macmillan Publishers Ltd., London (1983).

Building maintenance cost information service, 85–87, Clarence Street, Kingston-upon-Thames, Surrey.

Life cycle cost analysis: a guide for architects RIBA Publications, London.

The Property Services Agency Library Service publication – *Current information on maintenance – Part C Management and economics*, has 404 entries in this subject which include journal articles, reports, BRE and other publications.

Also published by the Property Services Agency is *Quality surveying development* which includes a section on Costs-in-Use.

3. Energy utilization, audits and management

Energy conservation and energy management in buildings (A. F. C. Sherratt), Applied Science Publishers, London (1976).

Fuel efficiency booklets. *1 and 2 Energy audits: 7 Degree days*, Department of Energy, London (1976–77).

The efficient use of energy (ed. I. G. C. Bryden) IPC Science and Technology Press Guildford, (1975).

4. Thermal standards, methods and problems

Building for energy conservation (P. W. O'Callaghan), Pergamon Press Ltd, Oxford (1978).

Computer programs for energy in buildings (K. S. Burgess) Evaluation report No 5 Design office Consortium.

Industrial energy conservation, 2nd edn (D. A. Reay) Pergamon Press Ltd, Oxford (1979).

5. Building materials and their maintenance

Chemical materials for construction: handbook of chemicals for concrete, flooring, caulks and sealants, epoxies, latex coatings and heavy construction specialities (P. Maslow), Structures Publishing Co., Farmington, Mich, USA (1974).

Composite materials and their use in structures (J. R. Vinson and Tsu-Wei Chou), Applied Science Publishers, London (1975).

Insect factor in wood decay: an account of wood-boring insects with particular reference to timber indoors (N. E. Hickin), 3rd Edition revised by R. Edwards. Associated Business Programmes, London (1975).

Materials for construction (R. C. Smith), McGraw-Hill, London (1973). (1979 3rd edn IP)

Paint film defects: their cause and cure (M. Hess), Chapman and Hall, London (1965). (3rd edn 1979 IP)

The Property Services Agency Library Service publication: *Current information on maintenance Part E – Deterioration and weathering of materials*, has 334 entries on this subject which includes journal articles, reports BRE and other publications.

6. Services design and maintenance

Corrosion and oxidation of metals: scientific principles and practical applications (U. R. Evans), Edward Arnold & Co. Ltd., London (1960).

Electrical maintenance and repairs (J. L. Watts), Macmillan, London (1964).

Maintenance engineering handbook (ed. L. C. Morrow), McGraw-Hill Book Co., New York (1987).

Preventive maintenance of electrical equipment (C. I. Hubert), McGraw-Hill Book Co., New York (1969).

Safety in sewers and at sewage works, The Institution of Civil Engineers in association with the Ministry of Housing and Local Government (1972).

Service Guide, The Institute of Plumbing.

The Property Services Agency Library Service publication: *Current information on maintenance Part D – Building services engineering*, has 220 entries on this subject which includes journal articles, reports, BRE and other publications.

7. The maintenance and design of security systems

A J Design Guide Section VIII *Security and fire*, The Architectural Press, London.

BRS Digests 122 (October 1970) and 132 (August 1971) Building Research Establishment, Watford.

Handbook of security, Kluwer Harrap, Peter Hamilton (joint editor).

Security is an attitude Peter Hamilton.

8. Maintenance of the building structure and fabric

Composite materials and their use in structures (J. R. Vinson and Tsu-Wei Chou), Applied Science Publishers, London (1975).

The Soiling of facades: a study carried out under the direction of the R.A.U.C. (C. Carrie and D. Morel, associated by Jean Fourquin), Editions Eyrolles (1975).

The Property Services Agency Library Services publication: *Current information on maintenance Part A – Cleaning buildings*, has 229 entries on this subject which includes, journal articles, reports, BRE and other publications.

Greater London Council Development Materials Bulletins:

   44. *Metals in the building industry – general.*

   47. *Metals in the building industry – lead.*

   51. *Metals in the building industry – cathodic protection.*

   52. *Metals in the building industry – anti-corrosive coating.*

   55. *Metals in the building industry – zinc.*

   88. *Metals in the building industry – copper.*

 107. *Priming paints for metal surfaces – guidance notes for steelwork.*

 110. *Aluminium fixing bolts – feedback on a failure.*

 112. *The use of overlaid plywood.*

 114. *Effect of dark colours on resin exudation from exterior joinery.*

 114. *Feedback – a failure of sheet lead in roof gutters.*

 115. *Failure of plywood fascia board.*

 116. *Need for the protection of lead built into concrete and cement mortars.*

 117. *Corrosion avoidance in domestic hot water storage cylinders and tanks – a first interim report.*

 117. *The corrosion of aluminium trim by run-off from a copper clad roof.*

 117. *Priming paints for use on rusty steel surfaces.*

 118. *Exterior painting of resinous timber.*

 118. *Corrosion control in central heating systems.*

British Standards Institution Standards and Codes of Practice:

CP3 Chapter 5 1972 *Codes of basic data for wind loads.*

CP112 *The structural use of timber.*

BS1186 Part 1: 1971 *Quality of timber and workmanship in joinery.*

TRL Note 12. *Flooring and joinery in new buildings* Building Research Establishment.

TRL Note 38. *The movement of timbers.* Building Research Establishment.

9. The spaces between and around buildings

Hard landscape in brick (C. Handisyd), RIBA Publications, London.

ILA basic plant list (The Landscape Institute), RIBA Publications, London.

Landscape handbook (ed. D. Lovejoy), E. & F. N. Spon Ltd., London (1986).

Landscape planning (B. Hackett), RIBA Publications, London.

Landscaping in the middle east, The Landscape Institute, published for the South East Chapter.

Plants in the landscape (P. Carpenter, T. D. Walker and N. O. Lanptiear), W. H. Freeman (1990).

British Standards Institution Standards.
BS 5236: 1975. *Cultivation of advanced nursery stock.*
BS 5696: 1979. *Childrens' playground equipment.*

10. Conservation: the maintenance of older buildings

Practical building conservation (J. Ashurst and N. Ashurst), Vols 1–5, Gower/English Heritage (1988).

Emergency repairs for historic buildings (E. Michell), English Heritage (1988).

Common defects in buildings (H. J. Eldridge), HMSO.

Conservation of brick buildings (T. G. Bidwell), Brick Development Association.

Guide to domestic building surveys (J. Bowyer), Butterworth Architecture, Oxford (1988).

Guide to recording Historic Buildings (ICOMOS). Butterworth Architecture, Oxford (1990).

Household insect pests (N. E. Hickin), Rentokil (1974).

Kinkell: The reconstruction of a Scottish castle (G. Laing), Ardullie House Publishers, XXX (1984).

Old church and modern craftsmenship (Alban D. R. Caroe), Oxford University Press (1949).

The pattern of English building (A. Clifton-Taylor), Faber and Faber, London (1987).

Series of leaflets and booklets on Care of Churches published by the Council for Places of Worship (1970 onwards).
How to look after your church.
Maintenance and repair of stone buildings.
Cure of woodworm in churches.
Birds, bats, bees, mice and moths.
Protection of stained glass.
Wall paintings.
Heating your church.
Lighting and wiring.
Redecorating your church.
Churches and archaeology.
Church organs.
Care and maintenance of church clocks.

Series of Technical Pamphlets published by the Society for the Protection of Ancient Buildings (1971 to 1977).
1. J. E. Macgregor. *Outward leaning walls.*
2. J. E. Macgregor. *Strengthening timber floors.*
3. G. B. A. Williams. *Chimney in old buildings.*
4. John Ashurst. *Cleaning stone and brick.*

Space heating and hot water: for conversion and rehabilitation (R. Charlton), National Building Agency.

Structural renovation of traditional buildings (Alan Baxter and Associates), CIRIA Report No. 111 (1986).

The woodworm problem (N. E. Hicken), Rentokil (1981).

Care and conservation of Georgian houses: a maintenance manual for the new town of Edinburgh (Edinburgh New Town Conservation Committee), Butterworth Architecture (1986).

The Property Services Agency Library Service publication – *Current information on maintenance Part F – Preservation and restoration of buildings* price £1, has 389 entries on this subject which includes journal articles, reports, BRE and other publications.

Building user guide Greater London Council (1977).

Decay and preservation of natural building stones (C. A. Price), Building Research Establishment (1974).

Timber decay and its control FPRL/BRE (1971).

Building Research Establishment Digests.
Nos. 63, 64 and 67. *Soils and foundations* (1972).
No. 75. *Cracking in buildings* 1966 (1972).
No. 139. *Control of lichens, moulds and similar growths* 1972.
BRE Technical Note No. 44 *Decay in buildings* 1975.

British Standards Institution.
BS CP. Part 1. *Code of practice for cleaning and surface repair of buildings. Part 1 natural and artificial stone and brickwork.* (In preparation).
DoE Circular 23/77. *Historic buildings and conservation areas: policy and procedure* HMSO (1977).
DoE Circular 14/75. *Housing, action areas, priority neighbourhoods and general development areas* HMSO (1975).

11. The preservation of modern buildings

Please also refer to Chapter 11.

Astragal remembers the Festival of Britain, *Architects Journal,* pp. 6–9, 1 May 1991.

Banham, M. and Hillier, B., *A Tonic to the Nation: The Festival of Britain 1951,* Thames and Hudson (1976).

Banham, R., *The New Brutalism,* Architectural Press (1966).

Beck, H. (ed.), The State of the Art – A Cultural History of British Architecture', UIA–*International Architect,* Issue 5 (1984).

Collins, M. (ed.), *Hampstead in the Thirties – A committed decade*, Hampstead Artists' Council Ltd/Arkwright Arts Trust, catalogue to Exhibition 1974–5.

Concrete Quarterly, 144, January–March (1985). Special issue for Cement and Concrete Assocaition's 50th Anniversary, 'Five Decades of Building', ed. George Perkins.

Dannatt, T., *Modern Architecture in Britain*, Batsford (1959).

Dean, D., *The Thirties: Recalling the English Architectural Scene*, Trefoil Books (1983).

Dunnett, J., London – Images from the Modern City leaflet accompanying exhibition of the same title, British Architectural Library (1986).

Esher, L., *A Broken Wave – The Rebuilding of England 1940–1980*, Allen Lane (1981).

Ford, E., *The Details of Modern Architecture*, M.I.T. Press (1990).

Frampton, K., *Modern Architecture – A critical history*, Thames and Hudson (1980).

Frampton, K., Stephen, D. and Carapetian, M., *British Buildings 1960–1964*, Black (1965).

Glendinning, M. and Muthesius, S., *Tower Block; Modern Public Housing in England, Scotland, Wales and Northern Ireland*, Yale University Press (1994).

Gould, J., *Modern Houses in Britain, 1919–1939*, Society of Architectural Historians of Great Britain (1977).

Jackson, A., *The Politics of Architecture – A History of Modern Architecture in Britain*, Architectural Press (1970).

Landau, R., *New Directions in British Architecture*, Studio Vista (1968).

Maxwell, R., *New British Architecture*, Thames and Hudson (1972).

McCallum, I., *A Pocket Guide to Modern Buildings in London*, Architectural Press (1951).

McKean, C. and Jestico, T., *Guide to Modern Buildings in London, 1965–75*, Warehouse Publishing (1976).

Mills, E., *The New Architecture in Great Britain 1946–53*, London (1953).

Murray, P. and Trombley, S., *Modern Architecture Guide*, ADT Press (1990).

Powers, A., *The DoCoMoMo-UK Register of Modern Buildings and Sites*, held at DoCoMoMo-UK Offices, Unpublished draft (1993).

Richards, J., *Introduction to Modern Architecture*, Penguin (1940).

Riseboro, B., *Modern Architecture and Design – An Alternative History*, MIT Press (1983).

Russell, B., *Building Systems, Industrialisation and Architecture*, John Wiley (1981).

Saint, A., *Towards a Social Architecture*, Yale (1987).

Sharp, D., *Twentieth Century Architecture – A visual history*, Lund Humphries (1991).

Strike, J., *Construction into Design*, Butterworth-Heinemann (1991).

Yorke, F. R. S., *The Modern House in England*, Architectural Press (1937).

Yorke, F. R. S., *The Modern Flat*, Architectural Press (1937).

12. Safety and security in accessibility for maintenance

Abrasure Wheels Regulations 1970.
Annual Report of HM Factory Inspectorate.
Construction Regulations 1961 and 1966.
Engineering Equipment Users Association Handbook No. 7.
The Factories Act 1961.
Greater London Building Bylaws.
Health and Safety at Work Act 1974.
Health and Safety at Work Booklet 6B Roofing.
Health and Safety at Work Booklet 6D Scaffolding.
Offices, Shops and Railway Premises Act 1963.
Safety of Scaffolding Joint Advisory Committee Report 1973.
The above are published by HMSO (London).
British Standards Institution: Standards and Codes of Practice.

CP 93. The use of safety nets.
CP 153. Part 1: 1969. Windows and rooflights.
CP 152: 1972. Glazing and fixing of glass for buildings.
CP 413: 1973. Ducts for building services.
1129: 1966. Timber ladders.
1139: 1964. Metal scaffolding.
1397: 1967. Industrial safety belts and harness.
2037: 1964. Aluminium ladders.
3913: 1973. Industrial safety nets.
4211: 1969. Access ladders.
5062: 1973. Fall arrest devices.
5534 Part 1: 1978. Slating/tiling.

13. Maintenance, planning and information feedback

Building: the process and the product (D. R. Harper), Chartered Institute of Building, London (1990).

Building Cost Information Service, 85–87, Clarence Street, Kingston-on-Thames, Surrey KY1 1RB.

Building maintenance management (R. Lee) Blackwell Scientific, Oxford (1987).

The care of building (Cluttons), 5 Great George Street, London SW1P 3SD.

Maintenance management. A guide to good practice The Chartered Institute of Building, London (1990).

Weathering and performance of building materials (J. W. Simpson and P. J. Horrobin), Mechanical and Technical Publishing Co., London.

14. Fire safety and means of escape

Automatic sprinkler systems for fire protection (P. Nash and R. A. Young), Paramount Publishers (1991).

Section 7 of The Property Services Agency Library Service Publication *Current information on maintenance Part D – building services engineering – fire protection services* includes references on Fire Protection.

The following organizations and departments provide advice and issue publications on special topics related to fire safety:

Building Research Establishment – Fire Research Station.

Steel Construction Institute.

The Loss Prevention Committee (Insurance).

Fire Protection Association – general advice and industrial hazards.

Timber Research and Development Association.

The Institution of Structural Engineers.

15. Rehabilitation and reuse of existing buildings

16. Euro Legislation

17. Statutory inspections and Spare Parts

Architects journal handbook on environmental powers (C. Whittaker, P. Brown, J. Monahan), Architect's Journal, London.

Encyclopedia of health & safety at work, law and practice (P. Allsop), Sweet and Maxwell, London.

Encyclopedia of housing law and practice (A. Arden), Sweet and Maxwell, London (1972).

Encyclopedia of public health law and practice (C. A. Cross), Sweet and Maxwell, London.

The law of dilapidations (W. A. West), Estates Gazette Ltd., London (1988).

Terotechnology handbook, DoI (1978).

Research Associations

Timber Research and Development Association.

Zinc Development Association

British Cement Association.

Copper Development Association.

18. Maintenance manuals and their use

19. Directory of journals and organizations concerned with building maintenance and conservation.

The Architects Journal, 33-39 Bowling Green Lane, London, EC1R 0DA.

Building Refurbishment, Morgan-Grampian (Construction Press) Ltd., Morgan-Grampian House, Calderwood Street, London, SE18 6QH.

Building, Builder House, 1 Mullharbour, London E14 9RA.

Building Design, Morgan-Grampian House, 30 Calderwood Street, London, SE18 6QH.

Building Maintenance, Amabane Ltd., 886, High Road, Finchley, London N12 9SB.

Building Products, 9 Queen Anne's Gate, London SW1H 9BX.

Building Services and Environmental Engineer, Batiste Publications Ltd, Pembroke House, Campsbourne Road, London N8 7BR.

Building Specification, Manning Rapley Publishing Ltd, Noah's Ark House, 100 Brighton Rd, Redhill, Surrey.

Building Today, International Thompson Business Publishing, 100 Avenue Rd, London NW3 3TP.

Built Environment, Kogan Page Ltd., 120, Pentonville Road, London N1 9JN.

Cathedrals Fabric Commission for England (CFCE), 83, London Wall, London EC2M 5NA.

Chartered Builder, Chartered Institute of Building, Englemere, King's Ride, Ascot, Berks. SL5 8BJ.

Chartered Surveyor Weekly, Builder Group Plc, 1 Millharbour, London E14 9RA.

Civic Trust, 17, Carlton House Terrace, London, SW1Y 5AW.

Cleaning Maintenance, Turret-Wheatland Ltd, 12 Greycaine Road, Watford, Herts WD2 4JP.

Construction News, International Thomson Business Publishing, 100 Avenue Road, London NW3 3TP.

Construction Weekly, Construction Press Ltd, 30 Calderwood Street, Woolwich, London SE18 6QH.

Contract Journal, Reed Business Publishing Ltd, Carew House, Quadrant House, The Quadrant, Sutton, Surrey SM2 5AS.

Development and Materials Bulletin, GLC Bookshop, County Hall, London SE1 7PB.

DoCoMoMo, The Building Centre, Store Street, London, WC1E 7BT.

English Heritage, Empress House, 23 Savile Row, London, W1X 1AB.

Georgian Group, 37 Spital Square, London E1 6DY.

Historic Buildings Council for Scotland, 20 Brandon Street, Edinburgh EH3 5RA.

Insulation Journal, Comprint Ltd., 177, Hagden Lane, Watford, Herts WD1 8LW.

Maintenance and Equipment News, Crown Wood Publications Ltd, P.O.Box 249, Ascot, Berkshire, 5L5 0BZ.

Municipal and Public Service Journal, The Municipal Journal Ltd., 32 Vauxhall Bridge Road, London SW2V 2SS.

Painting and Decorating, David Pescod Associates, 12A High St, Kings Langley, Herts, WD4 8RH.

Stone Industries, Ealing Publications Ltd, Weir Bank, Bray, Maidenhead, Berks SL6 2ED.

Timber Trades Journal, Benn Publications Ltd, Sovereign Way, Tonbridge, Kent, TN9 1RW.

Twentieth Century Society, 58 Crescent Lane, London SW4 9PU.

Victorian Society, 1 Priory Gardens, Bedford Park, London W4 1TT.

Index